This series is written on the basis of several difficult projects, aiming at complex geological conditions such as complex topography and landform types, complex geological structure and poor geotechnical engineering geology, systematically describes the Key Technologies for Tunnel Construction under Complex Geological and Environmental Conditions like mountains, underwater, soft and uneven strata, from the aspects of geological exploration, early warning and risk control, design, construction and digital application. It's a monograph that introduces innovative achievements and core technologies formed in the construction of complex geological tunnels in recent years. The series contains both texts and illustrations. It can be used as a reference for engineers, technicians, scientific researchers, teachers and students of related specialties in tunnels and underground engineering.

This series systematically describes the Key Technologies for Tunnel Construction under Complex Geological and Environmental Conditions like mountains, underwater, soft and uneven strata, from the aspects of geological exploration, early warning and risk control, design, construction and digital application.

Contents of the series are mainly from a number of key projects, such as Yuelongmen Tunnel of Chenglan Railway in Sichuan Province, Jinan Yellow River Tunnel Project, Qiyueshan Tunnel of West Hubei Expressway, Jiaozhou Bay Tunnel of Qingdao, Nanjing Yangtze River Tunnel and Wuhan River Crossing Tunnel, as well as the 973 Projects of the State-Mechanism, Prediction, Early Warning and Control of Major Water Inrush Disasters in Deep and Long Tunnels, Science and Technology Project by the Ministry of Transport "Study on Key Technologies of Underwater Large Section Shield Tunnel Construction under Complex Geological Conditions", National Defense Research Project "Three-Dimensional Advance Prediction Technology for National Defense Underground Engineering".

More information about this series at https://link.springer.com/bookseries/16378

Key Technologies for Tunnel Construction under Complex Geological and Environmental Conditions

Series Editor

Qihu Qian, Nanjing, China

Jian Chen · Fanlu Min · Shouhui Wang

Construction Technology of Large Diameter Underwater Shield Tunnel

SSTP

Springer

Jian Chen
China Railway 14th Construction
Bureau Co., Ltd.
Jinan, Shandong, China

Shouhui Wang
Qingdao Metro Group Co., Ltd.
Qingdao, Shandong, China

Fanlu Min
College of Civil and Transportation
Engineering
Hohai University
Nanjing, Jiangsu, China

ISSN 2662-2904 ISSN 2662-2912 (electronic)
Key Technologies for Tunnel Construction under Complex Geological and Environmental Conditions
ISBN 978-981-16-5895-2 ISBN 978-981-16-5896-9 (eBook)
https://doi.org/10.1007/978-981-16-5896-9

Jointly published with Shanghai Scientific and Technical Publishers
The print edition is not for sale in China (Mainland). Customers from China (Mainland) please order the print book from: Shanghai Scientific and Technical Publishers.

Translation from the *Chinese Simplified* language edition: *大直径水下盾构隧道施工技术* by Jian Chen, et al., © Shanghai Scientific & Technical Publishers 2019. Published by Shanghai Scientific & Technical Publishers. All Rights Reserved.

This Springer imprint is published by the registered company Springer Nature Singapore Pte Ltd.
The registered company address is: 152 Beach Road, #21-01/04 Gateway East, Singapore 189721, Singapore

Foreword by Jun Sun

In recent years, the development of infrastructure construction in China has been developing vigorously. The construction of "The Belt and Road" has been continuously promoted. Many major projects have mushroomed vigorously, such as the Three Gorges Project, the Qinghai-Tibet Railway, the South to North Water Diversion, the three vertical, and four horizontal high speed rail network, the Hong Kong–Zhuhai–Macao Bridge, Shanghai Center Tower, and the construction with China's aid of the Jakarta–Bandung high-speed railway, the China–Laos railway, the China–Thailand railway, the Gwadar Port, the Piraeus port, and so on. There is no doubt that China has become one of the countries with the largest number of major projects in the world. In terms of construction scale, technical difficulty, and capital investment, these major projects are among the top not only in China but also in the world, and even in the first place. A series of major key technical problems emerged in the construction process of these projects, many of which have been well optimized and solved through analysis, exploration, and innovation. Some even created new technical means and methods on the basis of the original theory and technology, and applied for a large number of technical patents. For example, Shanghai Center Tower, 632m, as the world's tallest green building, has achieved many scientific achievements in many aspects such as super high-rise design, green construction, construction supervision, and building information modeling (BIM) technology. It has applied for 8 invention patents and authorized 12 practical new technologies. Only in the aspect of structural engineering, it has been applied to many innovative technological innovations, such as the support technology of super deep foundation pit, the technology of super high pumping concrete, the installation technology of complex steel structure, and the technology of structural crack control, some of which have reached the world level. These optimizations, breakthroughs, and innovations will be very valuable references for Chinese engineers and technicians.

During the plenary session of the National People's Congress held in early March 2016, many representatives said that a great deal of technological innovation and development is a key driving factor for realizing China's macroeconomic strategic adjustment during the 13th Five-Year Plan period, and a fundamental support and key driving force for realizing the overall development of the country under the overall layout.

At the same time, in the face of the opportunity of a new round of scientific and technological revolution, only by making innovations and breakthroughs in key core technologies one by one can we realize the overall leap of social productivity, make the level and ability of our scientific research achievements and engineering technology control enter the ranks of developed countries as early as possible and as quickly as possible, so as to continuously improve the technological competitiveness in the world, and the national strength will be stronger! At present, many engineering technology innovation has been widely recognized, but there are still many problems in the promotion and application of innovation results. In the field of major engineering construction, after the key engineering technical problems have been broken through and solved in practice, it is necessary to further sort out and summarize the new theories or methods, and then widely apply them to production practice again, which in turn will promote further innovation and development of technology, which is a huge driving force for the sustainable development of technology. One of the most effective aspects is to systematically summarize the innovative achievements and publish a set of technical monographs. This is also the significance of publishing the series of "Research on Key Technologies of Major Engineering Construction". With the promotion of academic innovation as the main goal, the series of "Research on Key Technologies of Major Engineering Construction" has the following characteristics:

1. Focus on major projects and key projects. At present, China's infrastructure construction is booming in various fields, and various kinds of engineering projects continue to mount. From the perspective of project volume and technical difficulty, we have selected a number of major projects and key projects, based on which, we have summarized the professional theory and technology to make them into books. Due to the large number of fields and disciplines involved in various projects, and the intersection and integration of disciplines, it is difficult to set a series of books with a single discipline. Therefore, based on the main line of Engineering categories, a series of books have been written in eight fields, i.e., tunnel and underground engineering, bridge engineering, railway engineering, highway engineering, super high-rise and large-scale public buildings, water conservancy engineering, port engineering, urban planning and architecture, basically covering the main fields of engineering construction in China, with a view to providing professional and technical reference guidance for major engineering construction in the future. As there are many fields and specialties involved, there are similarities and differences between technologies. In the field of cross technology, they are dealt with according to the specific situation to avoid content duplication and disjunction.
2. Highlight common technology and innovation achievements, and focus on application technology theorization. Focusing on a series of key technical problems in major projects in recent years, and based on the innovation achievements and technological breakthroughs made by the project, the series of books sorts

out pertinently the technical experience and scientific research achievements of common, key or significant promotion value in each series, and makes in-depth, systematic and detailed analysis and elaboration from the perspective of technical methods and engineering practice experience, and provides practical theoretical basis and application reference for the solution of similar problems and the improvement of technology. In the "Series of Key Technologies for Tunnel Construction under Complex Geological and Environmental Conditions" (Academician Qihu Qian is the director of the editorial board), the key problems in the current tunnel and underground engineering construction are systematically described and the corresponding professional technical theory system is formed, including the prediction, warning and risk control of major water inrush disaster in deep and long tunnels, the treatment of soft and hard uneven stratum and extremely soft stratum in shield engineering, a sort of rectangular shield method, underwater shield tunnel, ground access shield method tunnel, extra-long highway tunnel, three-dimensional geological detection of tunnel, rapid detection of shield tunnel disease, digitalization of tunnel and underground engineering, new anchorage material for large deformation tunnel in soft rock, and so on, and the key problems are solved in different degrees in the research and effectively implemented in the follow-up project.

3. Pay attention to the practical value of the project. The technical achievements involved in the series have been adopted many times in China, which have been proved to be reliable and effective in practice, and some of them have obtained technical patents. The series emphasizes the theory as the guide, the application as the key, and the case as the illustration. All the technical achievements are required to take the engineering project as the background and the production practice as the support, so that the series not only has the academic connotation but also has the important engineering application value. For example, the "Series of Key Technologies for Long-span Bridge Construction and Maintenance" (academician Jielian Zheng is the director of the editorial board, academician Zhengqing Chen is the deputy director), focusing on major bridge projects such as long-span suspension bridges, long-span bridges across the sea, multi tower cable-stayed bridges, long-span concrete-filled steel tube arch bridges, long-span pedestrian bridges, large-scale spatial suspension bridges with variable width, etc., focuses on the design innovation theory, construction innovation technology, technical breakthrough of construction difficulties, bridge structure health monitoring and state assessment, maintenance and maintenance during operation period, etc. The main contents include the vacuum assisted pouring technology of large-scale concrete-filled steel tube structure, the spatial cable plane suspension bridge system with large proportion and variable width, the new eddy current damping vibration reduction technology, the cable hoisting and cable-stayed buckle construction of long and large bridges, the super large-scale deep-water foundation super high composite bridge tower, intelligent deformation monitoring, and the integration of construction and maintenance based on BIM, etc. These technologies are based on major construction projects, including the first bridge of Hejiang

Yangtze River, the second bridge of Hejiang Yangtze River, Wushan Yangtze River Bridge, Nanpanjiang bridge of Guangxi Guangzhou railway, Zhangjiajie Grand Canyon bridge, Xihoumen Bridge, Jiashao bridge, Hong Kong–Zhuhai–Macao Bridge, Humen second bridge, etc., in the book, the relevant contents of specific project cases are analyzed in detail, with good application reference Test the value.

4. Focus on the hot spot, focus on risk analysis, disaster prevention and mitigation, health detection, engineering digitization, and other emerging branches in recent years. Under the guidance of the principles of green and sustainable development, in recent years, technological innovation in the field of infrastructure construction has many achievements and examples in energy conservation and emission reduction, low-carbon environmental protection, green civil engineering, risk analysis, disaster prevention and mitigation, health detection (remote wireless video monitoring), safety and economy in the life cycle of the project, reliability and durability, construction technology organization and management, digitization, etc. Series of books are also reflected in these aspects, in order to give full play to the role of series books in promoting the long-term, green and sustainable development of major project construction.
5. Establish an open framework. As a result of the above characteristics, the progress of each volume of the series is different, so the open framework is adopted, and the flexible way of publishing in stages is adopted in the setting of each volume of the subsequent series.
6. The editor-in-chief has first-class academic level, which lays a solid foundation for the academic quality of the series. The chief editor of each series is the academic authority in this field, and has an important academic status and influence in this field. For example, Professor Zhengqing Chen, academician of Chinese Academy of engineering, chief scientist of Project 985, bridge structure and wind engineering expert, Prof. Jielian Zheng, academician of Chinese Academy of engineering, bridge design and construction expert, Prof. Qihu Qian, academician of Chinese Academy of engineering, protection and underground engineering expert, Prof. Zhiqiang Wu, academician of Chinese Academy of engineering, urban planning and construction expert, etc. The main writers who participated in the writing were all the front-line personnel active in the research, education and engineering of infrastructure construction in China, who had undertaken major engineering construction projects or national major scientific research projects, mainly from China Railway Tunnel Bureau Group Co., Ltd., CCCC Tunnel Engineering Bureau Co., Ltd., China Railway Construction Co., Ltd., CCCC highway planning and Design Institute Co., Ltd., and the army Institute of engineering design, Qingdao Metro Group Co., Ltd., Shanghai Urban Construction Group, China Merchants Chongqing Communications Research and Design Institute Co., Ltd, Tianjin Urban Construction Group Co., Ltd., Zhejiang Communications Planning and Design Institute, Tongji University, Hehai University, Southwest Jiaotong University, Hunan University, Shandong University, etc. In addition to undertaking the heavy tasks of engineering construction and scientific research and teaching, all experts have

contributed their wisdom, knowledge and sweat to the progress of engineering technology in China. On behalf of the General Editorial Committee of the series, I would like to express my heartfelt thanks and respect for your hard work.

At present, not only the domestic infrastructure projects are in the ascendant but also under the initiative of “The Belt and Road”, China’s major overseas projects are booming, and the demand for high-level engineering technology is becoming increasingly urgent. It is believed that the publication of the series of books can provide certain assistance for the development of major engineering construction and the progress of innovative technology in China.

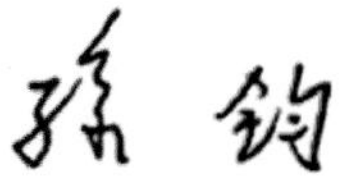

Shanghai, China
December 2017

Jun Sun

Mr. Jun Sun first-class Honorary Prof. of Tongji University, senior academician of the Chinese Academy of Sciences, well-known expert in geotechnical mechanics and engineering at home and abroad. Chief editor of the series “*Research on Key Technologies of Major Engineering Construction*”.

Foreword by Qihu Qian

Since entering the twenty-first century, with the continuous development of economy, the continuous improvement of comprehensive national strength, and the continuous application of high and new technology, China's tunnel and underground engineering have achieved unprecedented rapid development. China has become the country with the largest scale, the largest number, the most complex geological conditions and structural forms, and the fastest development of construction technology in the world. At the same time, with the continuous increase of urban subway construction, the number of cross river and cross sea tunnel projects continues to increase. National key construction projects such as long-distance water supply, underwater transportation, West to East Gas Transmission, and other projects will involve the problem of crossing rivers. The construction of railway, highway, municipal administration, water supply, gas supply, flood control, hydropower and other tunnel projects will greatly increase the number of tunnels.

In the aspect of tunnel construction technology, the technical system of high-speed railway tunnel has been basically formed; the construction technology of long large tunnels with complex geological conditions in difficult and dangerous mountainous areas has made continuous progress; the construction technology of large section soft surrounding rock tunnels has made great progress; the construction technology of urban large-span shallow buried tunnels and river crossing underwater tunnels has made breakthroughs; the research and development and manufacture of tunnel boring machines have made great progress, these marks the tunnel construction technology in China has reached a new level of development. In particular, China has a vast territory, complex geological conditions, and extremely complex geological conditions are the main factors restricting the safe and efficient construction of tunnels, which are recognized as the difficulties of tunnel construction. Relying on a large number of key and difficult projects, such as Guanjiao tunnel of Qinghai–Tibet railway, West Qinling Tunnel of Lanzhou–Chongqing railway, immersed tunnel of Hong Kong–Zhuhai–Macao Bridge, Gaoligongshan tunnel of Dali–Ruili railway, Yangtze River Tunnel of Sanyang road in Wuhan, etc., China has made great achievements and more innovative achievements in tunnel and underground engineering, especially in tunnel construction under complex geological and environmental conditions. Therefore, in view of the complex geological conditions such as complex terrain and landform type, complex geological structure and poor engineering geology of rock and soil mass, based on the breakthrough and innovation achievements of key and difficult projects, the editorial board of the series and Shanghai Science and Technology Press jointly planned the series.

From the perspective of geological exploration, early warning and risk control, design, construction and digital application, the series systematically sorts out the key technologies of tunnel construction under complex geological and environmental conditions, such as mountains, underwater, soft and hard uneven strata. In the form of academic monographs, it introduces the innovative achievements and core technologies formed in the process of complex geological tunnel construction in recent years. The innovative achievements and technologies involved and introduced in the series are at the leading level in China, some of which have formed core technologies with independent intellectual property rights, and have been applied in major projects, making the series have the characteristics of cutting-edge, original, innovative and leading. For example, Kairong Hong's *Shield Tunneling Technology in Hard-Soft Uneven Strata and Extremely-Soft Strata* expounds shield treatment technology of soft and hard uneven stratum and extremely soft stratum of underwater tunnel from theory, technology, and engineering cases, which represents the latest theory and practice of soft and hard uneven stratum tunnel and underground engineering; based on the scientific research project "National Defense Engineering Geological Prediction and Quality Nondestructive Testing System" presided over by the author, and combined with the research and application results of the team for many years, Guohou Cao and Hao Liu's *Three-Dimensional Exploration Technology of Tunnel Geology* comprehensively introduced the new theory, new method, and new technology of tunnel complex geological geophysical exploration; Shucai Li's *Monitoring Method and Early Warning Technology of Tunnel Water Inrush* relying on the national major scientific research instrument and equipment development project "Comprehensive Geophysical Exploration Instrument for Quantitative Advance Prediction of Tunnel Unfavorable Geology Used in the Roadheader Construction", the National Natural Science Foundation for outstanding young scientists project "Mechanism and Disaster Control of Tunnel Water Inrush and Mud Inrush" and other major scientific research projects, introduces the advanced prediction technology of water inrush and mud inrush disaster source in complex geological tunnels and its application in large tunnel engineering, etc. The application cases in the series also include Yuelongmen tunnel of Chengdu–Lanzhou Railway in Sichuan, Yellow River Tunnel Project in Jinan, Qiyueshan Tunnel of West Hubei expressway, Jiaozhou Bay Tunnel in Qingdao, Yangtze River Tunnel in Nanjing, and river crossing tunnel in Wuhan.

Under the strategic background of vigorously promoting the "One Belt and One Road" construction, implementing the innovation driven development strategy and building a powerful transportation nation, we hope that the publication of the series will not only better summarize the above technological achievements and promote the popularization and application of innovative technologies but also play a catalytic role in the breakthrough of basic theoretical research and common key technologies, and can play a positive role in the cultivation of technological innovation mode and professional talents. In the process of discussion, planning, organization, compilation, and review, the series has been greatly supported by relevant large enterprises, colleges and universities, research institutions, societies and associations. Many experts have put forward many very good suggestions and ideas for the series in their busy schedule. Thank you all.

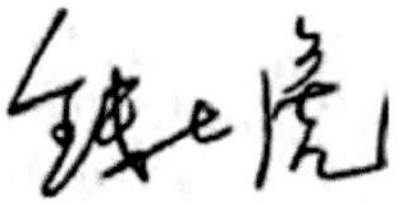

Beijing, China
August 2018

Qihu Qian

Mr. Qihu Qian Professor of the Army Engineering University of the Chinese people's Liberation Army and academician of the Chinese Academy of Engineering. Director of the editorial board of "Series of Key Technologies for Tunnel Construction under Complex Geological and Environmental Conditions".

Preface

Since the beginning of the twenty-first century, the construction scale and construction technology of underwater tunnels in China have developed rapidly with the construction of a large number of large river-crossing shield tunnels. At present, dozens of large-diameter underwater tunnels have been built in China, including Wuhan Yangtze River Tunnel, Shanghai Yangtze River Tunnel, Nanjing Yangtze River Tunnel, Guangshen Port Shiziyang Tunnel, Wuhan Sanyang Road Yangtze River Tunnel, Wuhan Metro Line 8 River-crossing Tunnel, etc. Slurry shield technology has become the main construction method of underwater large tunnel due to its excellent pressure control mode and safe and efficient advantages.

Due to China's vast territory and different geological conditions, the large-diameter underwater slurry shield projects that have been built and are under construction mostly pass through the composite strata of upper soft and lower hard, upper soil and lower rock, which will have a negative impact on the wear and replacement of cutters and the penetration of slurry on the excavation face. In addition, many unpredictable factors such as shallow overburden in the middle of the river and embankment during construction make the construction of large-diameter underwater shield tunnel face great risks, and serious deformation of stratum and instability of excavation face are easy to occur if the control is not good. Underwater tunnels in China are gradually developing towards "long distance", "large diameter", "large buried depth", and "high water pressure". At the same time, it also brings some urgent problems to be solved, such as initiation and arrival, crossing embankments, shallow overburden construction, crossing high water pressure permeable sandy cobble composite strata in the middle of the river, long distance tunneling and tool replacement, crossing the middle channel.

In order to make the construction personnel of the engineering front line better grasp the new technology of large-diameter underwater shield tunnel construction, based on the scientific research achievements and engineering experience of large-diameter underwater shield tunnel construction in the past ten years, and referring to the technical achievements of domestic relevant engineering and technical personnel of scientific research units, the author summarizes and compiles a book based on technical research and theoretical analysis, combined with the relevant typical engineering cases in China, and systematically introduces the new technology used in the construction of large underwater slurry shield under complex conditions. This book clarifies the basic principles, scope of application, construction technology and technical points of key technologies such as the initiation and arrival of shields, crossing shallow overburden in the middle of the river, crossing embankments, and replacing cutters with pressurized cabins, which can be used for reference by engineering technicians, researchers and teachers and students of relevant majors in colleges and universities.

All the contents of this book are compiled and finalized by Professor of Engineering Chen Jian, Deputy Chief Engineer of China Railway Fourteenth Bureau Group Co., Ltd., Professor Min Fan Road, Associate Professor of Tunnel and Underground Engineering Institute of Hehai University, and Professor of Engineering Wang Shouhui, Vice General Manager of Qingdao Metro Group Co., Ltd. after full discussion and consultation. In the process of writing this book, Xu Jingbo, Song Hangbiao, Bai Yuxin, Liu Jie, Du Jiarui, Yu Chaojie, Liu Hai, Wang Yucheng, Lv Huanjie, Wang Sheng, Wang Dengfeng, Xu Zhicheng, graduate students of the Institute of Tunnel and Underground Engineering of Hehai University, participated in the preparation of relevant chapters, data collation, chart drawing and revision, and so on. In the process of writing, this book also quotes some research results of scientific research cooperation units, and refers to some technical data and literature at home and abroad.

The book is jointly funded by the Enterprise Science and Technology R&D Project of China Railway Fourteenth Bureau Group (2017–07), the National Natural Science Foundation of China (51778213) and the Special Fund for Basic Research Services of Central Universities (2016B01214).

Due to the limited level of the author, there are inevitable deficiencies and errors in the book, please experts and readers to criticize and correct.

Jinan, China	Jian Chen
Nanjing, China	Fanlu Min
Qingdao, China	Shouhui Wang

Contents

Chapter 1
Introduction

The underwater tunnel crossing the underwater strata of rivers or straits has the advantages of less land occupation, less influence on navigation, stable traffic capacity and large carrying capacity, which is an important means of crossing the water area. China has a vast territory and a large area of rivers, lakes and seas, so it has excellent conditions for the construction of underwater tunnels.

1.1 Development and Application of Large-Diameter Underwater Shield Tunnel Technology

The idea of underwater tunnel construction was first proposed by British engineer Abel Mathieu in 1802, but the British-French war in 1803 failed to implement the plan. In the 1940s, French engineer Mare Isambard Brunel successfully constructed the first underwater pedestrian tunnel crossing the Thames River in London by using open manual shield in Britain, marking the birth of underwater shield tunnel. Since then, shield tunnel technology has gradually been widely used in underwater tunnel construction projects. Janes Heary Greathead first applied the compressed air method to the shield construction in 1886, and then the underwater tunnel of the Yibei River in Hamburg, Germany was successfully constructed by the compressed air shield in 1911. In the 1960s, the UK pioneered the slurry pressure shield, and in the 1970s, Japan made great progress in the slurry pressure shield. At the same time, a more novel earth pressure balance shield emerged, and the shield tunnel has been greatly developed in the stability control and excavation efficiency of the excavation face. At the same time, the emergence of new shield also makes underwater tunnel construction technology greatly enriched and improved.

In the mid-1990s, foreign underwater shield tunnels entered a period of rapid development and construction, and began to build strait tunnels. The five most representative underwater shield tunnels are the British-French Strait Tunnel completed in

J. Chen et al., *Construction Technology of Large Diameter Underwater Shield Tunnel*, Key Technologies for Tunnel Construction under Complex Geological and Environmental Conditions, https://doi.org/10.1007/978-981-16-5896-9_1

1993 (49.5 km long, 8.4 m in diameter), the Danish Stobel Strait Tunnel completed in 1995 (7.26 km long, 8.5 m in diameter), the Tokyo Bay Crossing Highway Tunnel completed in 1996 (9.6 km long, 13.9 m in diameter), the fourth Yibei River Tunnel completed in 2003 (2.56 km long, 13.8 m in diameter) and the Dutch Green Heart Tunnel completed in 2005 (7.16 km long, 14.5 m in diameter). Up to now, nearly one thousand underwater tunnels have been built for various purposes in the world, among which hundreds of large-diameter underwater shield tunnels (diameter >10 m) have been built, mainly in the United States, the United Kingdom, Japan, Denmark, Iceland and other countries.

In 1966 and 1984, Dapu Road cross-river tunnel with an outer diameter of 10.22 m and Yan'an East Road cross-river tunnel with an outer diameter of 11.30 m were built in Shanghai respectively, which opened the history of large underwater tunnel construction by shield method in China. In the 1990s, Shanghai built three cross-river tunnels with an outer diameter of 11.0 m, including the South Yan'an East Road Tunnel, the Dalian Road Tunnel and the Fuxing East Road Tunnel.

Since the twenty-first century, a large number of large-scale cross-river shield tunnels have been put into construction, covering railway, highway, urban subway and other fields of use. The construction scale and construction technology of underwater tunnels in China have been promoted to a new height. Some representative large-diameter underwater shield tunnels built in China are shown in Table 1.1. At present, dozens of large-diameter underwater tunnels have been built in Shanghai, Nanjing, Wuhan, Hangzhou, Guangzhou and other places in China, including Wuhan Yangtze River Tunnel (11.0 m in outer diameter, 3630 m in length, and two-way four lanes), Shanghai Yangtze River Tunnel (15 m in outer diameter, 8740 m in length, and two-way six lanes), Hangzhou Qiantang River Tunnel (15 m in outer diameter, 4250 m in length, and two-way six lanes), Nanjing Yangtze River Tunnel (14.5 m in outer diameter, 3905 m in length, and two-way six lanes), Guangzhou-Shenzhen Port Shiziyang Tunnel (10.8 m in outer diameter, 9340 m in length, and two-way single lanes), and Wuhan Sanyang Road Yangtze River Tunnel (15.2 m in outer diameter, and 2.59 m in length). Other representative tunnels that have been built are Qiantangjiang Tunnel, Qingchun Road, Hangzhou, which opened in December 2010 (11.3 m in diameter, 3765 m in length, four lanes in two directions), and Shouxihu Tunnel, Yangzhou, which opened in September 2014 (14.5 m in diameter, 2230 m in length, four lanes in two directions). Most of these projects are distributed in the middle and lower reaches of the Yangtze River. In the process of construction, they have passed through complex and changeable strata from muddy silty clay to silty sand and gravel, and then to the upper soil and lower rock, and the upper soft and lower hard composite strata. They have encountered various technical problems and accumulated many valuable key technologies.

Table 1.1 Some representative large diameter underwater tunnels built in China

Name of tunnel project	Tunnel outer diameter (m)	Tunnel length (m)	Tunnel use	Year of completion
Shanghai Yangtze River Tunnel	15.0	8740	Highway	2009
Hangzhou Qiantang River Tunnel	15.0	4250	Highway	2012
Shanghai Shangzhong Road Tunnel	14.5	2802	Highway	2008
Shanghai Yaohua Road Crossing Tunnel	14.5	1275	Magnetic suspension	2009
Nanjing Yangtze River Tunnel	14.5	3905	Highway	2010
Yangzhou thin West Lake Tunnel	14.5	2230	Highway	2014
Nanjing Weicheng third Road Crossing the Yangtze River	14.5	4134	Highway	2015
Wuhan Metro Line 8 Crossing Tunnel	12	3185	Railway	2017
Shanghai Yan'an East Road Crossing River Tunnel	11.3	2261	Highway	1984
Yujiang Tunnel, Qingchun Road, Hangzhou	11.3	3765	Highway	2010
Nanjing Metro Line 10 Crossing the Yangtze River	11.2	3600	Railway	2013
Wuhan Yangtze River Tunnel	11.0	9340	Highway	2008
Guangdong Guangshengang Lionyang Tunnel	10.8		Railway	2011

1.2 Common Technical Difficulties in Large Diameter Underwater Shield Tunnel Construction

Due to China's vast territory and large differences in geological conditions, in addition to the Shanghai area and individual tunnels crossing the muddy clay stratum, the large-diameter slurry shield projects that have been built and are under construction in China have almost encountered the composite stratum of fine sand gravel, upper soil and lower rock (Fig. 1.1 is the composite stratum map of Nanjing Weisan Road crossing the river channel) due to many unpredictable factors during the construction of crossing the river and the seabed. The construction of large-diameter underwater

Fig. 1.1 Composite stratum map of Nanjing Weisan Road crossing the river

shield tunnels is facing great risks. If the control is not good, serious deformation of the stratum and instability of the excavation face will occur, which will bring great risks to the project.

The construction of large diameter underwater shield tunnel technology in China is mainly faced with the problems of initiation and arrival, crossing the embankment, shallow overburden construction, crossing the high water pressure permeable composite stratum in the middle of the river, long distance tunneling and tool replacement, crossing the channel in the middle of the river. At present, China has become the country with the largest number of shields in the world, and the number of large section slurry shields is considerable. In the future, slurry shield construction method will be widely used in many major tunnel projects, especially in tunnel projects across the river and sea. Therefore, combined with the current slurry shield engineering examples in China, this book analyzes and summarizes the common problems, providing theoretical support for the formation of systematic, comprehensive and standardized construction technology methods.

1) Initiation and arrival

The initiation and arrival stage of shield mainly includes two parts: soil reinforcement and construction technology at the end of the tunnel. The problems in the construction are mainly manifested in the poor reinforcement effect of the shield in and out of the tunnel end stratum, the water gushing at the tunnel entrance when the shield enters and out of the tunnel, the difficulty of shield attitude control, the failure of establishing a good pressure balance as soon as possible, and the sealing failure of the tunnel entrance. Poor reinforcement effect of the portal will lead to a large amount of mud overflow and sand gushing. When the shield enters and exits the hole, due to the high pressure head of deep groundwater in the foundation pit, it is prone to inrush or piping problems. In the past, grouting method was the main method for stratum reinforcement at the initiation and arrival parts of shield tunnel. However, in recent years, with the increase of buried depth and section of shield tunnel, high-strength reinforcement is urgently needed. In addition, due to the time-consuming removal of diaphragm walls and other underground walls, the reinforced strata are required to remain stable for a long time.

The construction technology of shield initiation and arrival includes two parts. The starting section mainly includes four aspects: auxiliary engineering construction, negative ring segment and box culvert assembly, door breaking and starting tunneling parameter control. The arrival of shield has its particularity and importance relative to the construction of interval tunnel, mainly including: stratum reinforcement, shield positioning and receiving portal position review measurement, portal processing and installation of portal ring temporary sealing device, installation of receiving base, etc. Through the above construction technology, the deviation correction between the central axis of the shield machine and the design axis of the tunnel and the position of the receiving gate is gradually completed, and the tunneling is controlled to control the settlement and segment assembly, so as to prevent the displacement of the shield receiving base and the surface settlement caused by the breakdown of the shallow overburden layer.

2) Shallow overburden construction

When the tunnel passes through the shallow soil section of river bottom, due to the requirements of the tunnel design line, the overlying soil layer of the tunnel will be shallow. When the overlying soil weight and river pressure cannot resist the buoyancy of the tunnel caused by grouting during the excavation of the tunnel, the tunnel will float or roof fall, which brings immeasurable disasters to the project. There are many examples of similar problems encountered in shield tunnel construction at home and abroad. In the construction process, it is of great significance to solve the balance between the soil around the shield and the shield itself, which is of great research significance to the shallow overburden problem caused by the tunnel crossing the inland river.

There are three main risks and problems in the construction of shallow overburden of tunnels under water (river, lake and sea) by shield method: water flowing through the roof of shallow overburden, tunnel floating, sand flowing and piping. In the case of shallow overburden and high head pressure, the earth pressure balance in front of the large cutterhead is not easy to establish. Water sand often enters the shield through the opening of the large cutterhead and the tail of the shield from the cracks in the disturbed soil, causing the shield to be flooded, and such accidents of machine damage and casualties occur from time to time. The upper and lower forces of the shield advancing under the water area are not balanced, the shield posture is raised, the pressure slope is difficult, the tunnel is floating, and the axis is difficult to control. After the assembled tunnel ring is removed from the tail of the shield, the tunnel floats because the upper pressure load and the weight cannot resist the buoyancy caused by groundwater. If the corresponding reinforcement measures are not taken, it is easy to cause local tunnel cracking and water leakage. Construction in liquefiable soil layers such as sandy soil and sandy silt, due to shield driven cutter head rotary cutting extrusion disturbance, plus high head pressure (sometimes up to 0.3–0.5 MPa). Liquefied sand leaks into the tunnel along the shield tail and tunnel joint with groundwater. If measures are not taken in time, local foundation hollowing, tunnel subsidence, spiral fracture and tunnel failure may occur.

3) Crossing embankments and other buildings

When shield tunneling in the stratum, it will inevitably disturb the stratum, cause stratum settlement, and then affect the stability of the surrounding buildings, so that the adjacent buildings have settlement, inclination and even cracking, which has become one of the most noteworthy problems in the shield construction process. At present, relevant studies have shown that the main modes of building damage caused by shield tunnel construction include uneven settlement damage, damage caused by ground curvature, damage caused by inclination, and damage caused by horizontal deformation. Usually, the compression and negative curvature of the surface appear at the same time, and the tensile and positive curvature of the surface occur at the same time. In practice, the deformation of buildings is often a combination of various forms. Among them, the curvature, inclination and horizontal deformation caused by differential settlement are more obvious to the damage of buildings, which will appear distortion, subsidence, cracking and even collapse. In the process of shield tunneling, the influence of shield construction disturbance on buildings is restricted by many factors, including engineering geological conditions, the form of building foundation, the position relationship and relative distance between buildings and tunnels. In particular, when the large-diameter slurry shield is constructed in the urban river-crossing tunnel, the river bank of the city is often the central area, which is an important barrier for urban flood control and is related to the safety of people's lives and property. Therefore, it is essential to ensure the stability of the embankment during the shield construction.

Embankment has potential safety hazards such as ground subsidence, cracks, collapse and seepage in flood season. Its harm mainly has two aspects: one is that the looseness of embankment soil reduces the safety of seepage resistance, especially the voids and voids above the shield will form seepage channels, which are prone to piping. Second, the physical and mechanical indexes of soil are reduced due to soil loosening, which affects the anti-sliding stability of embankment slope and further affects the embankment safety. Therefore, the use of corresponding embankment reinforcement means is the key to safe construction.

4) High Water Pressure Permeable Composite Strata Crossing the River

In recent years, the common characteristics of large-scale cross-sea cross-river tunnel projects at home and abroad are large diameter, which often passes through coarse sand, gravel, sandy silt and other granular strata rich in groundwater. In the harsh working conditions, especially in a large number of water-bearing gravel layer, no cohesion, very unstable soil layer and shallow overburden tunnel engineering, as well as large diameter shield and ground deformation requirements are particularly high area construction, most of the slurry shield construction. Slurry shield is a kind of sealed mechanical shield. In the construction process, the slurry needs to meet the following three requirements: ① reliable mud film is formed on the excavation surface to stabilize the excavation surface. ② Slag soil was successfully transported to avoid sediment. ③ Slurry separation was completed by screening equipment. The stability of excavation face is maintained by the following three factors: ① Mud pressure balances soil pressure and water pressure. ② The impervious mud film is formed on the excavation surface, so that the mud pressure can effectively play a role. ③ The slurry penetrates into a certain stratum from the excavation surface, and the stratum on the excavation surface increases the cohesion. With the continuous excavation of the excavation surface, it is required that the mud pressure always acts on the excavation surface in the form of a surface. In this process, the formation of impervious cement film on the excavation surface is the core factor. The stratum in the middle section of the river crossing by slurry shield is generally complex. Therefore, according to different geological conditions, the corresponding construction technology is used to ensure the safe and efficient tunneling of shield.

In low permeability clay and silty clay strata, the formation pore is small, and the mud is easy to form mud film on the excavation surface. However, in the high permeability formation such as fine sand and gravel sand, mud easily penetrates the formation and directly seeps out, and cannot form mud film on the excavation surface, which will lead to a large number of mud filtration, mud pressure is difficult to maintain, thus inducing excavation surface instability and other accidents, the consequences will be very serious. At the same time, large size particles in the formation are unfavorable for carrying slag, but the separation of slurry is relatively easy. Therefore, how to form a good mud film on the surface of high permeability stratum is an urgent problem to be solved in the cross-river tunnel project.

5) Long distance tunneling and tool replacement

In most shield projects with complex geological conditions, tool wear occurs in different degrees. For example, when shield tunneling in sand formation, because of the characteristics of small cohesion, large internal friction angle, poor flow plasticity, large permeability coefficient, high quartz content, easy to appear cutter head, cutting tool, screw conveyor wear, cutter head torque and thrust increase, abnormal excavation parameters, difficult to unearth and difficult to control land subsidence and a series of risk problems, such as serious safety accidents will be caused by a series of risk problems, such as small cohesion, large internal friction angle, poor flow and plasticity, large permeability coefficient, high quartz content, wear, torque and thrust of cutter head, abnormal excavation parameters, difficult to unearth and difficult to control land subsidence, and so on. In recent years, more and more attention has been paid to the problems caused by shield tunneling in sand strata, especially in long distance excavation.

The normal wear of the tool is the result of the normal use of the tool. The wear is mainly due to the contact and sliding friction between the blade surface of the tool and the rock and soil body, which makes the blade surface of the tool lose continuously. The main phenomenon is the uniform wear of the alloy block. However, when the formation quartz content is high, the wear coefficient of the cutter will greatly increase, and the single cutter heading mileage and service life will be greatly reduced. In addition, in the composite strata of upper soft and lower hard, upper soil and lower rock, due to the unevenness of the section, the abnormal wear of the cutter such as tooth collapse, eccentric wear and tool fracture seriously affects the development of shield technology. On the other hand, due to the technical personnel is not clear about the cutting mechanism of the cutter and the matching law with the stratum, there is a serious situation of abnormal wear of the cutter due to improper configuration of the cutter, such as the hob configured in Guangfo subway cutting soft soil layer and the occurrence of eccentric wear accidents. The improper selection of cutting tools is due to the section shape of the cutting tool is not suitable for cutting strata, such as Nanjing Wei Qi Road Yangtze River Tunnel crossing sandy cobble strata, due to the sharp blade of the scraper, resulting in a large number of fracture of the alloy block of the cutting tool.

With the continuous application and development of shield technology in China, the river-crossing tunnel has the characteristics of large diameter, long distance, high water pressure and soft and hard excavation surface. Long-distance tunneling in upper soft and lower hard strata puts forward higher requirements for the life of the cutter. It is inevitable to repair or replace the cutter in the tunneling process. However, due to the complexity and variability of the geological conditions and the limitations of the surrounding conditions, it is often difficult to implement atmospheric cabin entry by reinforcing the soil in front of the excavation. With the characteristics of no need for stratum reinforcement, strong geological adaptability, low requirements for the surrounding environment and small influence, the technology of pressurized cabin entry solves the problem of tool change under harsh conditions. The so-called pressure into the cabin refers to the pressure of compressed gas to replace the mud pressure or soil pressure in the pressure cabin. According to the principle of compressed air construction, the excavation face is supported by air pressure, and the operator enters the pressure cabin under a certain air pressure environment for operation. Although there are many successful cases of large-diameter slurry shield construction in China, the research on the cutter replacement technology with pressure in ultra-high water pressure (above 0.68 MPa) and strong permeable strata is still in its infancy. There are still many difficulties in the operation of pressurized cabin, such as the stability control of excavation face under air pressure, personnel safety and welding repair operation.

6) Crossing the channel in the middle of the river

The slurry in shield construction is not only responsible for maintaining the stability of excavation face, but also responsible for carrying slag. The selection and control of slurry parameters are of great significance to the normal and safe construction of slurry shield. Especially in the construction of the punching section, due to the shallow overburden thickness, the slurry parameters must be strictly controlled to maintain the stability of the excavation face. When the shield punching section passes through the silty sand layer, due to the single particle size and poor gradation of the silty sand layer, the slightly disturbed stratum is easy to be liquefied. It is necessary to form a dense mud film on the excavation surface to maintain the stability of the excavation surface and ensure the safety of the excavation surface. However, the particle size of this kind of fine sand stratum is small, and the mud is not easy to penetrate into the formation to form a permeable zone, which can only form mud skin in the surface siltation. Since the mud is prone to damage under disturbance, in order to ensure the stability of excavation face in such strata, it is necessary to ensure the formation of dense mud with high strength and strong anti-disturbance ability on the surface of excavation face.

1.3 Significance and Application Prospect of Underwater Shield Tunnel Construction Technology

Nowadays slurry shield technology has been widely used in underwater tunnel construction in China for its safety, high efficiency, energy saving and environmental protection. With the gradual implementation of the national maritime strategy, regional economic integration and national large-scale channel construction plan, China's underwater tunnels have gradually developed to long distance, large diameter, large buried depth and high water pressure. The construction of underwater shield tunnels has good prospects and great potential. Table 1.2 is some representative large-diameter underwater tunnels under construction and planning in China.

Table 1.2 Some representative large-diameter underwater tunnels built or under construction in China

Name of tunnel project	Tunnel outer diameter (m)	Tunnel length (m)	Tunnel use	Year of completion
Chek Wat Kok Tunnel, Tuen Mun, Hong Kong	17.6	4200	Highway	Under construction
Wuhan Sanyang Road Tunnel	15.2	2590	Highway, railway	Under construction
Jiajiang Tunnel of Nanjing Yangtze River Bridge	15.0	1159	Highway	Under construction
Nanjing and Yan Road Tunnel across the Yangtze River	14.5	2976	Highway	Under construction
Wuhu Chengnan River Crossing Tunnel	14.5	4090	Highway	Under construction
Jinan Jiluo Road Yellow River Tunnel	14.5	4760	Highway, railway	Under construction
Shantou Soviet Tunnel	14.5	6400	Highway	Under construction
Nanjing Jianning West Crossing River Tunnel	14.5	2450	Highway	Planning
Suzhou Sutong GIL Pipe Corridor Tunnel	12.6	5468	Pipe gallery	Under construction
Hangzhou Wangjiang Road Tunnel	11.3	3240	Highway	Under construction

At present, the representative large-diameter underwater tunnels under construction mainly include Hong Kong Tunmen Chilajiao Tunnel (17.2 m in diameter, 4.2 km in length), Wuhan Sanyang Road Tunnel (15.2 m in diameter, 2.59 km in length), Nanjing Wuqiao Jiajiang Tunnel (15.0 m in diameter, 1.59 km in length), Shantou Su'e Tunnel (14.5 m in diameter, 6.4 km in length), Nanjing Heyan Road Crossing the Yangtze River Tunnel (14.5 m in diameter, 2.98 km in length), etc. In the next 10–30 years, China plans to build five world-class subsea tunnels and nearly 100 underwater tunnels. These subsea tunnels include the Bohai Sea subsea tunnel from Dalian, Liaoning Province to Yantai, Shandong Province, the Hangzhou Bay subsea tunnel from Shanghai to Ningbo, Zhejiang Province, the Lingdingyang subsea tunnel connecting Hong Kong and Macao with Guangzhou, Shenzhen and Zhuhai, the subsea tunnel connecting Guangdong and Hainan Provinces across the Qiongzhou Strait, and Fujian and Taiwan. A subsea tunnel across the Taiwan Strait. It is foreseeable that our country will be in the near future and for a long time to come.

Hong Kong Tuen Mun Chiwajiao tunnel project is under construction, which adopts slurry pressurized shield construction with a diameter of 17.6 m. It is the largest diameter shield tunnel in the world at present. Bohai Strait tunnel and Qiongzhou Strait tunnel under planning will also adopt or plan to adopt slurry pressurized shield construction method. This shows that China's shield technology will develop in the direction of high water pressure, long distance and complex strata in the future, and will also face severe challenges in complex working conditions such as high water pressure, long distance, strong permeability, upper soft and lower hard, upper soil and lower rock composite strata. Therefore, it is necessary to systematically study and analyze the key technologies in the construction of large-diameter underwater shield tunnels in China, which is of great significance to promote the construction of underwater shield tunnels in China.

Chapter 2
Large Diameter Shield Initiation and Arrival Technique

Shield initiation and arrival are the two processes most prone to accidents in shield tunneling construction, and they are also a very important link in shield tunneling construction. The shield machine type is different, the structure of the shaft wall starting mouth and arriving mouth is different, and the starting and arriving operations are also different. At present, many accidents of shield tunnel in China occur in the process of shield tunnel construction, and the reason is mostly soil instability. This is because during the start-up and arrival of shaft construction, pile driving and excavation disturb the surrounding strata, resulting in stress release of the section, that is, the soil is unstable. This chapter introduces the technology of shield initiation and arrival, including the construction method of shield initiation and arrival, the soil reinforcement technology at the end of the tunnel and the construction technology of shield initiation and arrival.

2.1 Shield Initiation and Arrival Construction Method

2.1.1 Main Construction Methods of Shield Initial Stage

Shield initiation refers to the use of temporary assembly segments and other reactive equipment to push the shield onto the launching platform, from the starting point into the stratum, starting a series of operations along the given line. The shield launching stage mainly includes two parts: the soil reinforcement at the end of the tunnel and the initiation construction technology. According to the different methods of removing temporary retaining wall and preventing stratum collapse of excavation face, the main construction methods in the initial stage are the following types (Fig. 2.1).

(1) The end soil reinforcement method is to use reinforcement measures to stabilize the excavation stratum, and then shield tunneling into the reinforced

J. Chen et al., *Construction Technology of Large Diameter Underwater Shield Tunnel*, Key Technologies for Tunnel Construction under Complex Geological and Environmental Conditions, https://doi.org/10.1007/978-981-16-5896-9_2

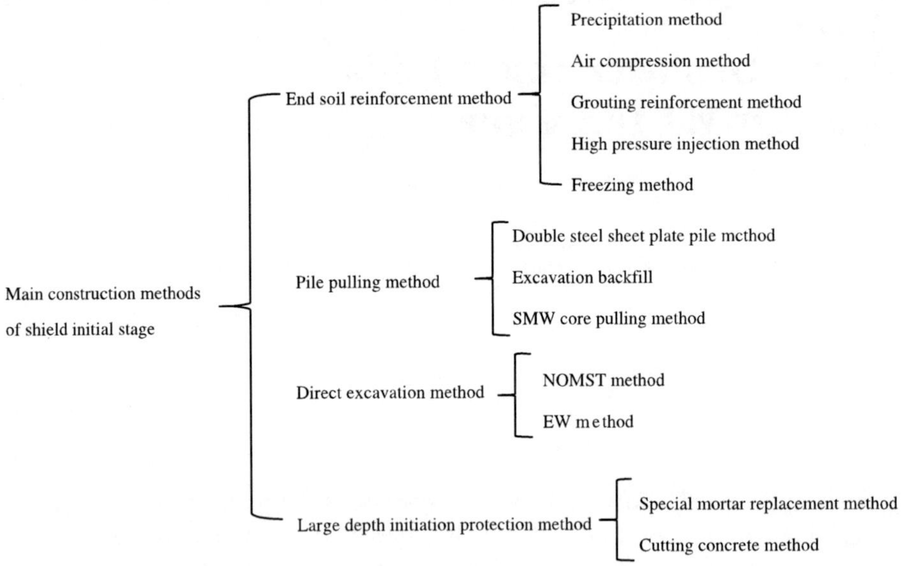

Fig. 2.1 Classification of main construction methods of shield initial stage

stable stratum. Reinforcement methods used in the past are more precipitation, air compression and chemical grouting reinforcement method, now gradually using high pressure injection method and freezing method more safe construction method.

(2) Pile pulling method is a process to overcome the pile side friction (the soil around the pile acts on the surface of the pile) and the self-quality of the pile depth, and pull the pile out of the surface, including the double steel sheet plate pile method, excavation backfill method and SMW core pulling method. Double steel sheet pile method is to start the shaft steel sheet pile retaining wall as two layers. The shield tunneling after pulling out the inner steel sheet pile can ensure that the outer soil will not collapse due to the retaining effect of the outer steel sheet pile, which ensures the stable tunneling of the shield. When the shield is pushed to the front of the lateral steel sheet pile, the lateral steel sheet pile is pulled out. Due to the stability of the soil between the internal and external steel sheet piles, the shield can be maintained to continue to advance after the lateral steel sheet pile is pulled out. Excavation backfill method is to use the starting shaft as a rectangle (longer than 2 times the length of the shield), set a partition in the middle of the shaft (or build two parallel shafts), half as the starting shield assembly, and backfill when the shield machine is pushed into the other half of the shaft. Due to the isolation supporting effect of backfill soil, the stratum can not collapse when the steel sheet pile of the end shaft wall is removed, which provides a reliable guarantee for the safe penetration of shield into the stratum. The SMW core pulling method is to use the SMW retaining wall as the lining of the starting wall of the shaft. Before

the shield starts, the I-steel core material is pulled out, and then the shield starts to excavate and cut the shaft wall without core material.

(3) Direct excavation method includes NOMST method and EW method. The characteristics of NOMST method is that the wall material of the starting port is special, and the tool can be directly excavated without damaging the tool. The starting operation of NOMST method is simple, without auxiliary method, and the safety and reliability are good. The principle of EW method is that I-steel, the core material of retaining wall, is corroded by means of electric erosion before shield launching, which brings convenience to shield excavation and cutting directly. The advantages of EW method are the same as those of NOMST method.

(4) Large-depth initiation protection method includes special mortar replacement method and cutting concrete method. The special mortar replacement method is to use the excavator to dig out the soil of the original protection part directly from the surface, and then pour the low strength (compressive strength at month age >10 MPa) without separating the mortar in water. After the completion of all the mortar conversion soil, the in-situ curing method of wall protection slurry is used above the mortar (until the ground). The cutting concrete method is to use PC steel wire to reinforce the starting wall constructed by limestone concrete. When the shield starts, the PC steel wire is pulled out to make the starting wall in a state of no reinforcement, and then the shield cutterhead is used to directly cut the starting wall.

For the starting construction of a specific shield tunnel, one of the methods in Fig. 2.1 can be used alone, and the combination method can be used. Details on the characteristics, application scope and construction points of various reinforcement methods are provided in Sect. 2.2. The specific selection of construction methods depends on factors such as geology, groundwater, overburden, shield diameter and construction environment, and the requirements of safety, construction, cost and progress should also be considered.

Taking the launching construction of large-diameter slurry shield in Nanjing Yangtze River Tunnel as an example, the main construction methods of soil reinforcement at the end of the launching stage are introduced. The soil layer of Nanjing Yangtze River Tunnel is silty clay, and the stratum is reinforced by high pressure jet grouting pile in the 0–18 m section ahead of the tunnel portal, with the reinforcement depth of 23 m and the width of 23 m. Due to the connection between groundwater and Yangtze River water, the high-pressure rotary jet grouting slurry is easy to be taken away by flowing water. The reinforcement effect of high-pressure rotary jet grouting pile by drilling core detection is not ideal, and it does not condense into a whole. There is a local weak zone that cannot seal sand water, which is prone to sand gushing and water gushing, affecting the normal exit of the broken portal and shield. In order to better enhance the strength and sealing of soil and ensure reliable water sealing of reinforced soil, a 1.6 m wide frozen soil wall is designed between the soil reinforced by high pressure jet grouting pile and the underground diaphragm wall at the end. The frozen soil wall formed by artificial refrigeration process cements the reinforced

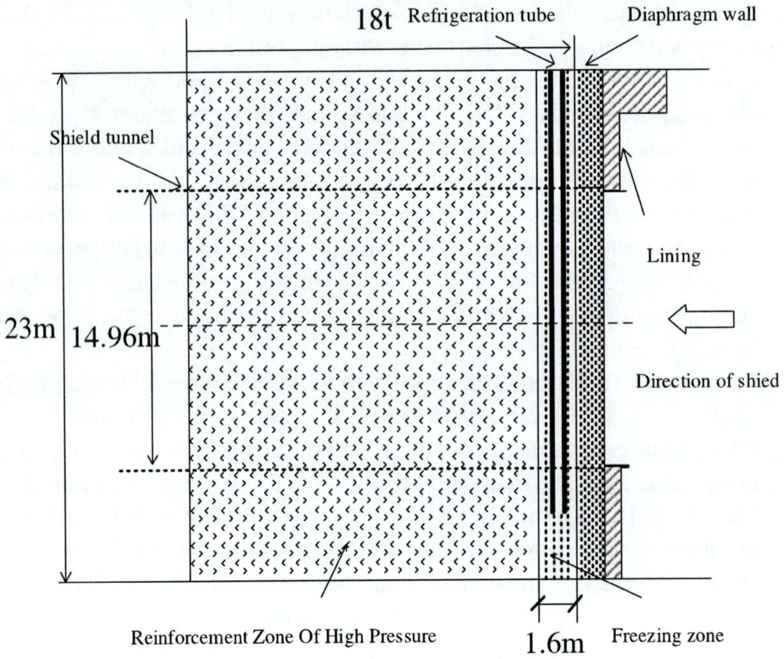

Fig. 2.2 Profile of soil reinforcement at shield initiation end

area of high pressure jet grouting pile and the underground diaphragm wall, so that the soil reinforced by jet grouting and the frozen wall jointly resist the water and soil pressure to ensure the safety of the portal breaking and shield launching. Figure 2.2 is the schematic diagram of the reinforcement area at the starting end of the shield.

2.1.2 Construction Method of Shield Arrival

The arrival construction of shield machine refers to the whole construction process from 50 m before shield tunneling reaches the receiving well to the tunnel entering the receiving well into the receiving well and being pushed onto the shield receiving pedestal (usually the initial base of shield machine). The construction of shield arrival includes the reinforcement of the reinforcement of shield to front end, and the shield arrival method and the shield receiving method. The specific scheme of each part is described as follows.

1) Soil reinforcement of shield to front end

With the change of stratum conditions, the soil at the end of the shield arrival section also has the soil stability problem, and the reinforcement method is referred to the reinforcement method of the shield arrival section.

2) Shield arrival method

Shield arrival construction refers to the whole construction process from 50 m before the shield tunneling to the receiving well to the shield tunneling through section tunnel into the receiving well is pushed to the shield receiving base (usually the shield arrival base).

According to the different time of removing the retaining wall, it is divided into two kinds: one is to remove the retaining wall reaching the shaft after the shield arrives, and then push it forward, the other is to remove the retaining wall in advance, and then push it to the designated position.

(1) The construction method of removing the retaining wall after the shield arrives.

The method is to push the shield outside the retaining wall reaching the shaft, stabilize the stratum by stratum reinforcement, remove the retaining wall, and then push the shield to the specified position. When the retaining wall is removed by this method, the shield is stopped in front of the open surrounding rock, and the gap between the cutter head and the shaft is small, so the stability is strong. However, it is widely used process and good construction, it is widely used. Because the stratum is prone to collapse when the shield is re-propelled, it is mostly used for small and medium section shield projects with good stratum stability.

(2) The method of removing the retaining wall before the shield arrives.

This construction method needs to remove the retaining wall in advance, so high-strength strata reinforcement should be carried out before removal, and steel partition wall with easy removal and good stress should be built in the well. Then, the retaining wall is removed from the bottom to the top, and the gap between the stratum and the reinforcement and the partition is filled sequentially with cement or poor ratio mortar. After completely replacing it with cement soil or poor ratio mortar, the shield is pushed to the front of the partition, and the partition is removed to complete the arrival process. Because the shield is not allowed to advance again, the construction method has the advantages of preventing stratum collapse and strong anti-seepage at the entrance of the tunnel, but the scale of stratum reinforcement increases, and the partition wall must be set up, so the scale of arrival preparation operation is expanded. This method is mostly used in large-section shield engineering.

3) Shield receiving method

In order to solve the engineering problem of poor reinforcement effect in some strata encountered in the field construction, the construction technology of water arrival and the receiving technology of steel sleeve have been developed in recent years, as shown in Figs. 2.3 and 2.4.

The construction technology of shield arriving in water refers to the construction technology that, in order to prevent or control the engineering danger caused by the large amount of groundwater and soil gushing out of the open tunnel circle during the shield arriving, the water and soil of the shield arriving well is backfilled actively

Fig. 2.3 Construction technology of arrival in water

Fig. 2.4 Steel sleeve receiving technology

by using the mechanism that the water and soil pressure balance inside and outside the arriving well can be controlled and the shield is pushed into the arriving well safely under the condition of water and soil pressure balance.

The steel sleeve receiving technology adopts the main design principle of the steel sleeve receiving shield, which is a steel sleeve structure that can completely cover the shield machine body in the shield well. The soil pressure condition of the normal excavation of the tunnel is simulated in the steel sleeve, and the normal tunneling pressure is established in the process of shield breaking the door, and the door ring is closed to the steel sleeve to prevent water and soil from pouring into the shield well during the hole exit, so as to ensure the safety of shield receiving. The construction method has the advantages of safe construction, short construction period and reusable, which has become an effective measure for shield safety reception.

2.2 Strata Reinforcement at Initiation and Arrival Sites

2.2.1 Soil Precipitation Technology

When the shield enters and exits the hole, due to the high pressure head of the deep groundwater in the foundation pit, it is prone to inrush or piping. In order to ensure the smooth progress of deep foundation pit excavation and underground construction, it is necessary to reduce the pressure water of the site, and when the foundation pit soil is saturated silty clay, it is not convenient for earthwork excavation and transportation. The precipitation technology should be used to drain the water in the silty clay stratum, so as to reduce the water content of groundwater in the excavation soil layer and meet the requirements of earthwork excavation and transportation.

Artificial lowering groundwater level refers to the construction technology of burying a certain number of filter pipes (wells) within the scope of construction, extracting water from wells with pumping equipment, and lowering groundwater level to the water level favorable for construction. In the process of construction, continuous pumping is still maintained, so that the soil of the working face is always dry, and the occurrence of flow sand is fundamentally prevented. At the same time, due to the extraction of water in the soil, the hydrodynamic pressure is reduced or eliminated, and the vertical surface of the upper body is more stable.

The use of drainage method generally requires ground downward drilling points, so its scope and area are limited. However, in the stage of shield construction, drainage method is often used as the main method. At present, the well point drainage method in foundation engineering is widely used in the soil drainage method of shield tunnel entrance and exit, and it is also the most commonly used method in foundation pit excavation drainage, so it is also called "Well point drainage method".

Taking the precipitation at the end of the shield arrival section of Nanjing Yangtze River Tunnel as an example, the precipitation at the end well point is used to further ensure the smooth arrival of the shield and prevent accidents such as sand flow and piping at the entrance of the tunnel. Before the shield reaches the reinforcement area, the groundwater level near the entrance of the tunnel is reduced to 1 m below the shield tunneling surface. Precipitation construction avoids freezing reinforcement and freezing stage to prevent water flow from affecting freezing effect. After the shield is completely separated from the reinforced soil and the door sealing grouting reinforcement is completed, the drainage operation is stopped. Figure 2.5 show the layout of precipitation wells.

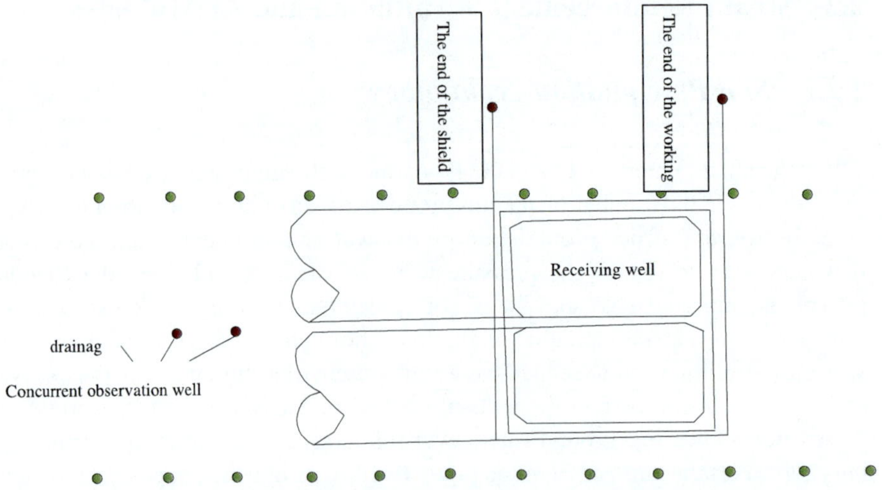

Fig. 2.5 The layout of drainage wells at the end of Nanjing Yangtze River Tunnel

At present, the commonly used well points are light well point, spray well point, electroosmotic well point, deep well point, etc. The specific method should be selected according to the permeability coefficient of soil, the depth of required water level reduction, engineering characteristics, equipment conditions and field construction conditions, as shown in Table 2.1.

Scope of application of drainage method: to prevent mud gushing or sand flowing in shield well construction. During the construction of shield tunnel, the soil at the excavation face is stabilized, and the mud leakage and water leakage at the shield tail are also prevented. The well points are arranged on both sides of the shield axis. Well point precipitation is particularly suitable for the construction of shield tunnels. In the use of building-intensive areas, its impact on surrounding buildings must be fully considered.

2.2.1.1 Light Well Point Drainage

Light well point drainage is now widely used in various projects, especially in coastal or high water level soft soil foundation. The use of light well point precipitation has good economic and technical effects, as shown in Fig. 2.6.

1) Characteristics

The light well points are flexible in layout, easy to use, fast in construction, and high in precipitation efficiency. Even if individual well pipes are damaged, the whole system will not be affected, and the project can adapt to the change of construction conditions. Light well point equipment can be used repeatedly with low construction cost.

Table 2.1 Common well point precipitation methods

Well point type	Range of application	Merits and drawbacks	Preparation note
Light well point (multi-stage light well point)	Suitable for foundation pit drainage with permeability coefficient of 10^{-2} – 10^{-5} cm/s and shallow excavation depth	While lowering water level, it can reinforce soil, Low cost, but not deep enough	Clay layer not applicable
Spray well point	Suitable for soil layer with high water content and permeability coefficient of 10^{-3} – 10^{-6} cm/s, especially sandy soil layer	Small amount of water, usually used to assist precipitation	The depth of the water level that may be lowered is 8–20 m
Electroosmotic well point	Suitable for silt and clay formation precipitation with rock permeability coefficient $<10^{-6}$ cm/s	More expensive and complex process, general foundation pit should not be used	The electroosmotic well point should be used in conjunction with other forms
Deep well precipitation	Suitable for high permeability formation with permeability coefficient $\geq 10^{-5}$ cm/s	The precipitation effect is significant, and deep precipitation can be carried out. However, the layout and construction of precipitation wells have large workload and long construction period	It can be used in silt layer and silty clay layer after vacuum negative pressure

2) Scope of application

Because the pumping unit of the light well is placed on the ground and the vacuum is produced on the ground, the precipitation depth is limited by the vacuum suction range. The buried depth of the first light well point drainage well is generally about 6 m. The soil layers suitable for light well sites are silty sand, sandy silt and clayey silt.

3) Process principle

Light well point drainage is that the well point branch pipe with filter is drilled into the soil layer, and sand is filled around the branch pipe, and then all the well point branches are connected to the pumping unit on the ground through the horizontal water collection pipe, and the groundwater is sucked to the ground by the pumping unit and excluded.

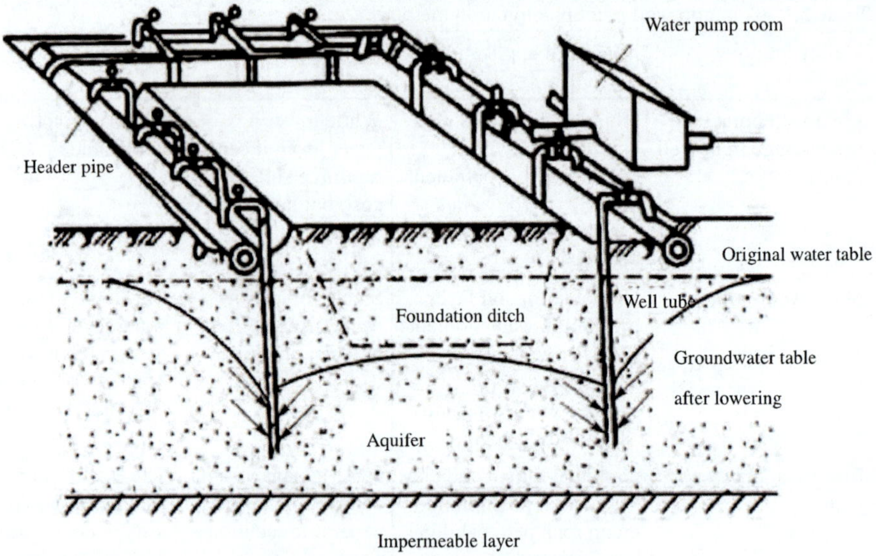

Fig. 2.6 Schematic diagram of light well point drainage

4) Technological process

Light well point drainage technological process as shown in Fig. 2.7.

2.2.1.2 Spray Well Point Drainage

1) Characteristics

The spray well point can effectively reduce the groundwater level of 8–20 m. Spray well point can meet the requirements of deep drainage during caisson and shield construction. The effect of spray well point precipitation is significant, and it has good economic effect in weak permeability strata, which has been widely used in underground deep foundation construction.

2) Scope of application

Because the spray well point uses high pressure water (or compressed air, water vapor) as the working medium, the vacuum is produced by jetting in the injector under the well tube, and the groundwater is inhaled into the well tube and the working medium, and then the groundwater is sent to the ground by the pressure of the working medium. Therefore, the water absorption capacity is large, and the precipitation depth depends on the pressure of the working medium and is not limited by the vacuum dust absorption. The precipitation depth can be increased by increasing the pressure of working medium. So the depth of primary drainage in general spray well point can reach about 20 m. The suitable soil layers for injection wells are silty sand, sandy silt, clayey silt and silty clay.

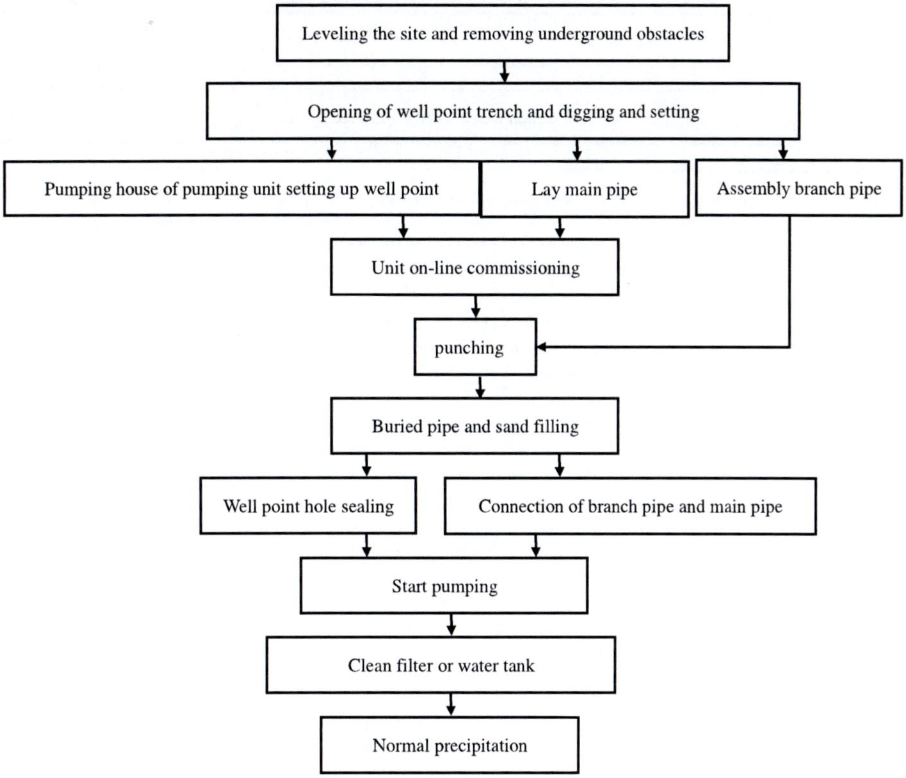

Fig. 2.7 Process flow of light well point drainage

3) Process principle

Spray well point drainage is to drill holes in the soil around the required precipitation position, put the double-layer well pipe with filter and ejector into it, fill sand around the well pipe, and connect it with the inlet and outlet main pipe, so that the high-pressure working water output by the unit (generally high-pressure water pump) enters the well pipe ejector. The vacuum produced by the ejector is inhaled by the filter and mixed with the working water into the water tank, and then the excess water is overflowed by the water tank to achieve the purpose of removing groundwater.

4) Technological process

The technological process of spray well point drainage is shown in Fig. 2.8.

5) Spray well point equipment.

Spray well point equipment mainly includes: high pressure centrifugal pump, well point pipe, filter pipe, circulating water tank, inlet pipe, drainage pipe and low pressure centrifugal pump.

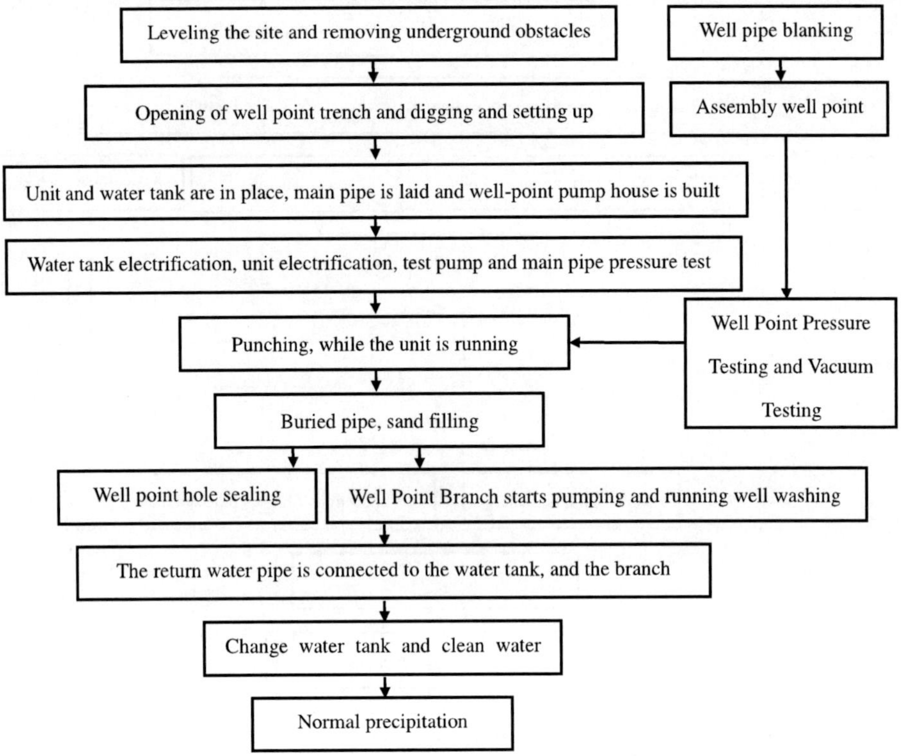

Fig. 2.8 Technological process of spray well point drainage

Well point pipe includes: inner pipe, outer pipe, jet water heater and filter pipe. Spray water lift is composed of inlet window, nozzle, mixing chamber (also known as throat), diffusion tube, etc.

The spray well point spacing is generally 2–3 m, and the well placement is similar to the light well, but the punching diameter should be 40–60 cm.

2.2.1.3 Electroosmotic Well Point Drainage

In saturated clay soil, it is difficult to discharge groundwater by general precipitation method because of poor water permeability (permeability coefficient <0.1 m/d), so it is necessary to discharge groundwater by electroosmosis well point.

The free water flowing in the soil layer under the action of gravity can be extracted from the well points except the electroosmotic well point. However, a large number of bonded water in the cohesive soil will not move under the action of gravity, so the free water is in a state of segmentation in the clay. Its flow is hindered by the bonded water, so the flow is also very slow. If two electrodes are inserted in saturated clay and direct current is applied, the water in the soil at the anode will move to the cathode, and the fine particles in the soil will move to the anode. The former is called

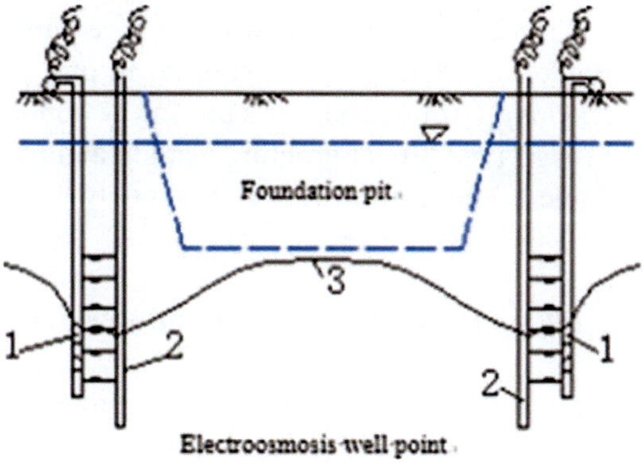

Fig. 2.9 Schematic diagram of drainage at electroosmotic well point. 1-Well point pipe; 2-Metal rod; 3-Underground water fall curve

electroosmotic phenomenon, and the latter is called electrophoresis phenomenon, as shown in Fig. 2.9.

The electroosmotic well point is to reduce the groundwater level in the clay layer by electroosmotic principle. Generally, the light well point is used as the cathode, and the corresponding anode is inserted next to it. The anode can be reinforced, with a diameter of 20 mm and a spacing of 0.8–1 m. Cathode is 20–40 cm deeper than anode, using no more than 60 V direct current as electroosmosis well point voltage, common is using welding machine instead of transformer (voltage 40 V, current 600 A).

2.2.1.4 Deep Well Point Drainage

The biggest feature of the deep well pump is to put the pump in the well pipe and rely on the head of the pump to send the deep groundwater to the ground, so the water level can be reduced to 30–40 m or more. Deep well point is one pump per well, working independently. There are two types of deep well pumps.

(1) The motor in the ground deep well pump, mainly consists of three parts: pump body is vertical multistage centrifugal pump, water pipe. The bottom is a filter. Others include water pipe, motor drill, transmission shaft and filter head.
(2) The shape and structure of deep well submersible pump are similar to those of common submersible pumps. It is mainly a submerged deep pump with large head and good insulation.

2.2.2 *Soil Reinforcement Method*

In the past, grouting method was the main method for stratum reinforcement at the starting and arriving parts of shield tunnel. However, in recent years, with the increase of buried depth and section of shield tunnel, high strength reinforcement is urgently needed. In addition, due to the time-consuming removal of underground walls such as continuous walls, it is required that the ground reinforcement can be stable for a long time. Therefore, in order to make the strata reach high strength and long-term stability, and be closely bonded with the retaining wall, the examples of high-pressure jet grouting method and freezing method are gradually increasing. Table 2.2 compares the four commonly used reinforcement methods.

2.2.2.1 Deep Mixing Pile Reinforcement Method

Deep mixing pile is one of the commonly used methods for soft soil foundation reinforcement and lateral support of deep foundation pit excavation. As early as in the 1970s, Japan pioneered and used in steel works port quay wall, highway engineering. Since 1977, China began to develop and trial, then in 1980 by the Ministry of Metallurgy presided over by the ministerial technical appraisal, applied to the foundation reinforcement project.

1) Technical characteristics

   (1) The solidified pile and the original foundation form a composite foundation to improve the bearing capacity and deformation modulus of the foundation.
   (2) Self-supporting retaining without support and anchor.
   (3) After the pile is connected to the wall, it has the effect of waterproof curtain.
   (4) There is no vibration, no noise and no pollution in the construction, which has little influence on the surrounding buildings and underground pipelines.
   (5) The construction equipment is simple, easy to operate and low cost, which creates better conditions for civilized construction, especially in small construction sites.

2) Scope of application

   (1) Soft soil foundation reinforcement, including shield tunnel soil reinforcement.
   (2) The lateral retaining structure has a good protective effect on adjacent buildings.
   (3) Curtain engineering of waterproof and anti-flow sand.

3) Process principle

Using deep mixing machinery, cement is used as curing agent and ground soil in-situ forced crushing and mixing. After curing, different shapes of piles, walls or blocks are formed. The calculation theory is calculated according to the gravity dam rigid retaining wall, and the deformation is checked according to the rigid retaining wall calculation method.

Table 2.2 Comparison of four common reinforcement methods

Reinforcement method	Reinforcement principle	Scope of application	Merits and drawbacks
Deep mixing pile reinforcement method	Using cement (or lime) and other materials as curing agents, through special mixing machinery, deep in the foundation The soft soil and curing agent (slurry or powder) are forced to mix locally to make the soft soil harden into cement reinforced soil with integrity, stability and certain strength, so as to improve the foundation strength and anti-seepage performance	It is suitable for dealing with silt, sandy soil, silty soil, peat soil and silt strata. It is commonly used in deep foundation pit reinforcement, deep hole reinforcement, engineering pile, bearing pile, uplift pile, foundation pit retaining structure, foundation pit seepage curtain, etc.	The construction is simple, the project cost is small, and the pile diameter is various, but the bearing capacity of the mixing pile is small, and the strength of the pile is low
High pressure jet grouting pile reinforcement method	After the grouting pipe with nozzle is drilled into the predetermined position of the soil layer by the drilling rig, the slurry is sprayed out from the nozzle by the high pressure equipment, punching, disturbing and damaging the soil. At the same time, the drill pipe is gradually raised at a certain speed, and the slurry is forced to mix with the soil particles. After the slurry is solidified, a cylindrical consolidation body (i.e. rotary jet pile) is formed in the soil to achieve the purpose of strengthening the foundation or preventing water and seepage	It is mainly applicable to the reinforcement of engineering properties of loose permeable strata such as muddy soil, silty clay, silt, sand, collapsible loess, gravel and pebble. Not suitable for strata with rich water and large groundwater flow	The equipment is simple, lightweight, simple construction, easy operation, strong adaptability to soil, high strength, stable pile quality. However, the construction efficiency is low, the cost is relatively high (about 6 times that of the mixing pile), and the consumption of cement is large

(continued)

Table 2.2 (continued)

Reinforcement method	Reinforcement principle	Scope of application	Merits and drawbacks
Chemical grouting reinforcement method	Forced filling of certain chemical solutions into gaps or crevices of surrounding rock, forming cementitious materials through chemical reactions or activating the surface of soil particles, cementing and curing at the contact point to enhance the bonding between soil particles and improve the mechanical strength of soil	It is mainly suitable for fine sand, silt, gravel and rock fissures and other strata	The stratum are highly stable and effective in a short period of time, and the cost is relatively expensive. Generally, it is used in emergency situations such as engineering emergencies
Freezing reinforcement method	Artificial refrigeration technology is used to freeze the water in the stratum, transform loose water-bearing rock and soil into frozen soil, increase its strength and stability, and isolate groundwater, so as to carry out excavation and masonry work of underground engineering under the protection of frozen wall	It is suitable for construction of various strata, especially in sections with difficult construction conditions of urban underground works, which is generally adopted when other construction methods are difficult to achieve the purpose of surrounding rock stability	It can effectively isolate groundwater, high strength of frozen soil, and no pollution to the surrounding environment. After freezing, the frozen wall melts and does not affect the underground structure around the building. But high cost, long construction period

4) Construction flow

The construction site should be smoothed, rolled or compacted to ensure the positioning and movement of machine and the vertical drilling.

(1) The deep mixer is in place.
(2) Stirring and sinking. Start the motor, determine the calculation rate according to the soil quality, relax the hoist to cut the soil from top to bottom and mix and sink until the design depth.
(3) Grouting mixing and lifting. When the cement slurry reaches the mixing head, lift the mixing head, grouting, mixing and lifting according to the speed required by the calculation, so that the cement slurry and the original foundation soil can be fully mixed until the pump is closed after it has been lifted to the design elevation of the pile top.
(4) Repeat stirring down. Sink the mixer to the design elevation again.
(5) Repeated stirring lifting (no grouting). When stirring and lifting to the natural ground, one pile is completed by closing the mixer.

5) Key points of construction

(1) All underground obstacles must be identified and removed before start-up, and the parts that need to be backfilled must be filled and tamped in batches to ensure the quality of the pile.
(2) The driving track and sleeper of the pile machine shall not sink, and the vertical deviation of the pile machine shall not be greater than 1%.
(3) The cement should be 32.5 grade ordinary Portland cement, the cement mixing ratio should be 8%–16%, and different types of admixtures can be added according to different geological conditions and construction period requirements.
(4) The amount of grouting and lifting speed must be strictly controlled to prevent the occurrence of sandwich layer or pulp breakage.
(5) The speed of the two elevations of the mixing head should be controlled at the outlet pressure of the 2.5–3 m/min, grouting pump in the range of 0.4–0.6 MPa.
(6) The following matters should be taken into account in the work to be overlapped with the pile:
① The overlapping time between piles shall not be more than 24 h. If it is more than 24 h, attention should be paid to increasing the grouting amount during the construction of the second pile, which can increase by 20% and slow down the lifting speed at the same time. ② If the interval time is too long, the second pile cannot be overlapped, local pile replacement or grouting measures shall be adopted with the approval of the design.
(7) The pre-mixing sinking of the pile machine shall ensure that the structure of undisturbed soil shall be fully crushed according to the condition of the original soil, so as to facilitate the uniform mixing of the same cement slurry.

(8) The standard water tank is adopted, the water-cement ratio is strictly controlled, the mixing time of cement slurry is not less than 2 × 3 min, the slurry is poured into the collecting tank after filtration, and then the cement is continuously stirred to prevent cement segregation. The grouting should continue and cannot be interrupted.
(9) Each class must be a group of test blocks (3 blocks). After 28 days, the unconfined compressive strength should be measured and the design mark should be reached.

6) Quality standards

In addition to referring to the relevant national standards, the following points should be noted:

(1) The vertical deviation of pile is not more than 1%, and the deviation of pile position arrangement is not more than 50 mm.
(2) The mixing pile should be mixed, the surface should be dense and smooth.
(3) Pile top elevation and pile depth should meet the design requirements.
(4) There is no abnormal and excessive amount of cement slurry filling.

2.2.2.2 High Pressure Jet Grouting Pile Reinforcement Method

High pressure jet grouting pile method has single pipe method, double pipe method, triple pipe method and multiple pipe method in recent years. It plays a good role in foundation reinforcement, improving foundation bearing capacity and improving soil quality for wall protection, retaining soil and water isolation. Figure 2.10 shows the high pressure jet grouting pile reinforcement site.

1) Technical characteristics

(1) The high pressure rotary jet grouting pile method can be used to reinforce a certain depth of strata.
(2) The soil which cannot be filled with slurry cannot be strengthened in the fine particle soil layer with small permeability coefficient, and the slurry filling is uniform and the range can be adjusted and controlled.
(3) The soil can be strengthened when the distance between the public pipelines is small or the structure has only a small slit.
(4) Combined with fixed spray method, vertical water partition wall, horizontal water partition wall or closed water insulation curtain can be effectively formed.
(5) It is convenient to use and flexible to move, which can not only form a single row pile body, but also form a multi-row pile body, and the pile diameter can be properly adjusted.
(6) The discharge mud can be recycled, the construction environment can be improved, and the transportation cost can be saved.

Fig. 2.10 Reinforcement by high pressure rotary jet grouting pile

2) Scope of application

Suitable for sand, clay, silt and artificial fill.

3) Process principle

The engineering drilling rig is drilled to the design depth, and a certain pressure of cement slurry and air are sprayed at the same time through the special nozzle on the end side, and forced to mix with the sprayed slurry to cement and hardening. At the same time, the grouting pipe is rotated and lifted at a certain speed, that is, the mixing and solidifying with obvious diameter is formed in the soil.

The pile superimposed to form a wall with water insulation and retaining soil.

4) Construction process

(1) Drilling rig in place. It mainly refers to that grouting pipe can be installed only after drilling. If the grouting pipe is directly penetrated or sunk, it is not necessary to drill, but its verticality must be ensured. Intubation must take place immediately after drilling, not too long interval.

(2) Rotating spray lifting. Both slurry and gas must be sprayed at the same time. The lifting speed is closely coordinated with the slurry flow rate, so as not to meet the design requirements of pile diameter and pile mass, and at the same time, the lifting speed must be matched with the rotating speed.

(3) Cleaning of machinery and tools. Be sure to wash the grouting pipe, when all completed or phased pause, to mix slurry, grouting equipment

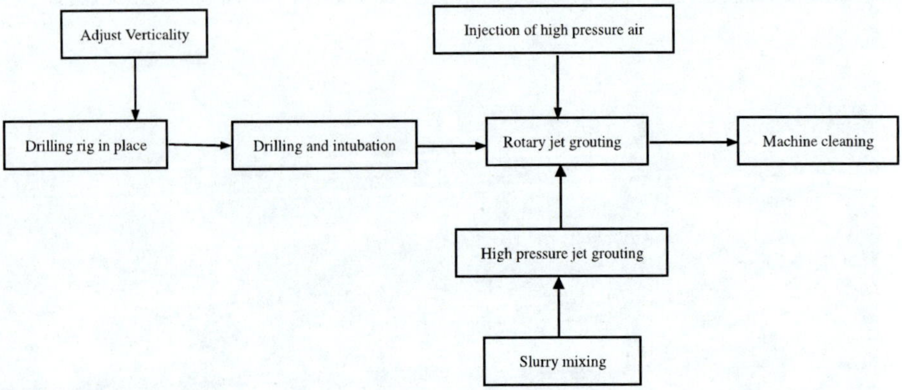

Fig. 2.11 Construction flow of high pressure jet grouting pile method

to do cleaning. The concrete construction flow of high pressure rotary jet grouting pile method is shown in Fig. 2.11.

5) Construction requirements

(1) Cement: the types of visual slurry (common type, early strength type, impermeable type, etc.) are selected, but slag cement is not required to be moisture-free, not caking, freshness and various indexes in accordance with national regulations, and slag cement is not suitable for impermeability requirements.
(2) Water for pulp mixing: clean, no deterioration, no corrosion, moderate acidity and alkalinity. Generally speaking, river water, well water and other clean water can be used.
(3) Admixture: in order to improve the performance of slurry, promote its early strength, improve impermeability and blowability. The admixtures available are as follows.
   ① Promoting coagulation, early strength, adding 2%–4% calcium chloride or 3%–5% triethanolamine, 1% salt and 2%–3% calcium chloride.
   ② To improve the impermeability, the initial setting is fast and the final setting time is long. 2%–4% sodium silicate and 10%–50% bentonite can be added.
(4) Slurry: the mixture is required to be uniform, the spray is good, and there should be no blockage of pipes and nozzles. To control the use within the initial setting time, do not time out.

6) Quality standards

(1) Excavation inspection: when the slurry reaches a certain strength, the soil can be excavated to check the verticality of the pile, the bonding condition between the piles and the diameter of the pile.

(2) Drilling inspection: the core samples were taken from the center of the pile body which had been rotated and consolidated, and the physical and mechanical properties of the standard specimens were made into standard specimens for indoor physical and mechanical properties test.
(3) The permeability coefficient of pile is measured by pressure water injection or pumping test.
(4) Standard penetration test: standard penetration test was carried out in the middle of rotary spray consolidation.
(5) Load test: do vertical or horizontal load test to determine its bearing capacity.

2.2.2.3 Chemical Grouting Reinforcement Method

Chemical grouting reinforcement is the chemical slurry with fluidity and gelation, according to a certain concentration, through the special grouting hole, compressed into the rock and soil. Slurry into the rock and soil cracks or pores, through chemical reaction to produce cementitious materials or make the surface of soil particles activated, cement solidification in the contact, in order to enhance the bonding between soil particles, improve the mechanical strength of soil, to play a reinforcement, seepage prevention, improve the physical and mechanical properties of foundation. The commonly used reinforcement methods include silicification reinforcement, alkali solution reinforcement, electrochemical reinforcement and polymer chemical reinforcement.

1) Silicification reinforcement method

By inserting a metal perfusion tube with holes, sodium silicate (commonly known as sodium silicate) solution was injected into the soil under certain pressure, or sodium silicate and calcium chloride solutions were successively injected into the soil. The former is called single liquid silicide and the latter is called double liquid silicide.

Single liquid silicification is suitable for strengthening collapsible loess with permeability coefficient 0.1–2.0 m/d and silt with permeability coefficient 0.3–5.0 m/d. When the collapsible loess is strengthened, the solution is composed of 2.5% sodium chloride mixed with 10 and 15% sodium silicate solution. After the solution is inserted into the soil, the sodium ion reacts with the calcium ion (mainly calcium sulfate) in the water-soluble salt of the soil, and forms silicic acid gel between the soil particles and their surface, which can make the unconfined ultimate compressive strength of loess reach 0.6–0.8 MPa. When the silt is strengthened, a certain amount of phosphate (specific gravity is 1.02) is added to the sodium silicate solution with lower concentration (specific gravity is 1.18–1.20). After uniform stirring and injection, the unconfined ultimate compressive strength can reach 0.4–0.5 MPa after chemical reaction.

Double liquid silicification is suitable for strengthening sandy soil with permeability coefficient of 2–8 m/d or for preventing seepage and stopping water to form impermeable curtain. The specific gravity of sodium silicate solution is 1.35–1.44, and the specific gravity of calcium chloride solution is 1.26–1.28. After contact with soil, the two solutions produce not only general chemical reactions, but also colloidal chemical reactions, which generate silica gel and calcium hydroxide. In the adjunctive reaction, the product can also enhance the adhesion between soil particles and fill the pores. The unconfined ultimate strength of sand soil can reach 1.5–6.0 MPa.

The reinforcement radius achieved by silicification method is related to the permeability coefficient of soil, perfusion pressure, filling time and viscosity of solution, which can be determined by single hole perfusion test. Each filling hole should be arranged according to the apex of equilateral triangle on the plane, and the hole distance can be 1.7 times of the radius of reinforced soil. The reinforcement depth is determined according to the soil condition and the requirements of the building, and is generally 4–5 m.

The modulus of sodium silicate is usually 2.6–3.3, and the impurity content insoluble in water does not exceed 2%. This method must consume sodium silicate or calcium chloride and other industrial raw materials, the cost is high. The advantages are that the deformation of the foundation can be suppressed quickly and the strength of the soil has been greatly improved, which is especially suitable for the reinforcement of the existing building foundation. However, it is not suitable to strengthen the foundation soil with pH value greater than 9 for petroleum products, gum, oil and groundwater.

2) Alkali reinforcement method

The reinforcement effect of lye on soil is different from that of other chemical reinforcement methods. It does not precipitate cementitious substances from the solution itself, but after the chemical reaction between lye and soil, the surface of soil particles is activated and self-bonded, so as to enhance the mechanical strength and water stability of soil. In order to promote the reaction process, the solution temperature can be raised to 80 °C and then injected into the soil. When strengthening collapsible loess foundation, the solution is generally infiltrated into the soil through the filling hole. The content of calcium and magnesium ion in loess is high, and good reinforcement effect can be obtained by using single liquid.

3) Electrochemical reinforcement method

A certain number of metal poles are introduced into the foundation soil, and the DC current is introduced through the electrode, so that the water is discharged from the cathode and the soil is consolidated. When the foundation is strengthened by electrochemical method, there are three main processes: ① Electroosmosis, after electroosmosis, a large number of soil is dehydrated and consolidated; ② Ion exchange, the adsorbed sodium and calcium are replaced by hydrogen and aluminum; ③ in the process of structure formation, the soil particle structure is formed by aluminum gel, and the soil can also be strengthened by the combination of current and chemical solution, that is, the chemical solution can be injected into the soil through the perfusion pipe network with holes, and the solution diffuses from anode to cathode with the movement of water after electrification, so as to improve the reinforcement effect.

Electrochemical method is generally used to strengthen silt foundation with permeability coefficient less than 0.1 m/d. However, this method is expensive and must be tested by special equipment and confirmed to be effective before it is adopted.

4) Polymer chemical reinforcement method

It is a method of ground treatment by pressing polymer chemical solution into soil. It is suitable for reinforcement of sand soil foundation, curtain grouting and water stop and leakage of underground works; the reinforcement of the mud interlayer and fault fracture zone of dam foundation works is also effective. For example, the compressive strength after the cyanide is poured into the sand soil can reach 10.0 MPa.

There are many kinds of polymer materials used in foundation reinforcement, such as urea–formaldehyde resin, acrylamide (also known as propioning), polyurethane (also known as polycarbamate or cyanide), among which polyurethane is better. Polyurethane is made of polyisocyanate and polyether resin as the main materials, and then adding Plasticizer, diluent, Surfactant, Catalyst and so on to form slurry. After pouring into the formation, the gel is formed when the water is encountered, which plays the role of strengthening the foundation and preventing seepage and plugging leakage. In the late 1960s, the TACSS grouting material first developed by Japan and the cyanide condensation material developed by China in the early 1970s were obtained by the reaction of excessive isocyanate with polyether, which was called prepolymer. The prepolymer contains a certain amount of free isocyanate (NCO) can react with water. When the slurry is poured into the soil, the NCO base is further polymerized and crosslinked under the action of catalyst, the viscosity of the reactant increases gradually and solidifies, and the polymer insoluble in water is formed to strengthen the foundation.

The characteristics of cyanide grouting: Since water is an integral part of the reaction, it is less likely that the slurry is diluted or lost by water. And the carbon dioxide gas released in the process of water reaction makes the slurry expand, diffuse around, and expand the scope of reinforcement. Polymer materials are expensive, limiting their use. Cyanide coagulation is highly toxic, anti-toxic measures should be taken in construction, and environmental pollution should be considered.

Polyurethane pulp is divided into water-soluble and insoluble, the former has low strength, but has good impermeability, the latter has high strength, good seepage prevention and widely used in engineering. Among them, the two-step method (also known as prepolymerization) is the best pulping method, which is to synthesize the main agent into polyurethane prepolymer first, and then to mix the prepolymer and admixture into slurry according to the need.

The characteristics of polyurethane slurry are as follows:

(1) The slurry has low viscosity, good filling property and high strength of stone, which can be combined with cement grouting to establish reliable anti-seepage curtain.
(2) After the slurry is treated with water, the viscosity of the slurry increases rapidly and will not be diluted and washed away. It can be used for waterproofing and leakage under the condition of moving water, blocking various forms of

underground, ground and pipeline leakage, sealing firmly, and stopping water quickly.

(3) Good durability, safety and reliability, no pollution to the environment.

(4) The operation is simple and the economic benefit is good.

5) Key points of construction

(1) Grouting method can be divided into single liquid and double liquid. The single liquid grouting method is to mix the components of the slurry evenly first, prepare them once, and then pump the grouting into the hole; The double liquid grouting method is to load the two kinds of slurry in their respective slurry grooves respectively, and then use the proportional pump or two grouting pumps to press the two kinds of slurry into the hole according to the proportion, so that the two liquids can be mixed in the mixer near the hole, and then sent to the hole section, or the double liquid will be uniformly and directly pressed into the hole and mixed in the hole.

(2) Grouting order: from the center to the outside, from the bottom up. The length of grouting section is generally about 5 m, if the size of cracks in grouting part is very uneven, the rock and soil are loose and heavy, and there are many cracks, then the length of the section is 1–3 m. The hole distance is generally 0.5–1.2 m, and the bottom end of the slurry pipe should be kept 0.3 m from the bottom of the hole. The grouting pressure is generally 0.05–0.2 MPa. For water curtain grouting, the surface section should not be less than 1–1.5 times water head, and the bottom should be 2 times 3 times water head.

(3) During grouting, attention shall be paid to stop grouting after grouting is no longer carried out, and the valve of the grouting pipe shall be closed to prevent backflow.

(4) After grouting, the pipe should be extubated in time, and the height of each extubation is 350 mm. Brush all equipment, pipe lines, and eliminate waste slurry pollution to the environment in order to keep it unobstructed and clean. Holes left by extubation are filled with cement mortar or earth.

6) Quality control

(1) Calculate the grouting quantity statistically and analyze the grouting effect by using the automatic flow and pressure curve in the grouting process.

(2) The mechanical index of soil before and after reinforcement is tested by static penetration test to understand the reinforcement effect.

(3) The field load test was carried out to determine the bearing capacity and deformation modulus of the strengthened soil.

(4) The mechanical energy of soil before and after reinforcement is measured in situ by standard penetration test or portable penetration test, or laboratory test is carried out to determine the physical and mechanical properties of soil before and after reinforcement, and to compare and judge the reinforcement effect.

(5) The test point is generally 2%–5% of the number of grouting holes. If the unqualified rate of the inspection point is greater than or equal to 20%, or less than 20%, but the average value of the inspection point does not meet the design requirements, after confirming that the design principle is correct, repeated grouting should be carried out in the unqualified grouting area.

2.2.2.4 Freezing Reinforcement Method

In subway construction, because of the abnormal congestion of the ground building, the complexity of the shallow pipe network on the surface, the soft and water-bearing strata of the subway tunnel, and the poor stability, it is necessary to study a reliable method of formation reinforcement. In subway construction, freezing reinforcement technology has achieved some results, which will be further popularized and applied.

Frozen reinforcement method (hereinafter referred to as freezing method) refers to a special method of geotechnical construction for underground engineering construction under the protection of the frozen soil wall by freezing the soil surrounding the underground space to be excavated into ice and cemented with the soil, forming a frozen soil wall or a closed frozen soil body according to the design profile, which is used to resist the earth pressure, isolating the groundwater, and carrying out the underground engineering construction under the protection of the frozen soil wall (Fig. 2.12).

The method was first used for gold mining in Russia, and then German engineers used for coal mine construction to obtain patented technology. It has been widely used in subway, deep foundation pit, mine construction and other projects. Its advantages are good reinforcement effect, and the frozen soil wall can also be controlled

(a) Frozen formation (b) Construction by freezing reinforcement method

Fig. 2.12 Construction site of freezing reinforcement method before opening of tunnel door by temperature, which can ensure long-term stability. The discussion on the reinforcement range is the same as that of the high pressure jet grouting method, which is calculated according to the general temporary buildings.

1) Basic principles

The principle of ground consolidation by freezing method is to use artificial refrigeration method to send cryogenic refrigerant into the stratum, freeze the stratum around the excavation body into a closed continuous frozen soil wall to resist pressure and isolate the connection between groundwater and excavation body, and then excavate and make permanent support under the protection of the closed continuous frozen soil wall. It is a special method of stratum reinforcement. The coolant entering the formation is connected with the formation through the inlet and return pipeline, and the heat exchange is carried out with the formation through the frozen pipe, and the cooling capacity is transferred to the surrounding formation, and the heat in the formation is taken away. As a result, the strata around the freezing pipe are cooled from near to far, and the water in the formation is gradually turned into ice, and the original loose or porous strata are glued together through ice to form impermeable permafrost columns. Several such frozen tubes are arranged and the refrigerants in the frozen tubes are recycled to freeze the soil around the frozen pipes into permafrost columns. With the continuous expansion of the radius of the permafrost column, the adjacent permafrost columns will be connected and tightly bonded with each other through ice to form a sealed continuous wall.

2) Refrigeration system

Compression refrigeration consists of three cycles: ammonia cycle, brine cycle and cooling water cycle.

(1) Ammonia circulation system.

Ammonia cycle plays a leading role in the refrigeration process. In order to transfer geothermal energy to cooling water and then release it to the atmosphere, saturated steam ammonia in evaporator must be turned into superheated steam at high pressure and high temperature, resulting in temperature difference with cooling water, heat transfer to cooling water in condenser, and superheated steam ammonia condensation into liquid ammonia to realize the transformation from gas to liquid. Liquid ammonia evaporates into evaporator through throttle valve, and then absorbs heat from brine around it to saturated steam ammonia. In this way, over and over again, constitute the ammonia cycle.

Ammonia circulating system equipment consists of evaporator, ammonia compressor, condenser and throttle valve.

(2) Brine circulation.

Brine cycle plays an important role in cooling capacity transfer. The circulation system is driven by the flow of salt water. The freezer is a heat exchanger in which low temperature brine exchanges heat with the formation. The faster the flow rate of brine is, the greater the heat transfer intensity is. The freezer consists of a freezing tube, a liquid supply tube and a return pipe. According to the needs of the project, two kinds of brine circulation system can be used, and the positive circulation can be used to supply liquid under normal conditions. During the positive freezing period, the temperature difference between the inlet and outlet of the freezer is generally 3–7 °C, and during the negative freezing period, the temperature difference between the inlet and export of the freezer is 1–3 °C. The evaporation temperature of ammonia in evaporator is 5–7 °C different from that of brine around it. The endothermic rate of the surface of the freezer is unit time, and the heat absorption per unit area is 263–292 W/m^2. In order to observe the leakage of the brine in the freezing pipe, the meter shall be installed at the inlet and outlet of the demineralized water main pipe and the freezing unit.

The brine cycle described above is referred to as a closed-water circulation system. A kind of open circuit return brine circulation is also used in foreign countries, the main characteristic point of which is no liquid collecting ring, and each return pipe returns liquid separately, which is convenient to observe whether each frozen pipe salt water is leaking or not. In this way, the amount of brine passing through the loop is large, and the closed circuit cycle is more complex. The brine pipeline should be treated strictly with heat preservation. In general, the heat loss of the brine pipeline accounts for about 25% of the total refrigerating capacity of the freezing station.

(3) Cooling water circulation.

The function of cooling water cycle in the refrigeration process is to cool the superheated steam of the compressor into liquid ammonia in order to re-evaporate into the evaporator. Cooling water releases heat from ammonia vapor to the atmosphere. The lower the cooling water temperature is, the higher the refrigeration coefficient is, but the cooling water temperature is generally 5–10 °C lower than that of ammonia. The cooling water is driven by the pump, heat exchange is carried out through the condenser, and then flows into the cooling tower pool. The circulating water after cooling should be replenished by groundwater at any time.

3) Construction characteristics of freezing method

Freezing method is mainly used in winter construction. In addition to freezing method, admixture method (salt mortar method) should be preferred in winter construction. When using freezing construction, the thickness of frozen soil wall, the spacing of frozen pipe group and the length of frozen pipe, freezing sequence and thawing sequence should be determined according to the construction progress, the strength requirement of frozen soil wall and the excavation sequence, etc., according to which the appropriate freezing equipment should be selected. In addition, it is necessary to measure the temperature of the representative parts in the construction, and according to the measured results, it is determined whether to maintain the freezing of the frozen soil wall in the form of continuous or intermittent cooling; At the same time, the expansion of formation freezing and the subsidence of thawing should be observed in time and the corresponding treatment measures should be taken.

Install refrigeration equipment on the ground, use ammonia (NH_3) as refrigerant, cool the solution of calcium chloride ($CaCl_2$) of refrigerant (commonly called brine) to −20 to −30 °C, and send the brine into the frozen pipe by circulating pump and PVC liquid supply pipe inserted into the depth of the frozen pipe. The heat outside the tube is continuously absorbed by low temperature brine for a long time to freeze the surrounding strata. After the brine absorbs the heat of the formation, the temperature rises, and under the action of the circulating pump, it returns to the refrigeration equipment and recools through the loop tube. The original liquid ammonia, after evaporating under the condition of decompression, absorbs the heat of brine, compresses and condenses it and liquidates it, circulates and flows in the pipeline, and reuses it. The diameter of frozen soil cylinders formed around each frozen pipe increases with time. These cylinders intersect into dense and closed frozen soil walls, which can withstand water, earth pressure and block groundwater, excavate strata and build lining under their protection.

Freezing construction has the following characteristics:

(1) Sealing water. There is free water (in general, the water content should be more than 10%, otherwise the auxiliary work method to increase the humidity of soil layer) can freeze into frozen soil and form frozen soil wall. Whether it is permeable layer or water insulation layer, frozen soil wall can block groundwater invasion and form a dry construction environment.
(2) High strength. It is generally considered that frozen soil is a kind of Visco-elastic–plastic material, its strength is related to soil quality, bulk density, moisture content, salt content and temperature, which can reach 2–10 MPa, which is much greater than the strength of molten soil, thus playing the role of structural support wall.
(3) Strong adaptability. It is suitable for all kinds of soil layer and many kinds of underground projects, especially for underground projects with large water content, weak strata and difficult or unable to be constructed by other construction methods.
(4) Reproducibility. At the end of the construction, the soil layer is restored to the original state, and the damage to the soil layer is very little. This is impossible to compare with other methods of work.
(5) Avoid obstacles. It has the ability to circumvent obstacles, freeze reinforcement and seal water.
(6) Pollution-free. Electricity energy in exchange for cold energy, does not pollute the atmospheric environment, no harmful substances discharge, no pollution to groundwater. In the high environmental protection requirements of the project, its advantages are particularly obvious.
(7) Controllability. Freezing duration, frozen wall thickness, frozen wall shape and so on can be controlled.
(8) Applicability. It can be constructed under dense building area and existing engineering building without foundation pit drainage, which can avoid the adverse effect of foundation settlement caused by pumping water on the surrounding buildings.

(9) The construction is convenient, no support, no anchor, open construction can be carried out and the construction area can be expanded, and the construction period can be shortened.

4) Freezing method is more mature technology in shield construction

(1) Selection of refrigerant. Calcium chloride solution with wide temperature range, large heat capacity, abundant sources and cheaper price is usually chosen. It is the most widely used refrigerant in refrigeration engineering.
(2) Selection of the refrigerant. Freon is used as a refrigerant, and liquid nitrogen or dry ice is also used as the refrigerant. Wherein the liquid nitrogen can reduce the temperature of the freezing tube to −190 °C, while the liquid of the dry ice and the alcohol is up to −70 °C, so that the freezing tube can be formed in a few hours.
(3) Selection of refrigeration equipment. At present, in shield engineering, whether in, out of the hole reinforcement, or side channel construction, refrigeration capacity is easier to meet the requirements.
(4) The arrangement of freezing pipe. Frozen soil walls formed vertically are usually rectangular and plum blossoms. The arrangement of plum blossoms may be more reasonable, but it should be arranged flexibly according to local conditions, depending on the needs of the project.
(5) Calculation of frozen soil strength. The main factors affecting the instantaneous strength of frozen soil are temperature, moisture content, particle composition and freezing speed of soil.
① The effect of temperature: the strength and cementitious ability of ice increase with the decrease of temperature, so the strength of frozen soil also increases with the decrease of temperature.
② The effect of water content: when the water content of soil reaches saturation, the strength of frozen soil increases with the increase of water content; after the water content reaches saturation, the strength of frozen soil decreases with the increase of water content due to the precipitation of ice; when the water content is much higher than the saturated water content, the strength of frozen soil is about the same as that of ice.
③ The influence of particle composition of soil: When other conditions are the same, the thicker the soil particles, the higher the strength of frozen soil, and vice versa. This is mainly caused by the difference of bound water in soil caused by different particle composition. Coarse soil such as coarse sand, gravel and gravel, in which there is almost no bound water, once the soil is frozen, there is no unfrozen water, so the strength is high; on the contrary, the particles of clay are fine and the total surface area is very large, which contains more adsorbed water and thin film water, but the adsorbed water is generally not frozen. The film water is frozen under certain negative temperature conditions, which increases the viscosity and activity of frozen soil and decreases the strength.

④ The influence of freezing speed: the freezing speed is fast, the fine particle ice in frozen soil is more, the strength of frozen soil is high, on the contrary, it is low.

The durable compressive strength of frozen soil is about 1/2–1/2.5 of instantaneous compressive strength, the durable shear strength of frozen soil is about 1/1.8–1/2.5 of instantaneous shear strength, and the durable tensile strength of frozen soil is about 1/12–1/16 of instantaneous tensile strength.

2.2.3 Scope of Soil Reinforcement

In this section, taking the deep mixing pile construction method as an example, the method to determine the range of foundation reinforcement area is described in detail. Table 2.3 and Figs. 2.13, 2.14 and 2.15 are the detailed forms of the three construction methods of deep mixing pile.

1) Reinforcement thickness

The opening part of the strengthened entity is regarded as a freely supported disc. Based on the checking calculation of the stress failure of the simply supported beam,

Table 2.3 Detailed forms of three construction methods of deep mixing pile

Classification	Type A	Type B	Type B
Application condition	1. Foundation reinforcement 2. Water seal 3. Basement reinforcement	1. Foundation reinforcement 2. No water seal 3. Foundation reinforcement 4. Preset ground beam in the middle	1. The defective part of the retaining soil 2. Anti-seepage curtain
Basic interval	1.As a standard, for lap configuration 2. Horizontal interval $l_1 = \frac{\sqrt{3}}{2}D$ 3. Longitudinal interval $l_2 = \frac{3}{4}D$	1. Configuration of contacts as a standard 2. Horizontal interval $l_1 = D$ 3. Vertical interval $l_2 = \frac{\sqrt{3}}{2}D$	1.As a standard, for lap configuration 2. Effective thickness $t_0 = \sqrt[2]{(\frac{D}{2}) + (\frac{l}{2})^2}$ 3. Lap length $h = D - \sqrt{D^2 + t_0^2}$ $(h \geq 0.2m)$
Suit	1. Water seal; 2. Surge; 3. Sand boiling	1. Anti-uplift of foundation soil; 2. Ground beam	1. Water seal; 2. Temporary wall material
Verticality	Accuracy 1/250	Accuracy 1/250	Accuracy 1/250

Note D-Effective diameter of pile; *Z*-Pile depth

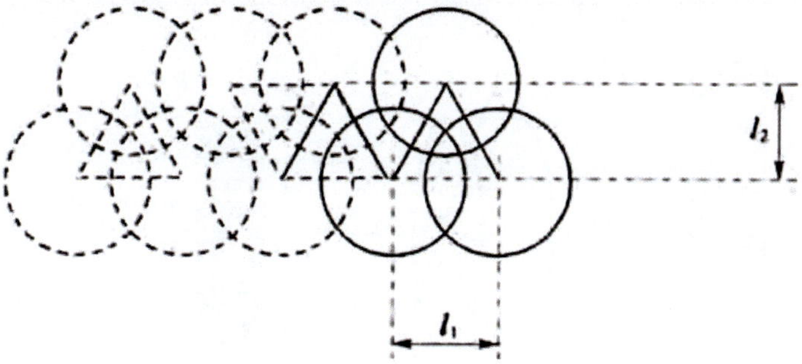

Fig. 2.13 Type A foundation reinforcement

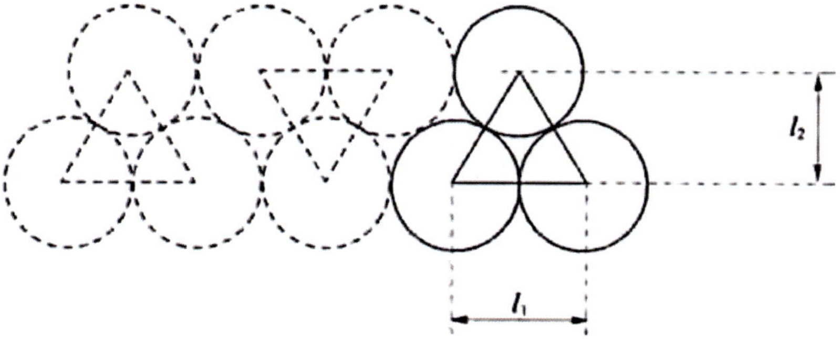

Fig. 2.14 Type B foundation reinforcement

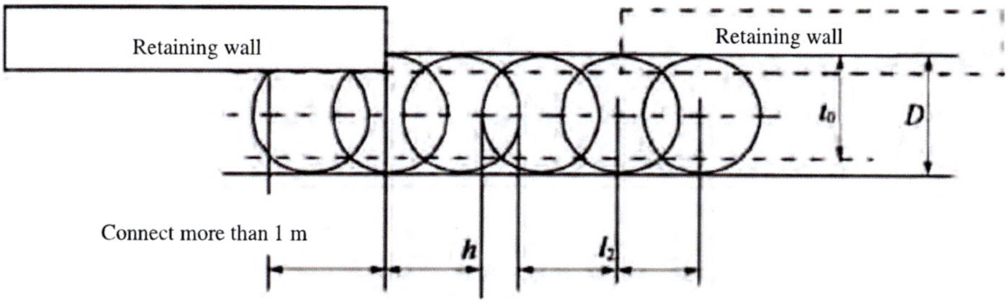

Fig. 2.15 Type C foundation reinforcement

the thickness which can resist the failure of the external force (active earth pressure + water pressure) on the back of the beam is calculated (Fig. 2.16).

$$t = F_s\sqrt{\frac{3W \cdot r^2}{\sigma_t}} \tag{2.1}$$

t—Reinforcement thickness (m);

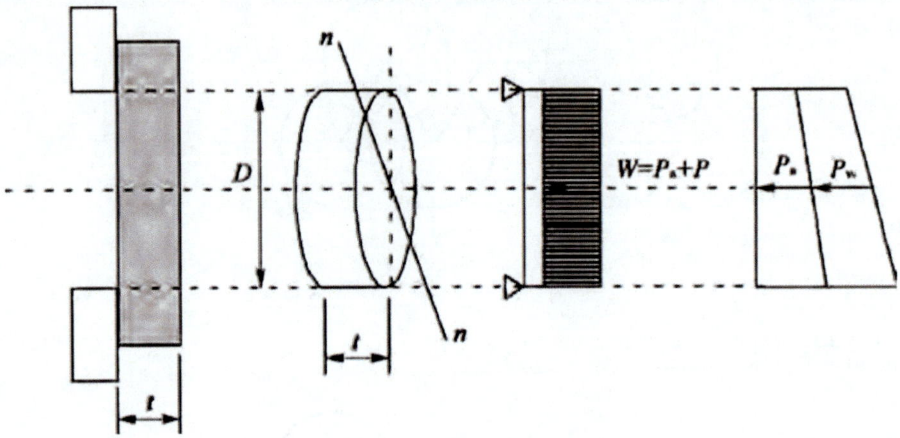

Fig. 2.16 Calculation model of reinforcement thickness

F_s—The safety factor is 1.2 in soil mechanics and 2.65 in non-seismic inspection of concrete structure;

W—External force (mainly active earth pressure and water pressure) (kN/m^2);

r—Shield excavation radius (m);

σ_t—The bending tensile strength of reinforced surrounding rock (kN/m^2).

2) Reinforcement range

(1) Thickness required for upper part.

The reinforcement thickness of the upper part of the shield is studied by using the method of deducing the additional stress generated by the surrounding when the tunnel is excavated.

$$t_{\mathrm{h}} = F_s\left\{\exp\left[\frac{(H-r)\gamma_t}{2C} + \lambda_n \cdot r\right] - r\right\} \text{ (When } t_{\mathrm{h}} \le 2.5\,\mathrm{m},\ t_{\mathrm{h}} = 2.5\,\mathrm{m}) \tag{2.2}$$

H—Depth to the center of the tunnel (m);

r—Shield open outer radius (m);

γ_{t}—The unit volume weight of surrounding rock after reinforcement (kN/m^3);

C—Cohesion of surrounding Rock after reinforcement (kPa);

F_{s}—The safety factor is 1.2 in soil mechanics and 2.65 in concrete structure.

The upper reinforcement range is shown in Fig. 2.17.

(2) Thickness required for lateral part.

The lateral part needs thickness to be the intersection range between the required thickness and the collapse angle ($\theta = \pi/4 + \varnothing/2$) obtained from the upper part (Fig. 2.18).

$$t_x = (r + t_h) \cdot \cos\beta - r \tag{2.3}$$

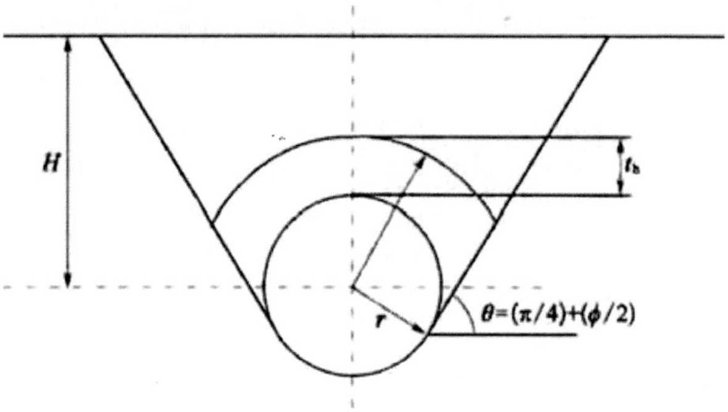

Fig. 2.17 Upper reinforcement range model

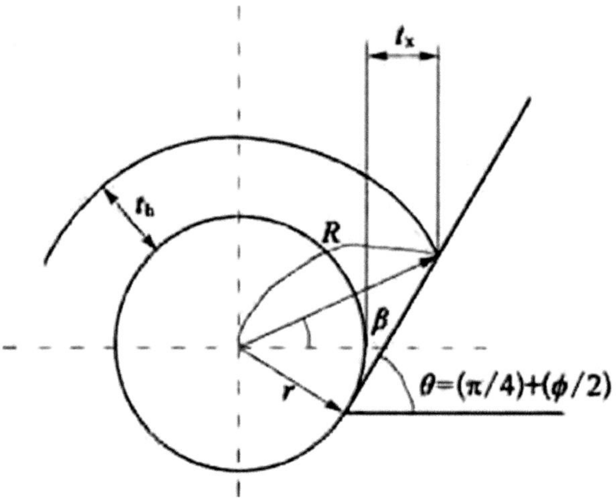

Fig. 2.18 Side reinforcement range model

$$\beta = \cos^{-1}\frac{r}{r + t_h} - \left(\frac{\pi}{4} - \frac{\varnothing}{2}\right) \quad (2.4)$$

(When tx 2.0 m, tx = 2.0 m)

r—Shield opening radius (m);
t_h—Upper reinforcement thickness (m);
$\varnothing$—The internal friction angle of surrounding rock;
t_x—Side reinforcement thickness (m).

(3) Thickness required for bottom part. Consider water seal to ensure minimum thickness.

To sum up, considering the actual value so far, the minimum thickness in the reinforcement range is 2.5 m in the upper part, 2.0 m in the side part and 1.5 m in the lower part as the standard value. Among them, the thickness of the lower part should be more than 60% of the upper thickness.

2.3 Shield Tunneling Initial Construction Technology

This section takes Nanjing Yangtze River Tunnel as an example to introduce the construction technology of the initial section of shield tunnel, mainly from four aspects: auxiliary engineering construction of the initial section, assembly of negative ring segment and box culvert, opening of tunnel door and control of initial excavation parameters.

2.3.1 Initial Construction of Auxiliary Engineering

1) Construction of initial reaction frame.

The reaction frame is a circular reinforced concrete structure, which mainly provides a solid force point for the origin of the shield machine, so that the shield Jack has enough reverse force to push the shield forward. The construction height of the reaction frame is large, so it needs to be poured in layers. The formwork is generally assembled with bamboo rubber board, the support system is full of red scaffolding, the scaffolding adopts steel pipe, and the square wood + arc I-shaped steel is used as the internal support.

The steel plate is embedded in the construction of the reaction frame, which is used to connect the steel pipe support of the steel pipe sheet (Fig. 2.19). During construction, the steel pipe piece shall be positioned first, and then the steel pipe is cut according to the measured distance between the steel tube sheet and the reinforced concrete reaction frame. One end of the steel pipe support is welded to the embedded steel plate of the reaction frame, and the other end is welded to the corresponding position of the steel pipe sheet. Steel tubes are filled with concrete of a certain strength before installation to enhance their stiffness.

2) Initial pedestal construction.

The initial pedestal is arc structure, the longitudinal two reserved grooves and the transverse one reserved grooves, and the grooves are reserved operating space for shield assembly and welding. Cast-in-place reinforced concrete structure is used in the initial pedestal and reaction frame.

(1) In order to facilitate the initiation of shield and offset the beginning of shield, the top surface of the structure is adjusted in the construction of initial pedestal. The size of the body is adjusted according to the center line of the shield design, and the adjustment of H-shaped steel is controlled according to the requirements in the construction of the height adjustment of the pedestal.
(2) In order to facilitate the welding of shield tail and middle shield during shield assembly, save the welding time of shield assembly, the reserved transverse welding operation groove structure of starting base shall be moved to a small distance of small mileage, and the length of reinforcement and position of embedded parts shall be changed accordingly.

Fig. 2.19 Construction of initial reaction frame

3) Installation of long guide track

After the installation of the initial pedestal, because there is a small distance between the front end of the reference guideway and the soil in front of the initial pedestal (that is, the thickness of the wall at the end of the shield working well and the gap left in order to facilitate the installation of the temporary sealing device for the door), in order to ensure the safety and accurate initiation of the shield, two long guide tracks are installed in the hole ring and in the corresponding position of the guiding track of the original pedestal, and the installation inclination angle position is consistent with the reference guiding track, The lower part of the guideway is supported by H-shaped steel, and the lower part of H-shaped steel is fixed by steel plate and expansion bolt (Fig. 2.20).

When the shield cutter head reaches the position of the hole door, there is a 20 cm gap between the cutter head and the bottom of the hole door. In order to prevent the cutter head from planting, fill the yellow sand and compaction in this position, pour the concrete guide slot or the embedded guide rail. The diaphragm wall position portal is broken according to the forward direction profile of the shield. No over-excavation is allowed to prevent the gap between the shield shell and the diaphragm wall from causing the shield head or segment to sink.

Fig. 2.20 Installation of starting base and sliding rail

4) Temporary sealing device for hole door

In order to prevent the mud from entering the shield working shaft from the annular building gap between the tunnel ring and the shield shell during the initial process of the shield, affecting the mud pressure of the shield excavation surface, the stability of the excavated surface soil mass and the construction in the shield, the sealing device with good performance must be provided at the tunnel portal before the shield starts (Fig. 2.21). According to the construction experience of shield machine in the past, the folding blade seal pressure plate has the advantages of good force, good sealing, simple operation, good stiffness, safety and reliability.

2.3.2 Installation of Negative Loop Pipe and Box Culvert

1) Negative loop pipe assembly

The positioning of the first ring of negative loop pipe is very important when the negative loop pipe is assembled, which plays the role of datum for the assembly of the rear segment. During the lining assembly of the first ring of negative loop pipe piece, the lining of the falling block shall be positioned horizontally at the lower part of the shield shell. The outer arc surface of the lining shall be reserved with the shell with the building gap, and the segment and the shield shell shall be temporarily welded and fixed with the connecting plate, and then the remaining segments shall be

Fig. 2.21 Initial door sealing curtain installation

assembled. Attention should be paid to the smoothness of the torus and the tightening of all circumferential bolts when assembling.

When the Negative loop pipe is assembled, its axis, plane elevation and torus verticality should be accurate and fixed. In order to ensure the smoothness of the torus, the opening Jack should be controlled when the lining is pushed out, the Jack should be selected to open the longitudinal joint of the lining, and the limit gasket should be set in the concrete position of the back seat.

When the shield tail comes out of the inner wall of the working well, the steel plate is immediately welded with the embedded iron of the hole ring. If the leakage phenomenon of the hole is serious, the chemical slurry is injected into the surrounding hole ring by the preset grouting pipe. In order to reduce the propulsion resistance at the beginning of the shield, butter is applied to the residual wall of the shield shell, guideway and continuous wall before the shield advance, and the shield tail grease must also be filled in the shield tail steel brush.

2) Box culvert assembly

The box culvert is initially fixed at the joint end wall, and the rest of the box culvert is connected with the adjacent box culvert through bolt holes. Concrete cast-in-place is used between the box culvert and the end wall of the starting shaft, as shown in Fig. 2.22.

The preparations for the culvert assembly of box culverts are as follows:

(a) Box culvert assembly

(b) Box culvert connection

Fig. 2.22 Box culvert assembly and connection

(1) driving the transport vehicle to the front end of the assembled box culvert, and after assembling a piece, the vehicle continues to move forward to the front end of the assembled box culvert waiting for the assembly of the second piece.
(2) Shield operator determines that there is enough space between the No. 2 trolley and the installed box culvert to install the box culvert (generally, two pieces are installed at one time).
(3) Before installation of box culvert, the shield machine captain and assembler shall determine the type of box culvert to be hoisted on site, inspect the appearance quality of box culvert, and allow assembly only after passing inspection.
(4) The box culvert installation group clears the garbage in the position of the pre-packing culvert, and takes the box culvert connector and the installed box culvert connection bolt and gasket to the front end of the installed box culvert.

The specific operation steps of box culvert assembly are as follows:

After preparation, the shield captain gives instructions, and after the box culvert assembler receives the assembly instruction, he organizes the personnel to carry out the box culvert assembly operation; in the course of operation, the shield driver is responsible for supervising the standardization of box culvert operator operation and monitoring the assembly quality, and the quality inspection engineer is responsible for checking the construction quality of box culvert.

(1) Cleaning up sundries. Before assembling the box culvert, the assembly position and the sundries at the bottom of the joint surface with the upper box culvert need to be removed.
(2) Treatment of box culvert. According to the requirements of the assembly scheme, the shield driver arranges the relevant personnel to paste 2–5 mm thick nitrile cork liner on the back face of the box culvert to adjust the direction and slope of the box culvert.

(3) Horizontal hoisting box culvert. The flat car drives to a suitable position and adjusts the height of the flat car so that the box culvert crane can grasp the box culvert; after the height adjustment is completed, the hoist is lowered and moved to the bottom of the box culvert, and after the initial alignment, the hoist is just stuck on the box culvert; when the switch is in place, the operator can clamp the fixture and lift the box culver, when the box culvert touches the limit switch, the box culvert can be shifted.
(4) Front moving box culvert crane. After grasping the box culvert, move the box culvert crane to the assembly position and confirm that there is enough space for the box culvert to be installed under the box culvert; Then the box culvert stops at a certain height horizontally, and the height of the box culvert and the distance between the box culvert and the upper box culvert can be adjusted at the same time, so that the box culvert can be roughly in place.
(5) Accurate positioning. Adjust the box culvert to the left and right according to the pre-lofting mark of the measuring engineer so that it is in line with the design azimuth base and close to the block box culvert at the same time. Adjust the box culvert so that it is closely affixed to the end face of the upper block box culvert; by accurately locating the box culvert by measuring the smoothness of the joint of the box culvert, it is required that there should be no error table larger than 5 mm between the box culvert. The gap between one side of the box culvert and the adjacent box culvert can be adjusted by the cushion cork liner, and the gap between the bottom contact surface of the box culvert and the segment can be filled with steel plate plug.

    Due to the segment construction error, it is possible to find a large dislocation (height or left and right position) with the upper box culvert in the positioning process. The leaders of the measurement team and the engineering department should be informed to analyze the reasons on the spot and formulate remedial measures.
(6) Install connections. After completion of positioning, install the connecting piece and the box culvert connected by the connector shall be removed and reused after the stress is stabilized.
(7) Restore the crane. The quality inspection engineer shall be notified for acceptance after the assembly is completed.

2.3.3 Broken of the Door of Tunnel

In order to ensure the stability of the retaining structure of the original well, the hole drilling is carried out in two stages in the process of opening the hole door. The first stage begins to be chiseled after the soil reinforcement inspection of the end well is passed, and the shield initiation facility is completed before going down the well; The second phase is carried out quickly after the assembly and debugging of the shield and other initial preparations are completed, as shown in Fig. 2.23.

(a Breaking of the opening of the starting section (b) Frozen excavation surface after fully breaking the portal

Fig. 2.23 Frozen excavation surface after door breaking and complete breakage in the initial section

Before chiseling, set up a double row scaffolding, from the top to the bottom layer chisel. In the first stage, the external concrete and steel bars are chiseled from top to bottom in five levels, and the inner steel bars are reserved, so as to protect the end stratum before starting. When chiseling, the excavated steel bar is first chiseled out and cut off by oxygen welding, and then it continues to be chiseled until the steel bar on the face of the earth is exposed. The order of concrete chiseling in each layer is first middle, then left and then right. When the shield cutter head reaches $0.5 \leq 1$ m in front of the enclosure structure, the excavation is stopped, then the remaining steel bar is cut off, the remaining continuous wall is pierced, and the rest of the continuous wall is checked to determine whether there is no steel bar.

The emergency measures for the chiseling process are as follows:

(1) abnormal conditions are found, which are quickly supported by planks and steel pipes to prevent the collapse of soil outside the continuous wall, and then grouting reinforcement is carried out from outside the retaining wall as soon as possible.
(2) when the pressure of rock and soil is high, the pre-made steel mesh is welded with the steel bar of the enclosure structure quickly and then supported and stabilized by plank and steel pipe, and then grouting reinforcement is carried out on the periphery of the envelope structure, and grouting reinforcement is carried out in the tunnel door at the same time.

2.3.4 *Control of Driving Parameters in the Initial Section*

1) Idling position

The cutter head does not need to rotate, no mud, Jack propulsion speed is 10 mm/min.

2) Strengthened area

(1) Keep the incision water pressure: because the shield is in the strengthened area, the incision water pressure should not be set too high. According to the actual working conditions, make appropriate adjustments during construction.

(2) controlling the propulsion speed: in order to ensure that the shield can grind the soil reinforcement area normally, prevent the suction of the mud discharge pipeline from clogging, control the propulsion axis and protect the cutter head, the propulsion speed should not be too fast in the actual crossing process, so that the shield machine can move forward slowly and steadily, the propulsion speed is controlled in 3 mm/min, and the change of cutter head torque is observed at any time, and the propulsion speed is adjusted in time.

(3) Controlling the tunnel axis: in the initial construction, because the synchronous grouting cannot be opened in time, the slurry on the front of the shield can easily run back, resulting in the "floating" of the shield, so the propulsion axis should be slightly lower than the design axis, which is generally suitable for −30 mm.

2.4 Construction Technology of Shield Arrival

Compared with the construction of interval tunnel, the arrival of shield has its particularity and importance. The work contents include: formation reinforcement, shield positioning and recheck and measurement of receiving tunnel position, portal treatment and installation of temporary sealing device of portal ring, installation of receiving base, etc. The specific construction flow chart is shown in Fig. 2.24. This section takes Nanjing Yangtze River Tunnel as an example and introduces the above aspects.

2.4.1 Installation and Construction of Receiving Pedestal

The structure of the receiving pedestal is the same as the original pedestal, and the receiving pedestal is installed after accurate measurement and positioning. The central axis should be consistent with the axis of the shield machine entering the receiving well, and the tunnel design axis should be taken into account (Fig. 2.25). The elevation of the rail surface of the receiving pedestal is suitable for the shield attitude. In order to ensure that the assembly segment has enough reaction force after the shield cutter head passes through, the rail surface slope of the receiving pedestal can be properly increased. After the receiving pedestal is positioned and placed, I-beam is used to reinforce the front and both sides of the receiving pedestal, so as

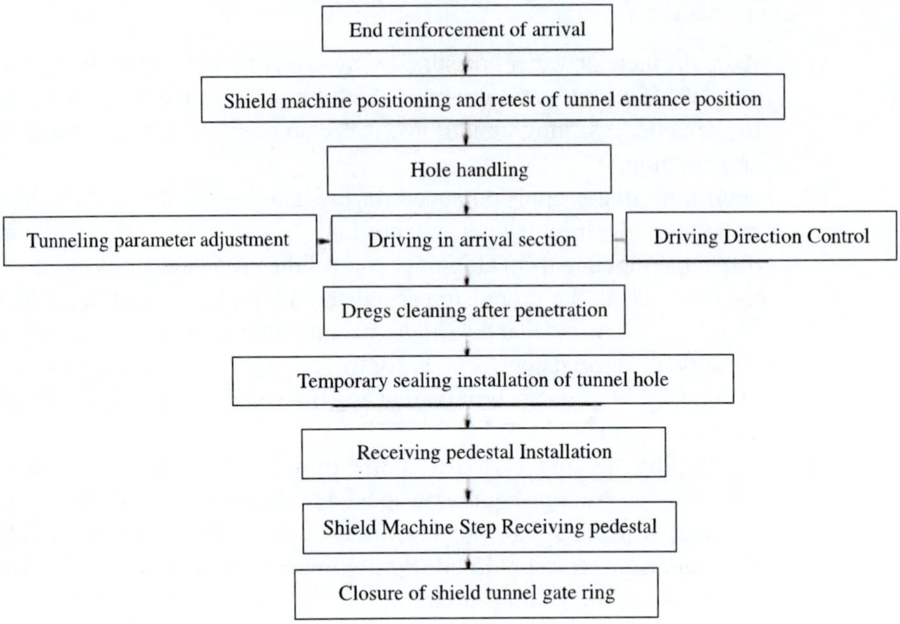

Fig. 2.24 Flow chart of shield arrival construction

Fig. 2.25 Receiving base installation

to prevent the shield machine from shifting the receiving pedestal in the process of pushing up the receiving pedestal, resulting in the failure of shield receiving.

The pedestal of the arrival section of Nanjing Yangtze River Tunnel is filled with M7.5 cement mortar. The length and width of the pedestal are the same as the size of the receiving well. In the working well, two forms of sections are adopted: within the range of 13.5 m from the small distance end, the base is a rectangular structure with full section. In order to ensure sufficient reaction support for the assembly of tube segments at the arrival stage, the bottom end of the shield is in the range of 2.3 m up, all of which are poured with M7.5 cement mortar. Within 4.5 m from the large mileage end, the pedestal is an arc structure, that is, the base is filled with yellow sand plug within the scope of shield excavation track line. Two grooves (cutting grooves) are reserved longitudinally in the pedestal structure, and one groove are reserved transversely. The grooves are used as the operating space when the shield is dismantled, and the grooves are filled with yellow sand in the arrival stage. The map of arriving at the pedestal and filling (soil) is shown in Fig. 2.26.

The foundation is constructed in two stages: the first stage is to construct the mortar base in the shaft, and the cutting groove and other parts are filled with yellow sand plug; in the second stage, the base shall be rapidly extended to the scope of the portal after the completion of the chiseling of the portal.

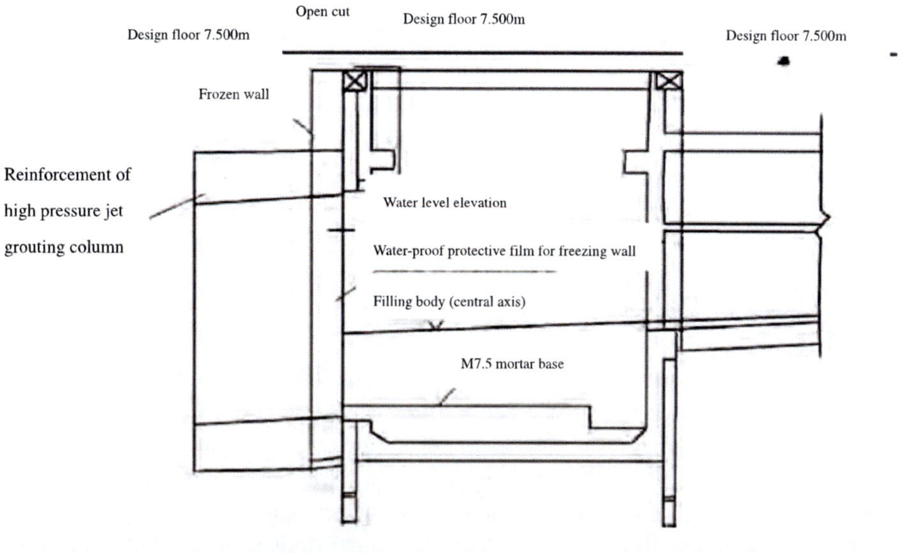

Fig. 2.26 Reaching the pedestal and irrigation (soil)

2.4.2 Breaking and Temporary Sealing of the Tunnel Door

1) Broken of tunnel door

When the shield reaches the back of the continuous wall, the concrete of the continuous wall in the tunnel door is broken by artificial air pick. The continuous wall chiseling of the tunnel door is carried out twice: For the first time, the palm surface of the hole door is divided into 9 pieces, and the reinforcement outside the continuous wall is cut off from top to bottom, and then the concrete of the main body of the continuous wall is chiseled out to retain the steel bar in the inner layer of the continuous wall. After the second time, the innermost steel bar is quickly removed after the shield opening chamber is checked for no leakage and the joint acceptance is completed. When the steel bar is removed for the second time, it is necessary to check the clearance of the shield to ensure that no steel bar invades the penetration range of the shield.

The continuous wall of the entrance of Nanjing Yangtze River Tunnel is chiseled three times: The outer concrete 10 cm is broken for the first time, and the steel bar in the inner layer of the underground continuous wall is removed. The second break is arranged after the freezing is completed, the concrete is broken 50 cm, and the temperature holes (a total of 35) are drilled on the door to detect whether the freezing has been completed. The third break is carried out in two stages. In the first stage, the thickness of concrete is 30 cm after the detection hole temperature (≤ -2 °C) meets the design requirements. In the second stage, the remaining 10 cm concrete and the outer steel bar of the underground continuous wall are broken within 2 days, and the exposure time of the frozen palm surface does not exceed 2 days.

2) Temporary sealing of tunnel door

After the shield tunneling, the small machine is used to clean up the mud during the penetration in time, and then the temporary sealing device of the tunnel door is installed. The temporary sealing device of the end tunnel door is similar to that of the beginning, the fixed bolt hole shall be welded outside the flip plate, and the temporary sealing device of the tunnel door shall be tightly affixed to the outer arc surface of the segment by tightening the wire rope pierced through the bolt hole.

After the receiving base is installed and fixed, the shield can push the receiving base slowly. When passing through the temporary sealing device of the tunnel door, butter is applied to the outer ring of the knife plate and the tool to prevent the shield cutter head and the tool from damaging the curtain rubber plate. When the shield is pushed on the receiving pedestal, the temporary sealing device of the tunnel door is tightened every time the shield is pushed forward 2 rings forward, and the outer annular gap of the pipe piece is filled with quick setting slurry through the synchronous grouting system, so as to ensure the correct attitude of the segment.

After the assembly of the last ring segment, the temporary sealing device of the tunnel door is tightened, so that the curtain rubber plate and the outer arc surface of the segment are tightly affixed, and the tunnel door ring is filled with grouting through the segment grouting hole. In the process of grouting, we should pay close attention to the situation of the cave door. As soon as the phenomenon of slurry leakage is found, the grouting should be stopped and the sealing treatment should be carried out to ensure that the grouting at the entrance is dense and the ring sealing of the hole door is tight.

2.4.3 *The Arrival Section of the Shield Machine is Excavated*

1) Shield positioning and compound measurement of the position of the entrance at the end of the shield.

When the shield is pushed to the construction range, the position of the shield and the survey control points of the shield tunnel should be accurately measured, and the relationship between the actual tunnel center axis and the tunnel design center axis should be clarified. At the same time, the tunnel door of the shield receiving well should be checked and measured, and the penetration attitude and driving deviation correction plan of the shield should be determined. Two points should be paid attention to when considering the penetrating attitude of shield: one is the deviation between the central axis of shield tunneling and the design axis of tunnel, and the other is the deviation of the position of receiving portal. These factors should be adjusted on the basis of tunnel design center axis, and the correction of deviation should be completed step by step.

In order to ensure the accuracy of arrival, the penetration survey is carried out 150 m before the tunnel penetration. The penetration survey includes the ground control network joint survey (plane and elevation) and the receiving well hole gate measurement (plane and elevation). The accuracy of height penetration survey is $\leq\pm25$ mm, and the measurement error of traverse penetration is $\leq\pm20$ mm.

(1) Ground control network joint survey. The GPS static positioning technology is used to measure all the control points of the receiving and starting control network. The instruments, methods and specifications used in the repeated survey are consistent with the timing of the establishment of the ground control network. According to the results of joint measurement, the coordinates of control points are checked to ensure that the control points are consistent with each other.
(2) Receiving well gate determination. By using the total station, a certain number of measuring points are measured on the reserved steel ring of the tunnel gate, and according to the results of the measurement, the actual central coordinates of the entrance door and the distance from each point to the center are calculated. According to the calculation results, it is judged whether it is necessary to adjust the digging axis to ensure the smooth arrival of the shield.

2) Shield machine arriving at excavation operation

Take the shield tunneling operation of Nanjing Yangtze River Tunnel as an example.

The driving speed of shield arrival section is relatively stable, the disturbance to soil is reduced, and the axis control of shield tunnel is strengthened. The amount of grouting and grouting pressure at the same step should be adjusted according to the propulsion speed and ballast discharge. Because the shield tail brush has been dug in such long distance, the synchronous grouting pressure should be controlled to prevent the leakage of slurry at the end of the shield.

After the irrigation is completed, the shield continues to push forward for 2 m, and stops at a position of 4 m from the freezer wall. It begins to pull out the frozen pipe and pull all the frozen pipes located in the tunnel propulsion range from the top of the tunnel 0.3 m, and then resume the freezing. After the shield passes through the frozen section, all the frozen pipes are pulled out and filled with yellow sand. After the frozen tube is pulled out in place, the shield is pushed smoothly to the door. The propulsion pressure is gradually reduced before shield penetration, until the incision pressure is 0, and the intake and exhaust valve of the system is closed after the cutter head breaks the wall.

When the shield cutter head breaks the freezer wall and enters the working well, the incision pressure disappears completely, which is consistent with the liquid level in the well. at this time, the flow rate of the inlet and outlet slurry is controlled within $100\text{m}^3/\text{h}$, and the flow rate in and out of the working well is kept unchanged (the liquid level in the working well remains 2.5–3.0 m below the top surface of the shield cutter head during the receiving process). After the shield machine is completely entered the receiving well by sealing grouting at the entrance of the tunnel, the debris in the well is removed, and the gap between the segment and the excavated surface (embedded steel ring) is filled with C20 jet coagulation soil, during which the precipitation has been running all the time. While the shield is disassembled, the temporary sealing tunnel door is used in time, and then the end ring segment is strengthened by grouting through the grouting hole reserved on the embedded steel ring. When the shield is dismantled, the permanent cast-in-place concrete tunnel door is applied under the condition of precipitation.

3) Shield arrival heading control

In addition to the purpose of correcting deviation according to the scheme of heading correction, special attention should be paid to the excavation control of the last 10 m section. Before the arrival of the shield, it is necessary to strengthen the observation and settlement monitoring of the entrance of the tunnel, and communicate with the main driver of the shield in time in order to control the excavation, such as the reinforcement of the strata, and so as to strengthen the observation of the working conditions and the monitoring of the settlement of the entrance after entering the reinforcement body.

According to the geological conditions of the entry section, the reasonable excavation parameters are determined and the written submission is made. The general requirements are: low speed, small thrust, reasonable slurry pressure and timely and full backfill grouting. In the assembly of the last 20 ring segments, the segments should be connected into a whole with a longitudinal pull rod in time to avoid loosening between the segments when the thrust is very small or there is no thrust.

Because the shield well enclosure structure adopts reinforced concrete at the shield gate, the shield must be broken before it arrives. When the shield is excavated in the last ring, it should be excavated with low speed, small thrust, small torque, reasonable slurry pressure, and through the interval tunnel after the synchronous grouting and condensation injected in the previous ring.

4) Key points and measures of shield arrival construction technology

(1) When the shield enters the arrival section, the workers shall specify the real-time mileage of the shield and the distance from the cutter head to the tunnel portal, and carry out the construction according to the determined construction technical scheme.
(2) Before the arrival of the shield, it should be checked to confirm whether the condition of the end formation meets the reinforcement requirements.
(3) Increasing the frequency of surface subsidence monitoring and feedback the monitoring results in time to guide the construction.
(4) Re-survey of conductors and elevations shall be carried out every 50 m before arriving at 200 m, and the clearance of the penetration point shall be measured to determine the position accurately.
(5) In the arrival section of shield machine, combined with the through position and referring to the design line, each ring must be carried out in strict accordance with the excavation plan.
(6) The mix proportion of grouting materials in the first six rings should be adjusted, and if necessary, special water stop materials can be injected into the outer part of the shield shell by drilling holes in the tail shell of the shield, so as to prevent the formation collapse caused by water gushing and mud gushing.
(7) In order to prevent the contact relaxation of the annular seam of the segment caused by the insufficient reaction force of the cutter head, the segment is tightened longitudinally along the tunnel with channel steel in the last 10 ring segments of the shield arriving section.
(8) Several ring segments installed after shield penetration must ensure that the grouting is full and dense, and prevent the segment from sinking and missetting.
(9) In the construction of shield arrival section, it is necessary to ensure the normal operation of slurry system and the amount of dry sand to be machined. the data of soil and sand quantity are positive, and the cut water pressure is reduced to the minimum.
(10) Closely observe the change of parameters such as the deformation of the tunnel portal and the torque of the cutter head. In case of any abnormality, the propulsion shall be stopped immediately and corresponding countermeasures shall be taken.
(11) when the shield incision is close to the shield working well wall protection 2–3 m, the viscosity and specific gravity index of slurry can be gradually reduced under the condition of determining the stability of the soil at the mouth of the tunnel, until it is replaced by clean water.

(12) When the cutter head excavates the shield working well to protect the wall, it should stop pushing immediately, replace the mud in the cut mud cabin with clean water, and clean the mud water pipeline at the same time.

2.4.4 Control of Driving Parameters in the Arrival Section

1) Slurry pressure

When calculating mud pressure, the groundwater level is 5 m. After entering the influence range of precipitation in the arrival stage, the groundwater level is adjusted by the measured groundwater level. The mud water pressure shall be strictly controlled according to the tunneling parameters, with the deviation range of ±10 kPa. The mud water pressure value shall be reduced by 10–15 kPa according to the calculated value.

2) Driving speed and cutter head speed

The driving speed and cutter head speed are set according to geological conditions and current construction experience, as shown in Table 2.4.

3) Backwall grouting

In order to prevent slurry leakage from shield tail, tunnel floatation and formation instability, it is necessary to strengthen the control of grouting behind the wall to ensure the quality of synchronous slurry. The volume mass of the slurry is 1.96 g/cm^3, and the slump of the slurry is controlled at 18–22 cm in this area. The grouting quantity is generally controlled at 25–30 m^3, and the grouting pressure is controlled to prevent the penetration of shallow overlying soil layer.

4) Setting and preparation of mud index

The volume mass of feed mud water is 1.23–1.26 g/cm^3, the pulping mud water is 1.30–1.45 g/cm^3, and the mud viscosity is controlled at 18–20 s.

Table 2.4 Shield tunneling parameters when Nanjing Yangtze River tunnel arrives

Stage mileage	Speed of cutter head (r/min)	Driving speed (mm/min)	Taper degree (mm/r)	Stratum feature	Excavation requirements
Arrival section K6 +470~+605	0.8–0.9	20–30	25–35	⑧-1 layer of fine sand, ⑦ layer fine powder sand	Low speed, small torque
Reach the reinforcement section K6+605~+622	0.6–0.7	10	10–14	Triaxial mixing to strengthen soil, Intensity 6–8 MPa	Low speed, small thrust

5) Ballast quantity control

During shield tunneling, the amount of ballast produced by each ring should be strictly controlled to prevent surface subsidence caused by over excavation. A small amount of under digging is allowed and over digging is not allowed to protect the stability of the palm surface.

6) Shield gesture control

According to the construction experience, the shield attitude and shaped segment are stable in the sand layer, easy to float in the shallow overlying soil layer, and easy to cause settlement deformation in the 10–20 ring near the tunnel gate. The shield attitude is strictly controlled when the shield is driven, and the horizontal and vertical deviation between the shield excavation and the design axis is controlled at −20 to +20 mm, and the shield tunneling attitude is in good agreement with the design axis.

Through the implementation of the above comprehensive construction measures, Nanjing Yangtze River Tunnel has safely and smoothly implemented the arrival and receiving of the shield on the left and right lines (Fig. 2.27). By canceling the sealing

Fig. 2.27 Nanjing Yangtze River Tunnel runs through the whole line

curtain of the tunnel door and receiving it by irrigation (soil) in the receiving well, the soil and water pressure of the longitudinal leakage channel is balanced. The tunnel door is temporarily blocked and sealed with arc steel plate, and then the 10 rings pipe piece of the hole mouth is filled with secondary grouting, and then the permanent sealing construction of the hole door is carried out under the condition of precipitation. The effect is very satisfactory, and there is no leakage trace of the hole door after construction. The use of cement mortar pedestal not only solves the problem of height difference received by guideway, but also provides reaction force for the assembly of several ring segments at the mouth of the hole, which is of great significance to improve the assembly quality of senior executives and prevent the occurrence of misalignment.

Chapter 3
The Key Technology of Large Diameter Shield Tunneling Shallow Overburden Construction

The construction of large diameter shield shallow overburden is one of the difficult problems in the construction process of river tunnel at present. Once something happens, the loss will be inestimable. Therefore, it is of great significance to improve and perfect the construction technology of large diameter shield shallow overburden.

This chapter first introduces the difficulties and technical status of large diameter shield shallow overlying soil construction, and focuses on the parameters and main technologies of shield shallow overlying soil excavation, including the construction technology of shield shallow overburden in the initial and arriving section, the shallow overburden shield construction technology through the bottom of the river, the incision water pressure control, the shield plane elevation attitude control, the mud water quality control and the synchronous grouting control behind the wall, etc. The slurry split (instability mechanism of excavation surface) is discussed, and finally, the three-dimensional visual simulation technology of shield shallow overburden excavation construction is introduced.

When the tunnel passes through the shallow overburden section at the bottom of the river, due to the requirements of the tunnel design line, the overlying soil layer of the tunnel will be shallow. When the overlying soil weight and river pressure cannot resist the floating force of the tunnel caused by grouting in the process of tunnel excavation, the tunnel will float or fall off the roof, which will bring inestimable disaster to the project. When the thickness of overlying soil layer is less than 1.0–1.5 (the outer diameter of excavation is the outer diameter of excavation), the problem of overlying soil in tunnel must be studied. At home and abroad, there are many examples of similar difficulties in the construction of shield tunnel. In the process of shallow overburden construction, it is of great significance to solve the balance problem between the surrounding soil of shield and the shell of shield itself, which is of great significance to the shallow overburden caused by tunnel crossing through inland rivers.

J. Chen et al., *Construction Technology of Large Diameter Underwater Shield Tunnel*, Key Technologies for Tunnel Construction under Complex Geological and Environmental Conditions, https://doi.org/10.1007/978-981-16-5896-9_3

3.1 Analysis of Construction Difficulties of Large Diameter Shield Shallow Overburden

3.1.1 Technical Difficulties

When shield method is used to construct shallow overburden tunnel in water area (river, lake, sea), it mainly faces the following risks and problems.

(1) Shallow overburden is easy to produce roof caving and permeable water flow. In order to reduce the slope of the line, the overlying soil in the bottom section of the water is very shallow. For example, in the north–south phase of Nanjing Metro, the minimum overlying soil thickness of the tunnel is only 0.97 m, which is very shallow overburden when the shield passes through the Qinhuai River; The diameter of the fourth Yibeihe tunnel in Germany is 12.33 m, and the thickness of the most difficult soil layer is only 7 m; The shallower overlying soil is only 5.5 m, less than half the diameter of shield machine, in Dalian Road Tunnel, Huxing East Road, Huangpu River, Shanghai. Under the condition of high head pressure and shallow overlying soil, the soil pressure balance in front of the knife plate is not easy to establish. Water sand often enters the shield machine through the opening of the cutter head and the tail of the shield from the cracks in the disturbed soil, resulting in flooding of the shield machine, resulting in machine damage accidents from time to time.

(2) Shallow overburden is easy to cause tunnel floatation. Under the water, the shield is subjected to uneven force, the attitude of the shield rises, the slope is difficult, the tunnel floats up, and the axis is difficult to control. After the assembled tunnel ring is removed from the tail of the shield, the tunnel floats because the upper ballast and self-weight cannot resist the buoyancy caused by groundwater. If the corresponding reinforcement measures are not taken, it is easy to cause local cracking and leakage of the tunnel.

(3) Shallow overburden is easy to cause sand flow and piping phenomenon. Construction in sand, sandy silt and other easily liquefied soil layer, due to shield propulsion cutter head rotating cutting extrusion disturbance, coupled with high head pressure (sometimes up to 0.3–0.5 MPa). Liquefied sand soil seeps into the tunnel along the shield tail and tunnel joint. If measures are not taken in time, local foundation emptying, tunnel subsidence, spiral fracture and tunnel failure may occur.

Taking the initial section of Nanjing Yangtze River Tunnel as an example, due to the influence of line control, the shield initiation section (K3+600~730) belongs to ultra-shallow buried (Fig. 3.1). The thickness of the shallower overlying soil is 5.5 m, which is only 0.37*D*. The main crossing strata are fluidizing silty clay strata, and the construction of shield in shallow overlying soil is easy to cause the following problems.

(1) Because of the small vertical pressure, it is difficult to control the attitude of shield propulsion.

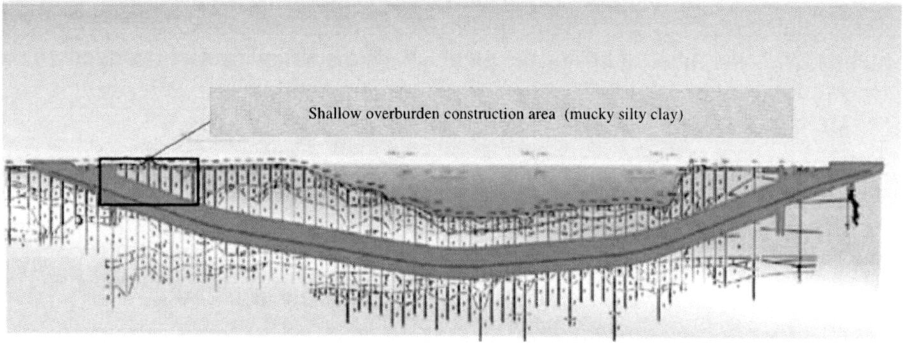

Fig. 3.1 Shallow overburden construction area in the initial section of Nanjing Yangtze River tunnel

(2) Because of the thin overlying soil layer, it is difficult to control the water pressure of the incision, and the mud water is easy to pop out of the ground, and the mud splits the ground layer and destroys the slurry balance.
(3) The soil is soft and the end soil is prone to instability, water and sand gushing, formation collapse and so on.

3.1.2 The Present Situation of Technology

1) Tunnel anti-floating technology

When the tunnel passes through the saturated upper layer, it will be subjected to the buoyancy of water. When the buoyancy exceeds the weight of the overlying soil on the tunnel and the self-weight in the tunnel and tunnel, the tunnel will float up. When the segment leaves the tail of the shield, the tunnel is surrounded by the grouting slurry behind the wall, and the buoyancy of the slurry is much greater than that of the saturated soil. At the same time, the earthwork excavated by the shield will lead to foundation unloading, and the assembled tunnel will deviate upward from the central axis by foundation rebound. Under the combined action of buoyancy and foundation spring back, the overburden of the tunnel produces uplift. If the maximum uplift is not effectively controlled, the overburden will be cracked, resulting in permeable cracks. The river flowing into the shield tail along the permeable crack will seriously affect the safety of the tunnel and tunnel construction.

Considering the formation loss caused by shield propulsion and the counteracting effect of foundation spring back on the tunnel, the anti-floating of the tunnel can be controlled by the following type:

$$p + q < F_y \tag{3.1}$$

q—Support pipe slice weight, $q = \pi\left(R_o^2 - R_i^2\right)\gamma_c$, R_o is the outer radius of the tunnel, R_i is the inner radius of the tunnel, γ_c is the bulk density of condensed soil;

F_y—Slurry buoyancy, $F_y = \gamma_y \pi R_o^2$, γ_y grouting bulk density after wall;

p—The overlying soil, $p = 2\ R_o h'$.

2) Underwater reinforcement measures for shallow overburden

When the thickness of shallow overburden is too small or the anti-floating setting cannot be taken, the reinforcement measures of underwater overburden should be taken. The commonly used reinforcement measures are as follows.

(1) Throwing soil under the water to increase the thickness of the overburden of the tunnel. This method is generally used in the reinforcement of deep-water river tunnel, including the famous Detian Yibeihe tunnel reinforcement. This method can not only meet the minimum thickness of overlying soil required by shield propulsion and tunnel safety, but also save cost, which is simple and easy to use.
(2) Grouting reinforcement at the bottom of the water. Due to the requirements of shipping and drainage, it is not suitable to strengthen the shallow overburden in shallow water channel by throwing soil by river channel. Therefore, underwater grouting reinforcement and anti-floating plate are the first choice of shallow water channel reinforcement. Through grouting, the overlying soil at the bottom of the river forms an impermeable hard shell, and the value of c, ∅ increases accordingly. When the overlying soil is impermeable, the increase of the value of c, ∅ can effectively reduce the calculated value of the minimum thickness of the overlying soil required for shield propulsion. The water pressure will effectively act on the overlying soil layer to prevent the tunnel from floating; at the same time, replacing the floating weight with the actual weight of the upper overlying soil can also increase the weight of the tunnel overlying soil, prevent the tunnel from floating, and reduce the calculated value of the minimum overlying soil thickness.
(3) Adding anti-floating plate to provide reaction force for shield propulsion and preventing tunnel floating. In the particularly shallow overlying area, the anti-floating structure, such as anti-uplift pile and anti-floating plate, should be added while grouting and reinforcement, which should be used to balance the lateral pressure produced by the soil during shield propulsion so that the shield can pass through the underwater dangerous area quickly with more appropriate propulsion force and propulsion speed.

Taking the first phase test section of Nanjing Metro North and South Line as an example, the reinforcement measures are introduced.

The test section of Phase I of Nanjing Metro North Line starts from Yingma Lane in the south and reaches Sanshan Street in the north, which consists of four parts. From south to north, it is divided into open section, buried section, shield working well and shield section. The shield section is 679.35 m long, in which crossing the inner Qinhuai River section is the key point of this section. The width of the shield crossing the Qinhuai River is 16.8 m, and the elevation at the bottom of the river is only 0.97 m from the top of the tunnel, which is less than 1/6 of the diameter of the tunnel. It is a typical construction case of shallow overlying tunnel at the bottom of the river.

Due to the requirements of dredging and waterway, the reinforcement method of adding uplift pile and anti-floating plate and grouting at the bottom of the water is adopted in this project. Steel reinforced concrete anti-floating slabs and anti-pull piles can not only prevent the tunnel from floating, but also provide a reaction force to prevent the shield from drifting when the shield is pushed forward, as shown in Fig. 3.2a. The grouting depth is 7 m below the bottom plate, AB liquid is used, A liquid is cement slurry with water-cement ratio of 0.7, B liquid is water glass, the ratio of A liquid to B liquid is 1: (0.8–1.0), and the initial setting time is controlled at about 90 s after double liquid mixing. The purpose of grouting is to prevent the river from pouring into the working face through the shallow overlying soil gap when the shield is pushed forward. In addition, a hard dielectric layer is formed between the anti-floating plate and the shield section, and the pressure measurement produced by the shield propulsion or the floating force on the tunnel is transferred to the anti-floating plate, and the grouting area is strengthened as shown in Fig. 3.2b.

3.2 Construction Technology and Control Technology of Shallow Overburden for Shield Tunneling

3.2.1 Main Technical Measures of Shallow Overburden Construction

1) Construction of shallow overlying soil with shield in arrival section

In the initiation and arrival areas of shield, due to shallow overburden and low vertical pressure, it is easy to "float upward" when shield is advancing, and mud is easy to leap out of the ground to "slurry" and destroy the balance of slurry, which makes it more difficult to control the cut water pressure. In addition, it is easy to cause the collapse of the top soil, so it is difficult to control the settlement of the ground surface.

It is difficult to control the shield axis and land subsidence in the shallow overburden section. It is necessary to take effective soil reinforcement measures and optimize the construction parameters to establish the mud water balance of the shield machine quickly and reliably in order to avoid the problem of arriving in the construction because of the initiation. The end is strengthened by high pressure rotary jet grouting pile, and the end layer is strengthened by freezing method. After the reinforcement construction of the end layer, the reinforcement area is coring vertically and several horizontal holes are arranged evenly at the entrance of the tunnel to detect the reinforcement effect. If there is any problem, continue to freeze until the freezing requirement is met to ensure the safety of shield.

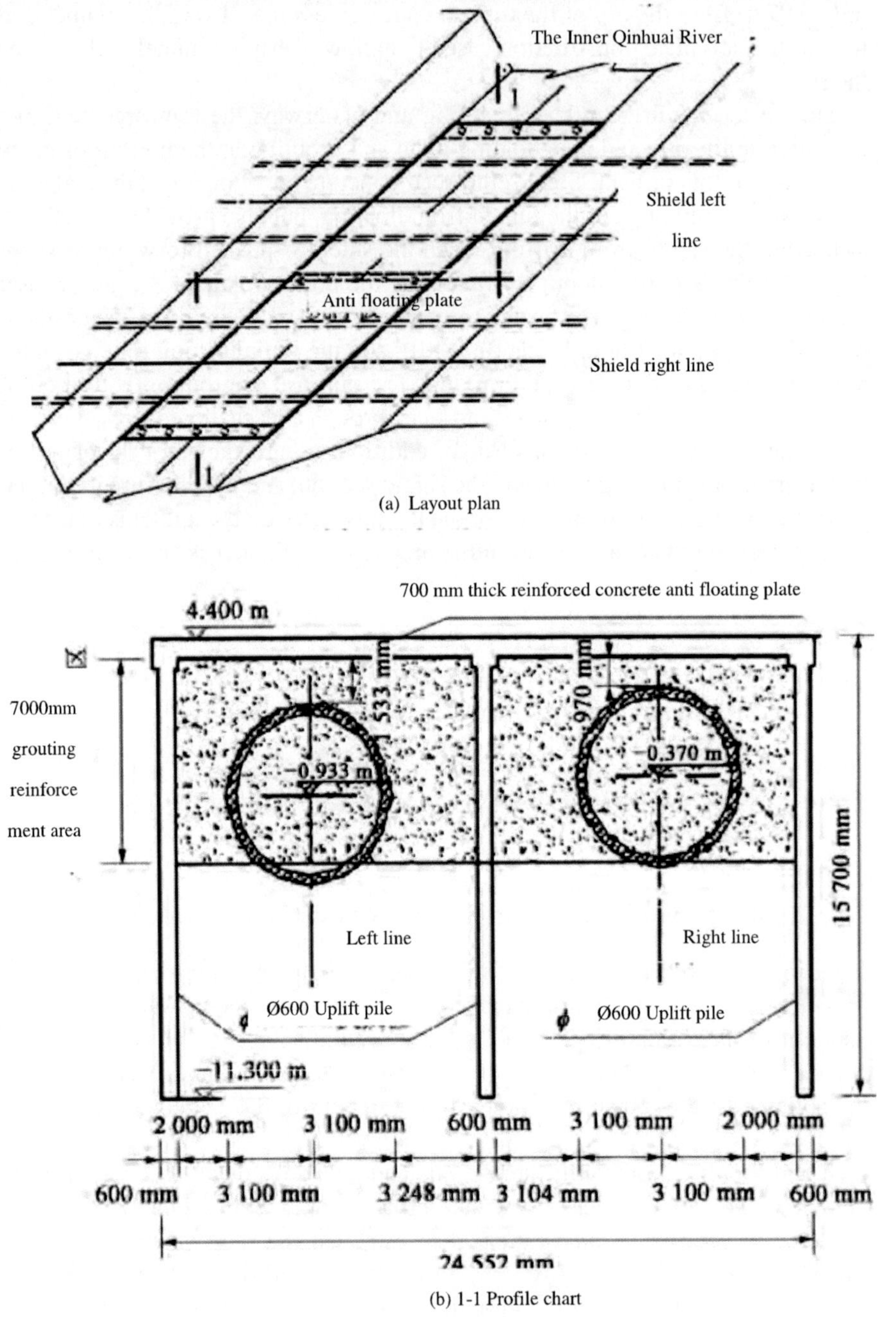

Fig. 3.2 Layout diagram of anti-floating plate, anti-uplift pile and grouting reinforcement area

2) Construction of shallow overlying soil shield across the bottom of the river

For the construction area where the overlying soil in the river is less than 1 times the diameter of the shield, it is easy to break through the overburden, resulting in mud leakage and even river water invading the inner part of the shield. Under this kind of working condition, the following four aspects are mainly adopted to ensure the smooth progress of the construction.

(1) Measures to prevent slurry leakage (leakage prevention) at the tail of the shield.

① Improve the quality of synchronous grouting. Before each ring is pushed forward, the grouting slurry must be sampled and tested to strictly control its initial setting time and slurry thickness. in the grouting process, the grouting pressure should be reasonably mastered so that the grouting quantity and grouting speed can match with the shield tunneling speed.

② Keep the incision water pressure stable. Keep the incision water pressure stable and make it slightly lower than the grouting pressure, beware that a large number of slurry forward channeling affects the grouting effect.

③ Water-stop sponge is pasted outside the segment. In order to prevent the leakage of the tail of the shield, the water-stopping sponge is glued to the outer arc surface of the segment, so that the tail of the shield can be blocked by water absorption and expansion when the slurry is leaking.

④ Uniform enough pressure injection of shield tail oil. The oil injection in the tail of shield was combined with automatic and manual mode, and the amount of grease in the tail of shield was increased appropriately, which was initially determined as 120 kg per ring. When the partial leakage of shield tail was carried out, manual mode was used to replenish the oil in the leakage part in time.

⑤ To assemble segments in the center to ensure uniform clearance at the end of the shield. The segment is centered to prevent the clearance between shield and segment from being too large and too small, resulting in uneven clearance of shield tail and reducing the sealing effect of shield tail, and even damaging the shield tail brush. The difference between shield tail clearance and shield tail clearance is controlled within 20 mm during construction.

(2) Prevent grouting (anti-spillage) measures at river bottom.

① Strictly control the fluctuation range of notch water pressure, which is controlled at -0.2 to $+0.2$ kg/cm^2.

② Strictly control the amount of soil produced, in principle, according to the theoretical unearthed amount unearthed, can properly under-excavate, keep the soil dense, so as not to infiltrate into the soil and enter the shield.

③ Strictly control synchronous grouting pressure and install safety valve in grouting pipeline to avoid the overlying soil due to high grouting pressure.

④ If mechanical failure or other causes cause shield to stop pushing, measures should be taken to prevent shield movement.

When pulping occurs at the bottom of the river, the main countermeasures taken are as follows:

When it is found that the grouting at the bottom of the river can be pushed forward without reducing the water pressure of the excavation surface, the propulsion speed can be properly accelerated, the assembly efficiency can be improved, and the shield machine can pass through the grouting area as soon as possible.

When the slurry is serious and cannot be pushed forward, the water pressure of the excavation surface is reduced to (soil pressure + water pressure) balance; increase the density and viscosity of slurry; in order to make the shield machine push forward, check the amount of sand drilled and confirm if there is any over-digging; after a certain distance of excavation, sufficient post-wall grouting is carried out; return the water pressure of the excavation surface to the normal state and carry on the normal excavation.

When it is found that the river water flows into the tunnel from the tail of the shield, the current situation should be analyzed first, and the strength should be organized to deal with the emergency. In the process of emergency, polyurethane is injected into the leakage part accordingly. At the same time, drainage work shall be arranged to ensure the smooth discharge of the river water entering the shield.

(3) Prevent the settlement of soil layer at the bottom of the river (prevent collapse).

① Set the cut water pressure according to the design value, and make corresponding adjustment according to the change of the water level of the Yangtze River at the propulsion time, strictly control the fluctuation value of the mud water pressure, so as to prevent the cut water pressure from supporting the soil on the excavated surface and cause the soil mass collapse; Prevent the high water pressure of the cut, and cause the soil mass collapse due to excessive disturbance to the soil mass.

② Strengthen the observation of soil settlement at the bottom of the river. To strengthen the settlement observation of soil at the bottom of the river, when it is found that the settlement of the bottom of the river is greater than 5 cm, the amount of synchronous grouting should be properly increased, and the grouting should be replenished if necessary.

③ Strengthen mud management to prevent over-digging. Strengthen mud quality control, mud delivery and discharge monitoring, properly increase mud specific weight and control its viscosity, ensure mud film quality, strengthen mud delivery and discharge monitoring, calculate the relationship between excavation speed and mud discharge in advance, find abnormal mud discharge, adjust parameters in time to prevent collapse caused by over-excavation.

④ The stability of the front soil is checked by using the soil detection device. During the advancing process, when the amount of dry sand is too large, the frequency of the formation detection device shall be increased, so as to master the collapse of the front soil mass of the cut in time, and the parameters shall be adjusted in time according to the specific construction conditions to make the data of dry sand close to the theoretical value and reduce the possibility of front soil collapse.

(4) Measures to prevent tunnel floating (anti-floating).

① During the construction period, the tunnel axis is strictly controlled, so that the shield can push along the design axis as far as possible, and each ring can correct the deviation evenly and reduce the disturbance to the soil.

② To improve the quality of synchronous grouting, it is required that the slurry has a short initial setting time, so that it does not crack after encountering mud water, and the slurry is required to have certain fluidity, can evenly fill the tunnel for one week and fill the building voids in time.

③ When it is found that the floating capacity of the tunnel is large and the spread range is long, it is necessary to immediately apply the grouting measures to the built tunnel in order to cut off the path of slurry loss.

④ On the basis of synchronous grouting, combined with polyurethane grouting, a hoop is formed around the tunnel, and a hoop is played every 10 m, so that the tunnel longitudinal water stop isolation zone is formed, so as to slow down and restrict the tunnel floatation, so as to control the deformation of the tunnel.

⑤ Strengthen the monitoring of longitudinal deformation of tunnel, and correct the grouting according to the monitoring results, such as adjusting the grouting position and grouting quantity, preparing fast setting and early strength slurry.

⑥ Strengthen the monitoring of topographic settlement at the bottom of the river.

3.2.2 Driving Parameters and Gesture Control of Shallow Overburden Construction

3.2.2.1 Water Pressure Control of Incision

Because of the shallow soil cover on the top of shield, it is difficult to control the incision water pressure. The fluctuation of the incision water pressure is too large, which will increase the disturbance of the frontal soil and lead to the loss of the frontal soil. Therefore, the incision water pressure fluctuation should be reduced as much as possible. Technically, the operator changes from automatic control to manual control, and the fluctuation value of notch water pressure is controlled at −20 to +20 kPa to ensure positive stability. In the process of construction, due to the high strength and good self-reliance of the strengthened strata, the actual slurry pressure is less than the design value. According to the monitoring information, the initial setting value of slurry pressure is 1 kg/m^2.

1) Theoretical arithmetic

(1) The upper limit value of incision water pressure.

$$P_U = P_1 + P_2 + P_3 = \gamma_w \cdot h + K_0 \cdot [(\gamma - \gamma_w) \cdot h + \gamma \cdot (H - h)] + 30 \tag{3.2}$$

P_U—Upper limit value of incision water pressure (kPa);
P_1—Groundwater pressure (kPa);
P_2—Static earth pressure (kPa);
P_3—When the earth pressure is changed, 30 kPa is generally taken;
γ_w—Bulk density of water (kN/m^3);
h—Tunnel depth below groundwater level (up to tunnel center, m);
K_0—Static earth pressure coefficient;
γ—Bulk density of soil (kN/m^3);
H—Tunnel depth (up to tunnel center, m).

(2) The lower limit of incision water pressure.

$$P_D = P_1 + P_2' + P_3 = \gamma_w \cdot h + Ka \cdot [(\gamma - \gamma w)h + \gamma(H - h)] - 2 \cdot C_u \cdot \mathrm{sqr}(K_a) + 30' \tag{3.3}$$

P_D—Incision water pressure lower limit (kPa);
$'$—Active earth pressure (kPa);
K_a—Active earth pressure coefficient;
C_u—Cohesion of soil (kPa).

2) Determination of incision water pressure during excavation

The mud water pressure of shield tunneling is between the upper and lower limits of the theoretical calculation value, and is adjusted according to the situation of the surface structure and geological conditions. In the process of reverse washing, because the mud discharge pipe in the mud tank or shield is in a blocked state, the mud discharge rate is increased during reverse washing, but the incision water pressure is not reduced. The difference of notch water pressure deviation between shield propulsion, reverse washing and bypass switching is controlled at −20 to + 20 kPa.

3.2.2.2 Strict Control of Main Tunneling Parameters

Total thrust, propulsion speed, mud discharge, reduce slurry pressure fluctuation, adopt low speed uniform excavation, avoid large disturbance to soil, strengthen mud management and unearthed monitoring, prevent over-digging and under-digging. The driving speed is controlled within 20 mm/min. Shield tunneling mainly controls the following parameters.

1) Excavation speed

Under the normal heading condition, the excavation speed is set to 20–40 mm/min; when the shield passes through gravel and gravel formation, and the excavation speed is controlled in 15–30 mm/min.

When setting the speed of shield tunneling, pay attention to the following points:

(1) When the shield is started, the shield driver checks whether the Jack is solid or not, and the speed should not be too fast before starting and ending the propulsion. At the beginning of each ring excavation, the excavation speed is increased step by step, and the starting speed is too large to disturb the formation.
(2) In the normal excavation process of each ring, the excavation speed value is kept constant as much as possible, and the fluctuation is reduced, so as to ensure the stability of incision water pressure and the smooth flow of mud discharge pipe. When adjusting the driving speed, it is adjusted step by step to avoid the impact disturbance caused by the sudden change of velocity to the formation and the excessive water pressure swing of the notch.
(3) The speed of propulsion must meet the requirements of grouting quantity of each ring, and ensure that the synchronous grouting system is always in a good working state.
(4) When selecting the excavation speed, we must pay attention to the matching with the geological conditions and the surface building conditions, so as to avoid the improper selection of the velocity to the shield cutter head, the abnormal damage caused by the cutting tool and the excessive disturbance of the soil around the tunnel.

2) Control of excavation quantity

The actual excavation quantity R of shield tunneling is calculated by the following formula.

$$V_R = (Q_1 - Q_0) \times t \tag{3.4}$$

V_R—Actual excavation quantity (m^3);
Q_1—Discharge flow rate (m^3/min);
Q_0—Mud flow rate (m^3/min);
t—Cutting time (min).

When it is found that the amount of excavation is too large, check the mud water density, viscosity and incision water pressure immediately. In addition, it can also be used to investigate the collapse of soil and adjust the relevant parameters in time to ensure the stability of excavation surface.

3.2.2.3 Shield Plane Elevation Gesture Control

The shield machine keeps moving smoothly, reduces the deviation correction, and reduces the disturbance to the positive soil. Considering that the overlying soil is shallow, the shield and segment may float because of the shallow overburden. In order to offset the influence of floating, the deviation between the shield center and the tunnel design elevation is controlled within -30 mm and the plane deviation is controlled within $\pm$ 30 mm.

Strengthen monitoring and measurement, strictly control settlement. In order to ensure the safe passage of shield through shallow overlying soil section and strengthen the information construction, through the test data provided by the monitoring system, the construction parameters in the process of shield crossing are adjusted and controlled in time, and the grouting and ground tracking measures are taken to reduce the influence of shield construction on the ground.

Nanjing Yangtze river tunnel large diameter slurry shield in the initial period of shallow overburden, adopt these measures, and take effective measures and optimize the construction parameters, soil strengthening quickly and reliably establish slurry balance shield machine, ensure the 5.5 m ultra-shallow overburden under the condition of shield construction smoothly tank and safe starting, setting a world large diameter shield the most shallow overburden soil safe starting record (Fig. 3.3).

Fig. 3.3 Safety initiation of shield tunneling through freezing wall 5.5 m super shallow overlying soil in Nanjing Yangtze River tunnel

3.2.3 *Slurry Quality and Grouting Control After Wall in the Construction of Shallow Overburden*

In the process of shallow overburden construction of shield tunnel, the quality of mud film formation in front of the cutter head is determined by the quality of mud water, and the grouting technology behind the wall is an important measure to ensure the overall stability of the tunnel and the stability of the surrounding soil.

3.2.3.1 Mud Water Quality Control

In order to strengthen the support of front soil mass and prevent grouting on the ground, heavy slurry shall be adopted for advancing. The density of slurry is 1.26–1.3 g/cm^3, and the viscosity is 25–35 s. In order to ensure the quality of slurry, in the process of advancing, increase the frequency of mud test, adjust the mud quality in time, and ensure the smooth progress of mud. The flow of slurry quality control is shown in Fig. 3.4.

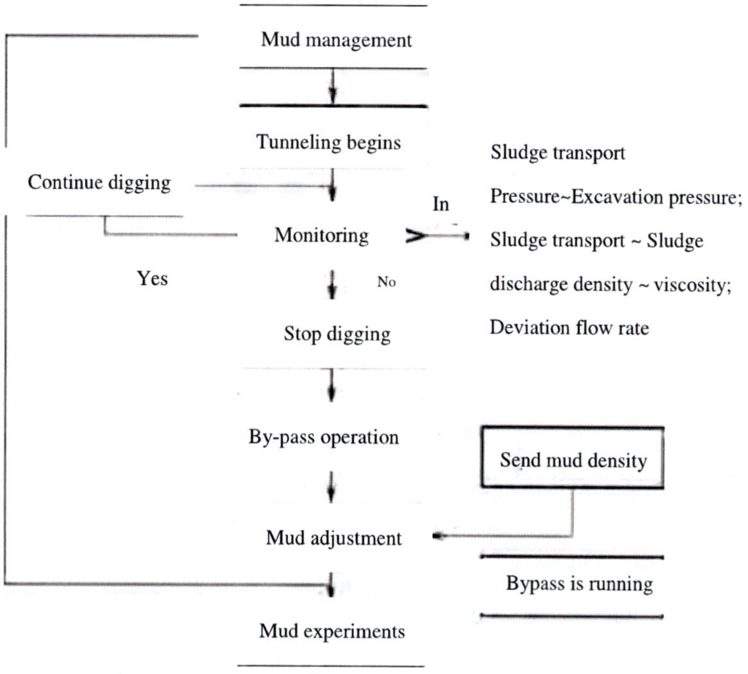

Fig. 3.4 Mud water quality control process

3.2.3.2 Strengthen Synchronous Grouting Control Behind Wall

Synchronous grouting is an important measure to prevent formation subsidence and control the synchronous grouting pressure and grouting quantity of shallow overburden section. The grouting capacity is controlled at 150%–250%. In order to prevent the grouting pressure from breaking the overburden, a pressure limit valve is set up in the control system of the grouting machine.

Taking the synchronous grouting construction behind the wall of Nanjing Yangtze River Tunnel as an example, this part introduces the slurry preparation and construction technology.

1) Grouting flow

The formation loss caused by shield construction and the reconsolidation of reconstructed soil disturbed or shear destroyed around shield tunnel and the infiltration of groundwater are the important reasons leading to the settlement of surface and buildings. In order to reduce and prevent settlement, in the process of shield tunneling, a sufficient amount of slurry material is injected synchronously behind the lining segment of the shield tail as soon as possible to fill the ring construction gap of the shield tail.

As a very important process in shield construction, grouting behind segment lining has the following three main purposes:

(1) Fill the gap of shield tail building in time, support the rock mass around the segment, and effectively control the surface subsidence.
(2) The condensed slurry will be used as the first waterproof barrier of shield tunnel to enhance the waterproofing ability of tunnel.
(3) Providing early stability for segment and integrating segment with surrounding rock mass is beneficial to the control of shield tunneling direction and can ensure the final stability of shield tunnel.

In general, three double piston grouting pumps with shield are injected at the end of the shield at the same time, which can fill the annular gap between the pipe and the formation in time, control the deformation of the formation, stabilize the segment structure, control the direction of shield tunneling, and strengthen the self-water prevention ability of the tunnel structure.

2) Slurry parameters and ratio

(1) Grouting ratio. Because most of the shield tunnel passes through fine sand and gravel layer, and the tunnel passes through sand layer and river water directly, combined with the construction case of large section mud water shield in China, cement mortar is used for synchronous grouting, and the material ratio is shown in Table 3.1.
(2) The main performance indexes of slurry.

Gelation time: Generally speaking, it is 3–10 h. According to the formation conditions and excavation speed, the gelation time is adjusted by adding accelerator and

Table 3.1 Preliminary proportion of synchronous grouting materials

Raw material	Stone ash	Flyash	Bentonite	Sand	Water	Additive
Allocation quantity	80	300	50	1180	300	3

changing the ratio in the field test. For strong permeable strata and areas requiring high early strength by grouting, the mix ratio is further adjusted and early strength agent is added through field tests to further shorten the gelation time, obtain early strength and ensure good grouting effect.

The strength of solidified body is not less than 0.2 MPa (equivalent to unconfined compressive strength of soft rock) and not less than 2.5 MPa (slightly higher than the natural compressive strength of strong weathered rock).

The rate of slurry stone formation is: >95%, that is, the consolidation shrinkage is <5%.

Slurry consistency: 12 cm ± 2 cm.

Slurry stability: the pour rate (the ratio of floating water volume to total volume after static precipitation) is less than 5%.

3) Synchronous grouting management

(1) Grouting mode. Automatic control or manual control is selected according to the need of grouting. The automatic control mode is to set the grouting pressure in advance, adjust the grouting speed automatically by the control program, and stop the grouting when the grouting pressure reaches the set value. The manual control mode is to adjust the grouting flow rate at any time according to the driving situation, so as to prevent the grouting speed from being too fast and affect the grouting effect. Generally do not grouting from the reserved grouting hole, in order to greatly reduce the possibility of leakage from the segment.

(2) Preparation of grouting equipment. Mortar is prepared by mortar mixing station and transported to shield by special mortar tank of segment carrier. Each line is equipped with two special moving mortar tanks with a single tank volume of 12.5 m^3 and a fixed mortar tank for shield machine with a volume of 25 m^3.

(3) Main parameters.

① The pressure of grouting. Synchronous grouting requires that the slurry pressure in the formation is greater than the sum of static water pressure and earth pressure at this point, so that it can be filled as far as possible and should not be split. If the grouting pressure is too large, the tunnel will be disturbed by slurry, resulting in later formation settlement and tunnel itself settlement, and easy to cause slurry running; but grouting pressure is too small, slurry filling speed is too slow, filling is not enough, will increase the surface deformation, and synchronous grouting pressure should be basically balanced with soil cabin mud water pressure, to avoid mud or mud forward channeling. The synchronous grouting pressure of this project is set to 0.3–0.6 MPa, and the appropriate adjustment is made according to the monitoring and measurement results.

② The amount of grouting. The synchronous grouting amount is 150–250% of the building gap, that is, 33.3–55.5 m^3/ring.

③ The time and speed of grouting. At the same time, synchronous grouting is carried out, the speed of synchronous grouting is matched with the propulsion speed of shield machine, and the excavation and grouting end at the same time.

④ Grouting sequence. Six grouting holes are used to inject at the same time, and a pressure detector is set up at the exit of each grouting hole in order to detect and control the grouting pressure and grouting quantity of each grouting hole, so as to realize the symmetrical and uniform injection behind the segment.

(4) The end standard of grouting and the inspection of grouting effect. The grouting pressure standard is adopted, that is, when the grouting pressure reaches the design pressure, it can be considered that the grouting meets the design requirements, and the injection can be stopped at this position. The grouting effect inspection mainly adopts the analysis method, that is, according to the P-Q-t curve, combined with the excavation speed and lining, the deformation measurement results of the surface and the surrounding buildings are comprehensively analyzed and judged. If necessary, nondestructive testing is used to check the effect, and the field radar detection is shown in Fig. 3.5.

Because the strata of shield tunnel are mainly silty sand layer and soil layer, the permeability of fine sand layer is large and the permeability is strong, which is suitable for the need of shield tunneling through this section of strata. The shield synchronous grouting system has the function of injecting double liquid slurry.

Fig. 3.5 Detection of synchronous grouting effect by radar detection technology

4) Secondary grouting management

After synchronous grouting, the annular gap behind the segment is filled, and the formation deformation and settlement in most areas are controlled. In the local area, affected by groundwater, there may be some problems in the solidification process of synchronous slurry, such as local inhomogeneity, solidification shrinkage of slurry and dilution loss of slurry. In order to improve the waterproof and compactness of the backlining grouting layer and effectively fill the annular gap after the segment, according to the monitoring results, the secondary reinforcement grouting is carried out if necessary. In the construction, the feedback of surface subsidence monitoring information is adopted, and combined with the subsidence and leakage of pipe pieces in the tunnel, it is comprehensively judged whether it is necessary to carry out secondary reinforcement grouting. The secondary reinforcement grouting material is mainly cement slurry and sodium silicate, and the slurry ratio (1 m^2) is cement slurry (water cement ratio 1:1): water glass (Baume degree 35, modulus 2.4) = 1:0.5.

KBY 50/70 grouting pump is used for secondary reinforcement grouting. The secondary reinforcement grouting pipeline is self-made, which can realize the function of quick connection and unloading and sealing and non-leakage, and equipped with slurry stop valve. The grouting pressure of secondary reinforcement grouting is 0.6–1 MPa, the gel time of slurry is adjusted to 4–15 min, and the amount of water glass is added if it's necessary to shorten the setting time. The grouting quantity is determined according to the monitored gap and the results of monitoring measurement. Grouting pressure is mainly used to control grouting.

3.3 The Instability Mechanism of Slurry Splitting Excavation Face in Shield Shallow Overburden Excavation

3.3.1 Split Extension Phenomenon of Slurry

In order to ensure the quality and safety of the shield, the most important problem is to ensure the stability of the excavation surface in the construction of a large cross-river and sea soft soil tunnel. When shield tunneling is carried out in the middle part of the river and sea, because the thickness of the overlying soil is generally small and the water pressure is large, the stability of the cutting surface is difficult to control. If there is carelessness, it is possible to split the formation with slurry to the bottom of the river (sea), resulting in major engineering accidents such as collapse and reverse irrigation of river (sea) water.

The problem of slurry eruption in the process of slurry balance shield tunneling is hydraulic fracturing in terms of physical phenomena. The study of hydraulic fracturing is still a relatively new topic in the field of soil mechanics and foundation, but it has been studied for a long time in the field of rock mechanics. Some research results are used in oil and natural gas exploitation, such as applying water pressure to drill holes to crack rock production in order to improve oil production efficiency. When the earth dam core wall is strengthened by split grouting, because the slurry pressure is too large, there will be leakage phenomenon, which also belongs to this category. In the field of grouting reinforcement, split grouting technology is used to strengthen soil in order to achieve the effect of improvement. The research in this field mainly focuses on the reinforcement range and reinforcement effect of split grouting.

There are few literatures on the split phenomenon of clay soil, and some existing studies have not been able to analyze the factors affecting the fracturing compression force, the direction of split and the extension of split. Many scholars in our country are also engaged in the study of fracturing phenomenon, mainly focusing on rock hydraulic fracturing and so on. The split study of slurry shield began with the Tokyo Bay Submarine Highway Tunnel in Japan. Li Yuan and Fu et al. studied the possible phenomenon of slurry split eruption in slurry shield construction by shield molding machine. In sandy soil layer, there is little possibility of slurry eruption except for the special condition that overburden is too thin, but in soft clay layer, the slurry pressure caused by split eruption is similar to the slurry pressure needed to maintain the stability of cutting surface, and an empirical formula of split eruption is given. However, the pressurized liquid used in the experiment at that time was water and did not take into account the effect of slurry viscosity. In addition, because it is a small model test, it is difficult to evaluate the phenomenon of splitting eruptions in the project without considering the effect of splitting extension. Morilin and Tamura Masahitoshi carried out laboratory experimental study on the split phenomenon of shield tail grouting, and studied the influence of various viscous slurry on the split pressure, and obtained the relationship between the split pressure and the clay and soil strength of slurry. the results showed that the split pressure was related to the viscosity of slurry, and the relationship between the split pressure and the viscosity of slurry and the strength of soil layer was obtained.

$$P_f = \sigma_3 + \alpha q_u \tag{3.5}$$

σ_3—Minimum principal stress;
q_u—Unconfined compressive strength;
α—Coefficient.

At the same time, Morilin and Tamura Masahitoshi pointed out that the coefficient α is related to the viscosity of slurry. When the viscosity is 25–28 s (the viscosity range of slurry used), $\alpha \approx 0.8$.

The research on splitting extension is very low at home and abroad.

The actual split phenomenon is different from the indoor small triaxial split test, and there will be an extension process after the split occurs. Yuan Dajun has studied and summarized the clay formation mud water split and extension phenomenon, especially through a large number of field fracturing tests (clay formation, depth 15–21 m, multi-point mud water field fracturing test). The empirical relationship between the split extension flow rate and the split extension pressure of slurry viscosity under the condition of 24–27 s is summarized:

$$q_c = 1.572\mathrm{P}_v \frac{A}{a_0} \tag{3.6}$$

q_c—The split continues to extend the flow of slurry;

P_p—The effective slurry pressure of split extension is defined as the difference between the split pressure and the minimum principal stress where the split occurs.

a_0—Field test hole split mud water inflow area (13.2 cm^2);

A—Shield machine and other split mud water inflow area.

Formula (3.6) cannot calculate and explain the occurrence and extension of slurry shield to reach slurry eruption, so it is necessary to study its characteristics according to the characteristics of slurry shield itself, so as to serve for the project.

3.3.2 *Model Test of Slurry Shield Excavation Mud Split*

Taking Nanjing Yangtze River Tunnel as an example, this section introduces the model test and split failure process of slurry shield excavation.

Figure 3.6 shows the diagram of the shield model developed by ourselves. The proportion and strength of artificial gypsum clay used are shown in Table 3.2. The thickness of the overburden is 1.75*D* (*D* is the outer diameter of the shield). The unconfined compressive strength of soil layer is determined by adjusting the proportion of water, and the bentonite with 10% slurry is mixed. In order to observe the split surface, the mud water is colored red. The viscosity of slurry is 25–29 s. The

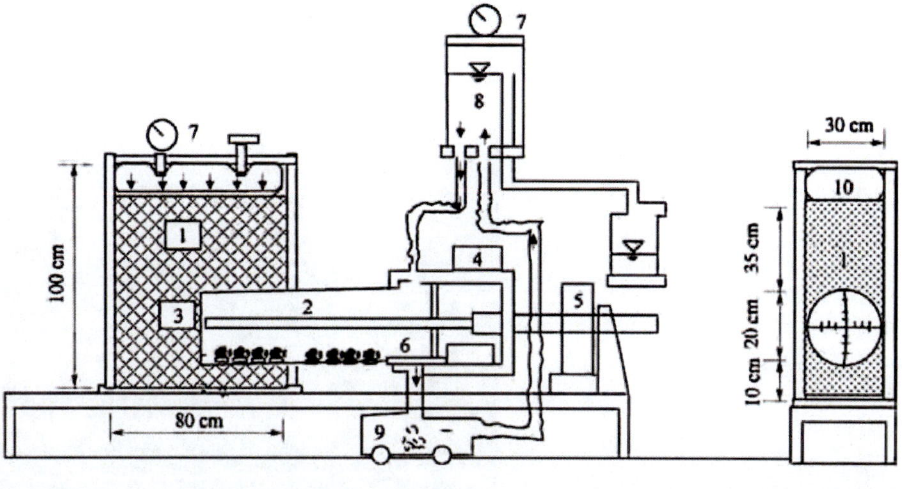

Fig. 3.6 Shield model test device. 1-Soil trough; 2-tool; 3-cutter head; 4-cutter head motor; 5-roadheader motor; 6-drain; 7-pressure gauge; 8-mud tank; 9-drain tank; 10-upper load

Table 3.2 Proportion and strength of artificial gypsum clay in soil trough

Artificial gypsum clay ratio				q_k (kPa)
Bolus alba	Gypsum	Silty clay	Water	
3000	2000	2000	4000–6000	10–140

test is carried out in two forms: one is in the case of upper load to simulate the large overburden, the other is the absence of upper load to simulate the thinner overburden. The shield modeling machine is driven by 1 mm/min and splits the formation with a fixed rate of slurry pressure (slurry eruption). The test is also carried out after driving 20 cm, and the formation is split and destroyed at a fixed rate of slurry pressure.

1) Splitting direction and width

Figure 3.7 shows the split condition of the upper load under the condition of 10 kPa. The split is radial pattern, it occurs on the vertical and parallel planes directly intersecting with the horizontal axis of the shield. In the case of upper load, the difference between lead straight earth pressure σ_1 and lateral pressure σ_3 (σ_1-σ_3) is larger, and the split should occur on the surface perpendicular to the minimum principal stress σ_3 in the horizontal direction, and the difference between the lead straight earth pressure and the lateral pressure is larger than that of the lateral pressure, and the split should occur on the surface perpendicular to the minimum principal stress in the horizontal direction. This is the same as the results of indoor split triaxial test, that is, the split occurs on a plane perpendicular to σ_3. Figure 3.8 shows the split condition without upper load.

Splitting occurs obliquely upward from the cutting surface. In the case of no upper load, it can almost be considered that $\sigma_1 = \sigma_3$, and the splitting occurs at any angle from the vertical direction to the horizontal direction. Therefore, without upper load, the split extends obliquely towards the only free surface, the ground surface. The angle of oblique upward cracking, as shown in Fig. 3.9, is 30°–45° in various cases. Table 3.3 is the results of shield model test.

Fig. 3.7 Split condition of upper load

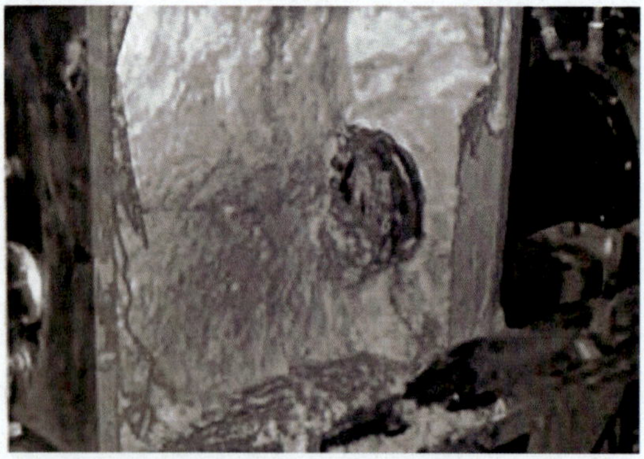

Fig. 3.8 Split condition without upper load

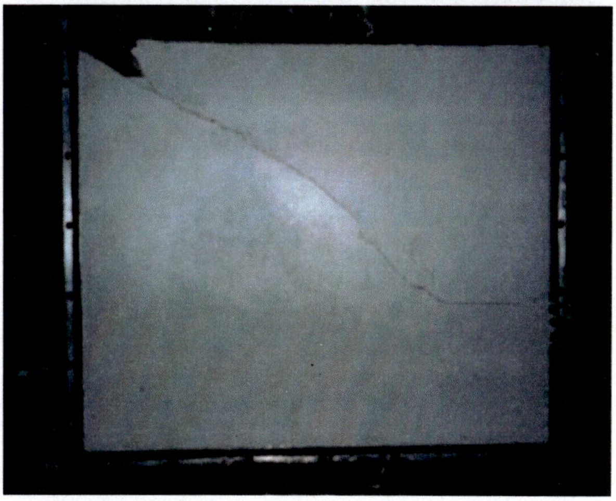

Fig. 3.9 The angle of oblique upward cracking

Table 3.3 Results of shield model test

Test number	Unconfined compressive strength q_u (kPa)	Upper load (kPa)	Split generating pressure (kPa)	Split angle (°)
1	69.5	0	71.3	45
2	114.0	0	92.3	38
3	132.1	0	108.5	30
4	134.3	0	190	35
5	101.2	10	175	90
6	78.3	10	148.1	90
7	60.5	10	126	90
8	68.0	10	157.6	90
9	19.0	10	55.7	90

Splitting occurs from occurrence to eruption, with each test taking approximately 2 s. In addition, the split width is 2.10–2.90 mm, with an average of 2.58 mm, as shown in Fig. 3.9.

2) Splitting process of shield excavation face

Figure 3.10 is the result of the 9th test of slurry pressure and shield propulsion distance. After the test, the soil layer passed by the shield model machine was excavated. The observation results showed that there was no abnormality in section A in the figure, and the cracks before splitting were found in section B, as shown in Fig. 3.11 (since the colored dye is added to the slurry, the cracks can be clearly identified). These cracks are all radially extended from the wall, and the closer the splitting depth is to the slurry eruption, the deeper the splitting depth is. It can also be seen from Fig. 3.10 that the mud pressure when cracks occur is much smaller

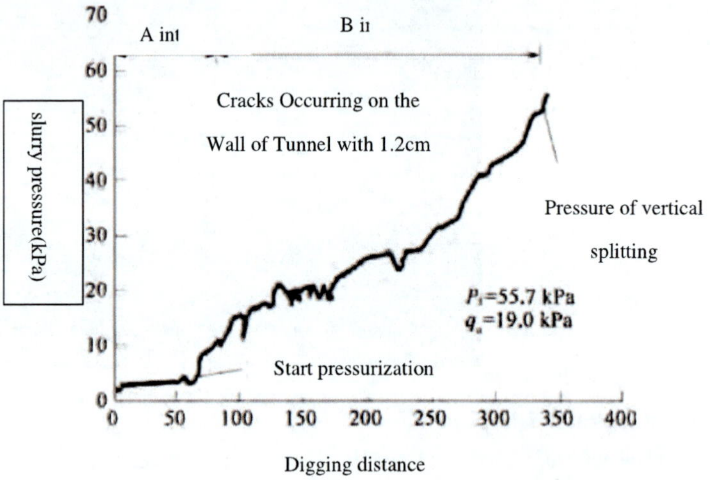

Fig. 3.10 During the 9th test excavation process

Fig. 3.11 Crack and split condition of cracks and splits

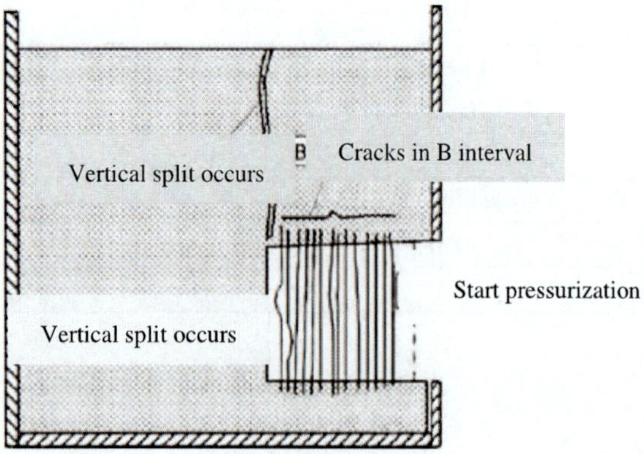

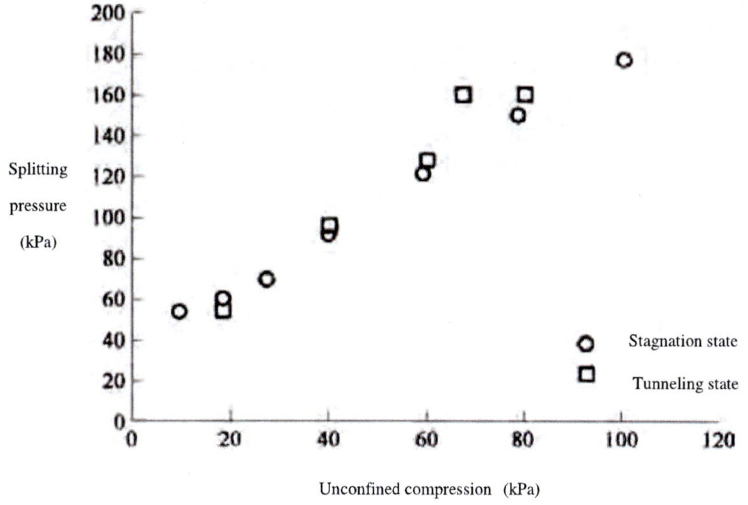

Fig. 3.12 Influence of heading state on split pressure

than the mud eruption pressure (splitting). Figure 3.11 shows the crack spacing and shield cutter.

The distance from the front end to the front end of the shield shell is almost equal, about 1.2 cm. When the actual slurry pressure is less than the slurry eruption pressure, the cracks also occur, but the length is very small, about 2 mm. At this time, the shield continues to dig 1.2 cm, the cracks have been blocked by the shield shell, slurry pressure can no longer act on the occurrence of cracks. The new excavation surface produces new cracks, so the cracks follow the ring. Figure 3.12 shows the effect of the tunneling state on the pressure of the split.

It can be seen from Fig. 3.12 that under the condition that the unconfined compressive strength of the soil layer is equal, the pressure of the split in the shield tunneling and stopping state is almost equal.

3) Split occurrence time and extension speed

From the beginning of reaching the splitting pressure to the time of splitting occurrence of mud eruption, the time of splitting occurrence can be counted by videotape for about 2 s, and the preliminary calculation of the splitting extension speed is 20–30 cm/s. This split extension velocity is the key to the occurrence of slurry eruption in actual shield excavation. If within a certain period of time, the split extends to the ground and causes slurry eruptions, it will lead to major accidents, which may usually occur when the overburden is thin or shut down for a long time. Generally speaking, the split extension velocity is related to the area of the split orifice and the total length of the split front, which can be expressed as the following formula.

$$V_f = \frac{\Delta A}{L_t} = \frac{q_c}{bL_s} \tag{3.7}$$

V_f—Split extension speed;

ΔA—Increase in area per unit time of split front end;

L_t—The circumference of the split orifice;

q_c—The flow rate of mud water extended by splitting is directly proportional to the inflow area of mud water at the crack occurrence of the excavated surface;

b—Split width.

3.3.3 Determination of the Eruption of Slurry

The actual split of slurry shield occurs when the pressure of slurry is greater than that of formation. From the above results of shield model test, it can be seen that: ① in the case of upper load on the surface, the split occurs in the vertical direction; ② When there is no upper load, the split occurs oblique upward. The former is equivalent to the case of $\sigma_1 > \sigma_3$, that is to say, it is equivalent to the case of thicker overburden; the latter is equivalent to the case of $\sigma_1 < \sigma_3$, that is to say, the overburden is thinner. According to the diameter of shield, it can be considered that the thickness of overburden is 1D or less than 1D.

When the splitting occurs on the cutting surface, whether the splitting crack tip stops in overburden or spreads to the surface to produce mud eruption depends on the relationship between the time t_s when the splitting reaches the surface and the time t_0 when the mud pressure acts on the splitting opening under a certain splitting extension speed. Time t_0 is closely related to the speed of shield tunneling, that is, if the split occurs in shield tunneling, with the continuous excavation of shield machine, the shell of shield body reaches the split hole, blocks the split hole, and the slurry pressure cannot continue to act, and the split extension will stop. The longest distance from the cutting surface to the front of the shield tunneling (Fig. 3.13), that is, the maximum distance L (m) is the distance from the front end of the tool to the front end of the shield shell. Therefore, the operating time of the slurry pressure at the opening of the split can be expressed as:

$$t_0 = \frac{L}{V_s} \tag{3.8}$$

V_s—Driving speed of shield (cm/s).

On the other hand, there is

$$L_t = \sqrt{D^2 + \frac{4q_c t \tan \varnothing}{b}} \tag{3.9}$$

t—Split continuation time;

$\varnothing$—Fig. 3.14 shows the angle of oblique upward split extension.

L_t increases with time, and become L_s when it reaches the surface. At this time, mud water erupts. If it is assumed that the time for the fracture to extend to the ground

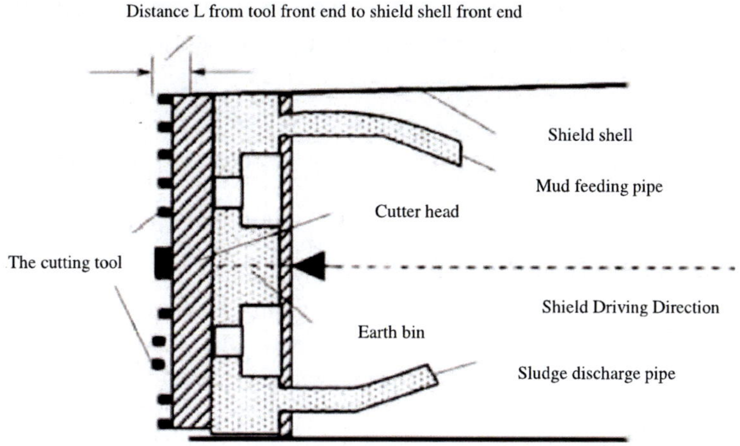

Fig. 3.13 Schematic diagram of the distance from the front end of the tool to the front end of the shield shell

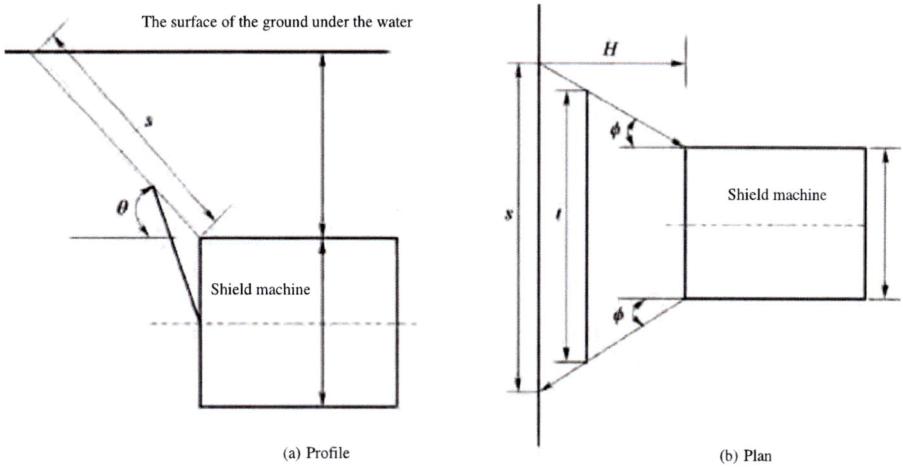

Fig. 3.14 Oblique upward split shape

is t, then L_s can be expressed as

$$L_s = \sqrt{D^2 + \frac{4q_c t_s \tan \varnothing}{b}} \tag{3.10}$$

From the formula (3.10), we can see that the expression of (t_s) is

$$t_s = \frac{(L_s - D^2)b}{4q_c \tan \varnothing} \tag{3.11}$$

By comparing t_s and t_0, we can judge whether there can be slurry eruptions: $t_s < t_0$, slurry erupts; $t_s > t_0$, slurry eruptions will not occur.

3.3.4 Prevention and Control Measures of Slurry Split

During the mud water shield tunneling, measures to prevent mud water splitting include: Controlling slurry pressure, increasing the strength of overburden, increasing the thickness of tunnel overlying soil and increasing the viscosity and specific gravity of slurry are the most important measures to control the pressure of slurry. Especially in the process of crossing the river and crossing the sea, improper setting of slurry pressure is easy to occur slurry split, erupt to the bottom of the river to lead river water recharge and other engineering accidents, such as the second tunnel of Heinenoord. In order to prevent slurry split in shield tunneling, it is necessary to study the mechanical properties of slurry split and put forward the control index to prevent split, and then to control slurry pressure in engineering, and to put forward the setting principle of slurry pressure under different excavation conditions, which is of great engineering significance.

Mud pressure is an important tunneling parameter in slurry shield construction, and its setting value is also difficult to choose. At present, most of the shield tunneling sections are set, such as 100 m before shield tunneling, and then the mud support pressure suitable for the current geological and topographic strip is selected by trial and error method.

By summing up the construction experience, this section divides the setting of slurry pressure into two situations: ① General shield driving section; ② Special shield driving section. The following is a classified discussion of these two situations.

1) Setting of slurry pressure in general shield excavation section

The general shield driving section refers to the situation in which the tunnel passes through the soil layer relatively single, the terrain is flat or the slope is flat, the fluctuation is small, and the tunnel line is smooth. Generally speaking, the shield driving section has the characteristics of simple operation and low risk of engineering accident, and its slurry pressure setting also has a lot of engineering experience, and the setting is relatively simple. Generally speaking, the slurry pressure in the center of shield machine is between its active earth pressure and static earth pressure:

$$P_{\min} = \gamma' h \cdot \tan^2\left(\frac{\pi}{4} - \frac{\varphi'}{2}\right) - 2c \cdot \tan\left(\frac{\pi}{4} - \frac{\varphi'}{2}\right) + \gamma_w H \quad (3.12)$$

γ'—Floating bulk density;
h—The vertical distance from the surface of the soil to the center of the shield;
φ'—Effective value of friction angle of soil;
γ_w—bulk density of water;

H—The vertical distance from the horizontal surface to the center of the shield.
c—For non-clay soil, $c = 0$.

$$P_{\max} = \gamma' h \cdot (1 - \sin\varphi') + \gamma_w H \tag{3.13}$$

For a specific tunnel crossing the river, the slurry pressure in the center of the shield is P_0. According to the existing engineering experience, there are:

$$P_{\min} \leq P_0 \leq P_{\max} \tag{3.14}$$

In general, the setting value of slurry pressure has a certain range of selection, which will be beneficial to the construction unit to adjust according to the specific situation, such as strict control of settlement, appropriate increase of slurry pressure, and so on. In fact, the setting of slurry pressure is also affected by its own performance parameters, such as slurry bulk density and so on, which need to be set in combination with slurry characteristics and formation characteristics.

2) Setting of slurry pressure in the excavation section of special shield

The special excavation section refers to the continuous change of geological environment in the process of shield tunneling, such as passing through buildings, or the shield body is in an abnormal working state, such as shield with pressure to change knives and so on. In the process of shield tunneling, it often crosses the building. The existence of building objects makes the setting of shield tunneling parameters different from that of general excavation section: ① Due to the overload of buildings, greater slurry support pressure is needed to maintain the stability of the excavation surface. ② Due to the limited area of overload, overload can not effectively increase the formation split resistance. On the one hand, in order to maintain the stability of the excavation surface, the influence of overload should be taken into account when setting the slurry pressure; on the other hand, the setting of the slurry pressure must prevent the formation from being split because of the excessive pressure.

To sum up, the slurry pressure determines the maximum support pressure according to the formation split resistance, and the minimum support pressure according to the active earth pressure considering overload, see formula (3.15), formula (3.16).

$$P_{\min} = (\gamma' h + q) \cdot \tan^2\left(\frac{\pi}{4} - \frac{\varphi'}{2}\right) - 2c \cdot \tan\left(\frac{\pi}{4} - \frac{\varphi'}{2}\right) + \gamma_w H \tag{3.15}$$

$$P_{\max} = \sigma_3 \cdot (1 + \sin\varphi) + c \cdot \cos\varphi + \gamma_f \cdot D/2 \tag{3.16}$$

γ_f.—Bulk density of slurry;
D—The diameter of the shield.
Then the setting value of mud water pressure of shield is satisfied:

$$P_{\min} \leq P_0 \leq P_{\max} \tag{3.17}$$

In addition, from the point of view of engineering application, controlling slurry pressure is not the only way to prevent slurry split. It is also an effective method to prevent the crack of slurry by increasing the strength of overburden, increasing the thickness of tunnel overlying soil and increasing the viscosity and specific gravity of slurry. The reinforcement of the overburden of the tunnel can be constructed in advance, but the cost of the project is high, and priority can be given to the projects with high split risk; increasing the thickness of tunnel overlying soil is affected by tunnel route selection, which may lead to higher cost; to improve the viscosity and specific gravity of slurry is a temporary measure to prevent the split of slurry in the construction process, and its engineering effect is also relatively weak. The control of slurry pressure has the characteristics of real-time, effective and economical, and can be carried out smoothly and effectively in most cases, which has become a better method to prevent and control accident risk in engineering construction.

3.4 Three-Dimensional Visualization and Simulation Technology of Shield Tunneling in Shallow Overburden

With the development of supercomputer technology and the maturity of numerical simulation technology, more and more computer aided technologies are applied to the field of engineering, which provides strong technical support for the preliminary design and demonstration of the project. Through numerical simulation, the essential understanding of various problems in engineering field is deepened, and a very powerful tool is provided for solving problems and further revealing the laws of problems. At present, the maturity of parallel algorithm theory and the wide application of supercomputer make it possible to use numerical simulation method to solve complex problems in large-scale engineering field. Taking Nanjing Yangtze River Tunnel as an example, based on the advanced high-rise technical resources of Shanghai Super calculation Center in China, this section solves the key technical problems for the construction of shallow overburden section of Nanjing Tunnel Project by using advanced means, and provides an example for the construction of super large diameter shield in the future.

3.4.1 Summary of Engineering Problems

There is a section about 150 m long in the middle section of Nanjing Yangtze River Tunnel, and the thickness of the overburden is about 11 m. The main overburden in this section is permeable silt, which is about 20 m underwater. The shield begins to be excavated from a certain distance from the shallower overlying soil layer, and gradually passes through the shallow overburden. When the shield passes through this section of overburden, it will cause great disturbance to the strata. At this time,

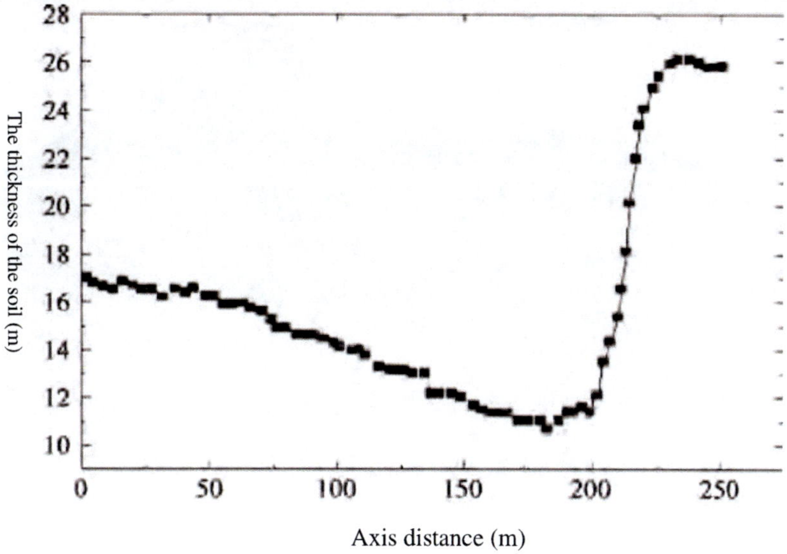

Fig. 3.15 Distribution of the thickness of the shallower overlying soil

whether the shallow overburden can remain stable, restrain grouting and settle too much when the shield passes through has an important influence on the safety of tunnel construction. Therefore, it is necessary to simulate and predict the settlement law, shield attitude change law and excavation surface stability in the process of shield crossing in shallow overburden section.

According to the thickness of tunnel overlying soil, the shallow overburden section is divided into four working conditions for comparative analysis:

① KLK5+900~LK5+975 section; ② KLK5+975-LK6+025 section; ③ KLK6+025-LK6+075 section; ④ KLK6+075-LK6+150 section.

The main contents are divided into three types, namely, surface subsidence, excavation surface stability and shield attitude. The total length of the selected section is 250 m, and the distribution of overlying soil thickness is shown in Fig. 3.15.

3.4.2 Finite Element Mode

Figure 3.16 shows a 3D reconstruction CAD wireframe model based on two dimensional geological exploration data, where the coordinate position of this section is LK5+900~LK6+150.

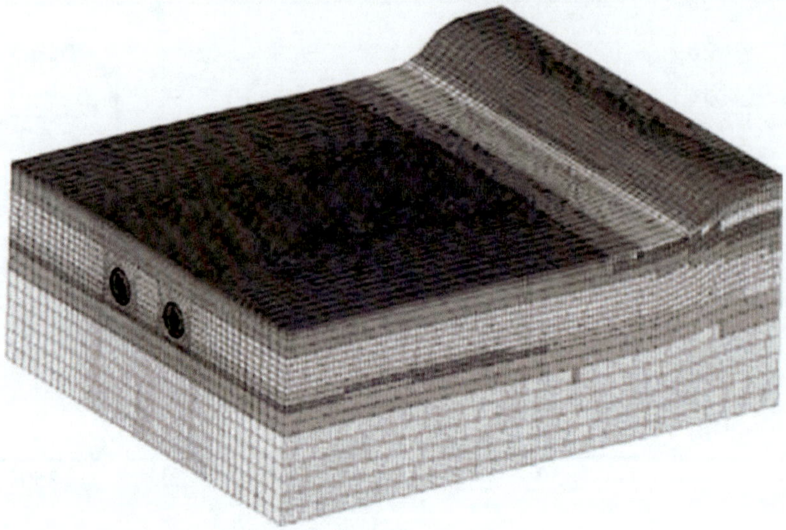

Fig. 3.16 Three-dimensional nonlinear finite element model

3.4.3 Analysis of Calculation Results

1) Surface subsidence analysis.

The variation of the maximum settlement value of each section under different working conditions is shown in Fig. 3.17.

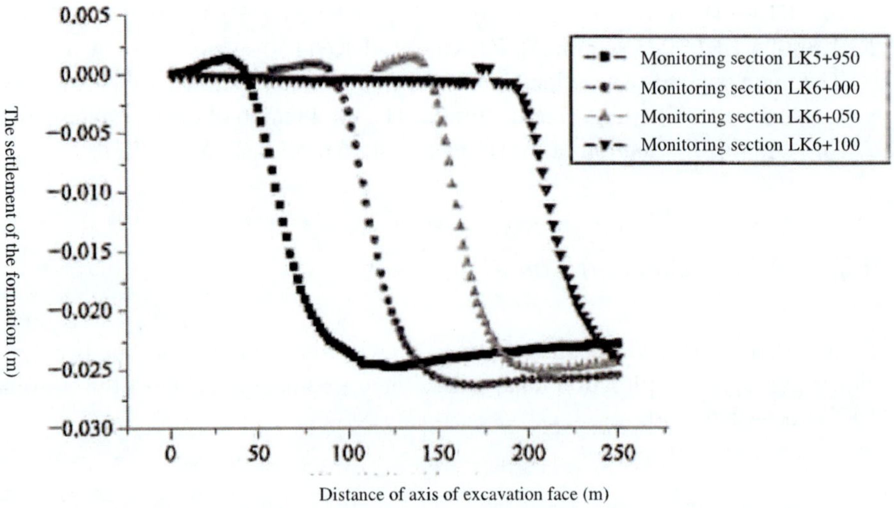

Fig. 3.17 Comparison of the variation of maximum settlement value in different working conditions

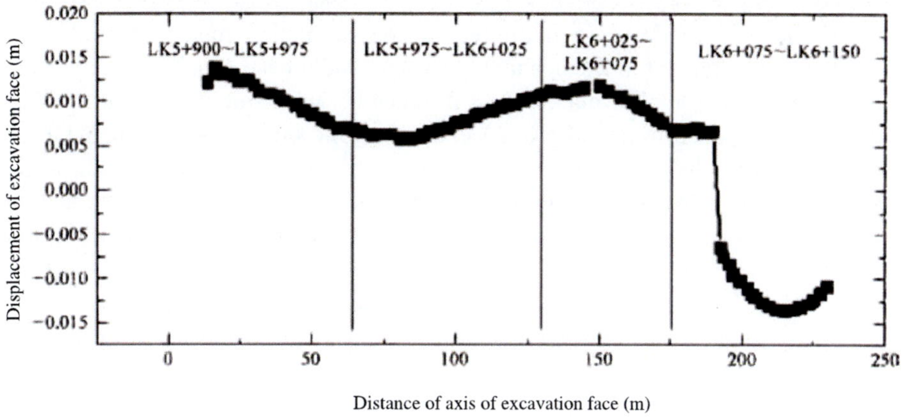

Fig. 3.18 Variation of soil displacement at shallow overburden excavation surface

With the gradual approaching of the excavation surface of the shield, the monitoring section shows a small amount of uplift, and with the shield excavation surface crossing the monitoring section, the settlement increases gradually, and finally reaches the stability. From the point of view of the maximum settlement, the second section (LK5+975~LK6+025) is the largest, the fourth section (LK6+075~LK6+150) still has a further settlement trend, and the first section (LK5+900~LK5+975) is close to the third section (LK6+025 and LK6+075). From the point of view of the overall numerical value, the final subsidence value of the four sections is close to each other.

2) Stability analysis of excavation surface

From the point of view of numerical simulation, the stability analysis of excavation surface is mainly represented by the displacement of soil body at the excavation surface. The numerical results are shown in Fig. 3.18.

When the shield excavation face is located in the first, second and third sections, the displacement of the excavation face gushes forward to the shield. At this time, the slurry pressure is slightly larger than the lateral pressure of soil and groundwater pressure. When the excavation face enters the fourth section, the displacement of the excavation face jumps, and the displacement of the excavation face changes from outward to inward. This is mainly due to the great changes in the overlying soil layer in the fourth section. Therefore, when the shield excavation face passes through the fourth section, the slurry pressure should be adjusted in time according to the situation of overlying soil, so as to maintain the excavation face within a certain stable range.

3) Shield gesture analysis

When the shield passes through the shallowest soil layer, its attitude is inevitably affected by the shallow soil. Therefore, the attitude of the shield machine crossing the shallowest soil layer is monitored by the simulation model. Attitude mainly has two measurement indexes, namely vertical shield head, shield tail displacement difference. Horizontal displacement difference between shield head and shield tail. The displacement difference between shield head and shield tail is shown in Fig. 3.19.

In the figure, the positive value represents the shield head forward, and the negative value represents the shield head forward. From Fig. 3.19a, it can be seen that when the shield passes through the shallowest soil layer, it mainly advances in a bump way.

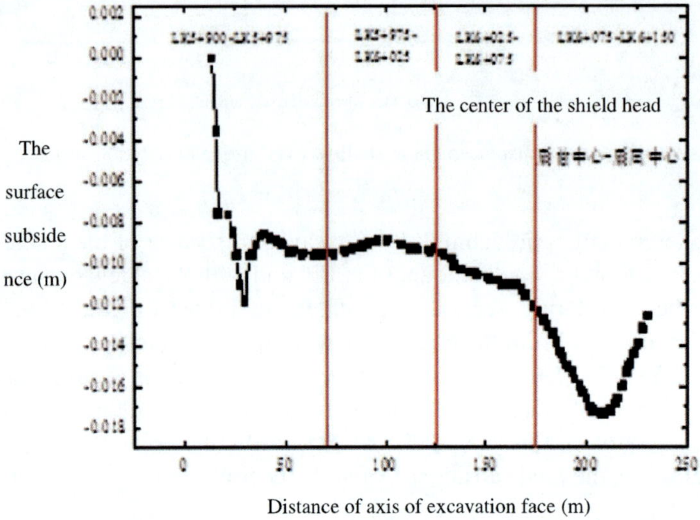

(a) Center displacement difference between vertical shield head and tail

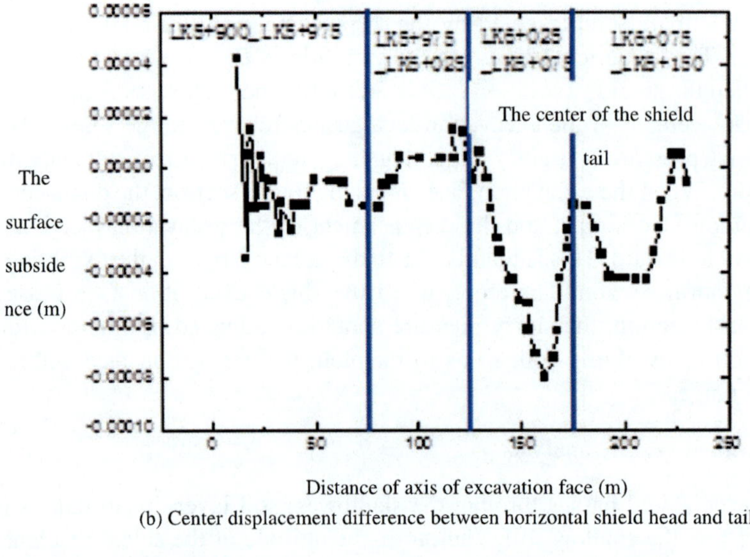

(b) Center displacement difference between horizontal shield head and tail

Fig. 3.19 Shield attitude diagram

With the decrease of overburden thickness, the amount of bump increases gradually. When located in the fourth section, the amount of bump begins to decrease due to the rapid increase of overburden thickness. Figure 3.19b shows that according to the boundary conditions of the simulation model, the attitude of the shield in the horizontal direction is basically unchanged.

Chapter 4
The Key Technology of Large-Diameter Shield Tunneling Through the Embankment

At present, the difficulty of large diameter shield tunnel crossing embankment construction is mainly manifested as stratum settlement, collapse, endangering existing buildings around. At the same time, when large-diameter slurry shield is constructed in urban river-crossing tunnels, the river bank of the city is often the central area, and it is also an important barrier for urban flood control, which is related to the safety of people's lives and property. Therefore, it is very important to ensure the stability of the embankment during shield construction. This chapter first introduces the influence on the stability of the embankment when the shield passes through the embankment construction, and focuses on the construction control technology of the shield crossing the embankment, including the control of the tunneling parameters, the grouting management behind the wall and the mud management, and introduces the deformation monitoring and reinforcement technology of the embankment.

4.1 Surface Subsidence and Hazard Analysis of Shield Tunnels

4.1.1 The Analysis of the Causes and Processes of Formation Settlement Caused by the Construction of Shields

1) Analysis of the causes of formation settlement caused by shield construction

There are many factors of ground deformation caused by shield construction, and settlement results are often caused by multiple factors. Based on the existing research at home and abroad, the main causes of stratum deformation caused by shield construction are: ① stratum loss caused by shield construction. ② deformation of

J. Chen et al., *Construction Technology of Large Diameter Underwater Shield Tunnel*, Key Technologies for Tunnel Construction under Complex Geological and Environmental Conditions, https://doi.org/10.1007/978-981-16-5896-9_4

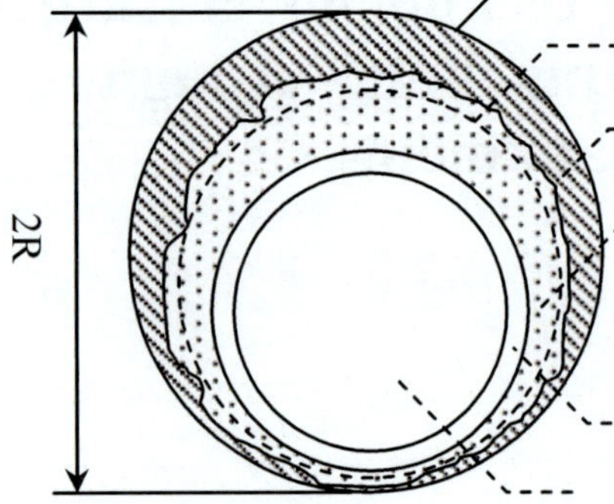

Fig. 4.1 Schematic diagram of soil loss

segment lining structure. ③ Consolidation and secondary consolidation of soil around the tunnel.

The stratum loss caused by shield tunneling is the main reason for stratum settlement deformation. The stratum loss is equal to the volume of soil excavated during shield construction minus the volume occupied after the completion of the tunnel (the volume occupied after the completion of the tunnel includes the volume of the internal space of the tunnel, the lining structure and the volume of the post-wall grouting slurry wrapped on the outer surface), as shown in Fig. 4.1. Due to over-excavation of shield cutterhead, thickness of shield body and assembly clearance, the excavated soil volume during shield construction is often much larger than the excavation volume (theoretical excavation volume) calculated according to the designed section area of tunnel.

The formation loss rate is represented by the percentage of formation loss in the theoretical excavation volume. Stratum loss can usually be divided into three categories: ① Normal ground loss in the case of a person's operation process in full compliance with the operating rules, which is all attributable to the objective conditions of the site. ② Abnormal stratum loss caused by improper operation of people in the construction process, such as shield machine body retreat, shield over-excavation, insufficient support pressure and insufficient grouting amount. ③ Hazardous stratum loss caused by special geological conditions or unpredictable disaster reasons, such as sudden large amount of soil flow in front of shield excavation.

Lining deformation is also one of the reasons for ground settlement. The lining segments of the shield tunnel are assembled by multiple blocks. The stiffness of the segment lining block, the block type of the lining ring, the structural type of the indirect head of the ring and the fineness of the segment assembly will affect the stress and deformation of the structure. When the shield passes through, because the original stratum is replaced by segment lining and tunnel, the surrounding stratum squeezes the lining, and the lining is subjected to stratum soil pressure and deforms under force, resulting in a certain displacement of the stratum. However, in practical engineering, as the main structure of tunnel stress, the segment needs to meet the specific strength and stiffness requirements in design, and the allowable deformation range is small. Therefore, the influence of lining deformation is often not considered in the analysis of stratum settlement.

The consolidation and secondary consolidation of soil around the tunnel are also important causes of formation deformation. The over-excavation, pushing, friction, shear and post-wall grouting of shield in the process of propulsion will disturb the surrounding soil, so that the surrounding strata will produce excess pore water pressure. When the shield tail passes through, the excess pore water pressure reaches the peak value, and then dissipates. The consolidation of the soil causes the formation settlement, which is called consolidation settlement. The excess pore water pressure generated by shield construction and its dissipation distribution are shown in Fig. 4.2. In addition, the soil skeleton around the tunnel will also generate creep due to shield disturbance, which is called secondary consolidation settlement. Generally speaking, the consolidation settlement caused by shield construction accounts for the main part, and its duration is relatively long. The secondary consolidation settlement is relatively small in value, and it occurs when the consolidation settlement occurs.

2) Analysis of formation settlement caused by shield construction

During shield construction, stratum will inevitably be disturbed. With the advance of shield, the settlement of stratum will gradually increase. Based on the relevant literature at home and abroad, the stratum settlement caused by shield construction is divided into five stages shown in Fig. 4.3.

(1) Pre-settlement: the settlement of the soil outside the sliding surface in front of the shield before the excavation surface reaches, and the settlement range depends on the formation conditions. For sandy soil, the main reason is that shield tunneling decreases the groundwater level and increases the effective stress of the stratum, which leads to the consolidation settlement of the stratum. For ultra-soft clayey soil, the pre-settlement is caused by over-excavation of excavation face. In general, the initial settlement is less than 5%–10% of the total settlement.
(2) The settlement in front of excavation: refers to the surface subsidence before the shield cutterhead is about to arrive. It generally refers to the settlement generated during the whole process from the observation point several meters away from the excavation face to the shield excavation face reaching this point. The main reason is the imbalance of water and soil pressure on the excavation

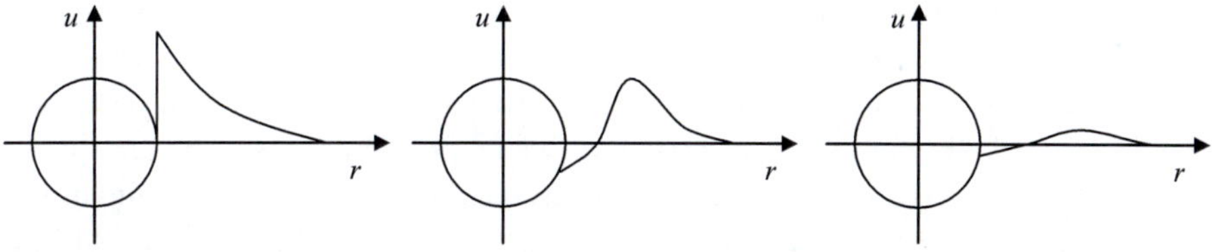

(a) After the shield enters the formation (b) After the shield tail leaves the formation (c) Shield leaves the formation for a while

Fig. 4.2 Schematic diagram of excess pore water pressure distribution

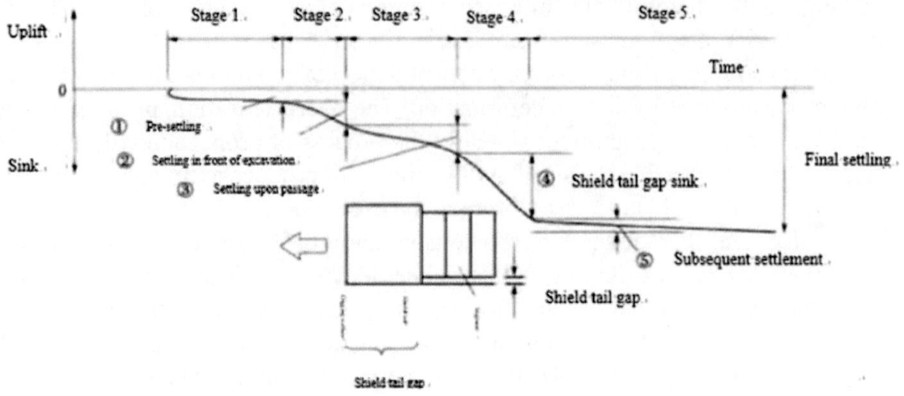

Fig. 4.3 Ground deformation law caused by shield tunnel

face. If the shield support pressure is too small, the soil in front of the excavation unloading deformation, resulting in settlement. If the supporting pressure is too large, the stress state of the soil in front of the excavation will change, resulting in surface uplift. This part of the settlement accounts for 15%–20% of the total settlement.

(3) Settlement generated when passing: that is, stratum settlement generated when passing from the incision position to the shield tail. The main reason is the friction between the outer surface of the shield and the stratum, or the formation disturbance caused by over-excavation, and the weakening of the three-dimensional support effect, resulting in the formation stress release. In addition, over-excavation will inevitably occur when the shield curve advances, resulting in ground settlement. This part of the settlement accounts for 15%–25% of the total settlement.

(4) Settlement when the tail of shield is out: when the tail of shield is out, the gap between the outer diameter of shield and the outer diameter of tunnel lining causes formation loss, ground stress release and elastic–plastic deformation of the stratum due to the delayed grouting behind the wall or insufficient grouting amount. When the lining segment is separated from the tail of the shield, the earth water pressure acting on the segment will squeeze the lining ring and cause the deformation of the lining ring, which will also lead to small ground settlement. This part of settlement deformation accounts for 30%–40% of the total settlement deformation of shield construction, which is the most important settlement during shield construction.

(5) Subsequent consolidation settlement: after the shield tail passes, due to the gradual dissipation of excess pore water pressure caused by the previous shield disturbance, the settlement caused by the continuous consolidation and secondary consolidation of soil mass accounts for 5–25% of the total settlement of shield construction. The settlement at this stage depends on the stratum

properties and the disturbance degree of shield to the surrounding soil, which is more obvious in clay stratum than in sand stratum.

Shield construction inevitably disturbs the surrounding soil, causing the movement of stratum and surface settlement. The stratum is more complex when large-diameter slurry shield is constructed at the bottom of the river, and the embankment structure also has a certain impact on tunnel construction, thereby increasing the construction difficulty. If the construction control is improper, it will cause excessive ground displacement and endanger the stability of embankment structure. During the construction of many cross-river tunnels in China, there have been major risks such as deformation of embankment gates, sand gushing and seepage of flood control walls. Therefore, it is very important to control construction to reduce the influence of shield on embankment.

4.1.2 The Hazard of Shield Construction to the Surface Structure

Shield construction inevitably disturbs the surrounding soil and causes surface settlement. When the local surface settlement is too large, it usually causes the settlement, inclination and even cracking of adjacent buildings (structures), which has become one of the most noteworthy problems in the shield construction process. The research on the influence of surface deformation on buildings (structures) was first carried out in the coal mining area. Many scholars have made in-depth studies on the deformation damage of buildings (structures) in the coal mining area, and have formed the damage evaluation standard of buildings (structures) in the mining area. However, the damage evaluation standard of buildings (structures) in the tunnel construction area has not been established.

At present, relevant studies have shown that the main damage modes of buildings (structures) caused by shield tunnel construction include uniform settlement damage, damage caused by ground curvature, damage caused by inclination, and damage caused by horizontal deformation. Usually, the compression and negative curvature of the surface appear at the same time, and the tensile and positive curvature of the surface occur at the same time. In practice, the deformation of buildings (structures) is often a comprehensive manifestation of various forms. Among them, the curvature, inclination and horizontal deformation caused by differential settlement damage to buildings (structures) is more obvious, and distortion, subsidence, cracking and even collapse will occur. In the process of shield tunneling, the influence of shield construction disturbance on buildings (structures) is restricted by many factors, among which the most important are the engineering geological conditions, the foundation form of buildings (structures), the position relationship between buildings (structures) and tunnels and the relative distance.

In 1956, Skempton et al. combined with nearly 100 tunnel construction examples, determined the allowable total surface settlement and differential settlement of the project, and summarized that the main reason for the cracking and failure of the building structure during tunnel construction was the excessive curvature radius of the settlement trough. However, compared with the curvature radius of the settlement trough which is difficult to measure, the measurement of the deformation angle is much easier, so the measured deformation angle is usually used as the main criterion of the deformation of the building structure.

In 2010, Zhang Zhongmiao and others based on Hangzhou Qingchun Road cross-river tunnel, combined with surface subsidence monitoring data, the Peck formula was used to fit the surface subsidence curve caused by shield construction, and the surface subsidence of the dam section and other sections were compared and analyzed. It is proposed that due to the complexity of the embankment structure, the disturbance of shield construction, the cyclic load of the vehicle on the top of the embankment and the combined action of rainfall, the settlement of the embankment is aggravated, which leads to the surface settlement caused by shield crossing the embankment is larger than that of other sections.

In 2014, Zhouyunjiang and Jieyu established a three-dimensional numerical analysis model for Shanghai Yangtze River shield tunnel crossing the levee section. The numerical calculation results were compared with the measured data to analyze the rationality of the numerical calculation. On this basis, the settlement law of the tunnel and the dam under different cross angles and different buried depths was analyzed. The results show that the larger the cross angle is, the smaller the maximum settlement value is, and the greater the uneven settlement is. The greater the buried depth, the smaller the settlement, the smaller the uneven settlement.

In summary, the formation loss caused by shield construction, the deformation of shield lining and the consolidation of soil after disturbance are the internal causes of land subsidence. There is a regular process of settlement, and the settlement of each stage is closely related to the state and location of the shield. The influence of shield construction disturbance on buildings (structures) is controlled by many factors, including the basic form of buildings (structures), engineering geological conditions and the position relationship between buildings (structures) and tunnels. In practical engineering, due to the influence of buildings (structures), especially its self-weight and structural stiffness, the deformation curve of buildings (structures) is very different from that of free ground. The existing research methods are mainly measured data regression method, indoor model test method, numerical analysis method and intelligent decision method. Through certain assumptions and simplification, the model is established by numerical analysis, and the relevant laws of shield construction are summarized. Combined with indoor test and field measured data, the correction and analysis are carried out, which has important guiding significance for practical engineering and theoretical research.

4.2 Study on the Effects of Large Diameter Shield Crossing on Embankment Subsidence and Stability

In this section, the shield tunneling of Line S of Nanjing Wei San Road Crossing River Channel Project through the bank section of the north bank of the Yangtze River is taken as the background, and the finite difference software FLAC^{3D} is used to numerically simulate the construction process of the shield. The model verification is carried out according to the previous settlement monitoring results, and the influence of shield tunneling on the surface deformation and stability of the bank is analyzed. The differences strata deformation and overall stability under different grouting effects, bank slope conditions and tunnel buried depths are compared.

4.2.1 An Overview of the Project of the Three-Way River Channel in Nanjing and the Geological Conditions of the Embankment Section

Nanjing Wei San Road Crossing Channel is located between Nanjing Yangtze River Tunnel (Nanjing Wei Qi Road Crossing Tunnel) and Nanjing Yangtze River Bridge. It is the traffic artery connecting Pukou New Town and Nanjing Main Town, and its location is shown in Fig. 4.4. Starting from the intersection of directional river and Puzhu road, the project sets the main tunnel toll station along the directional river south to Fengzi river road, and then converts the horizontal and vertical alignment in the open cut section, and introduces the vehicle into the double-layer tunnel of the

Fig. 4.4 Location of the Nanjing Wei San Road Crossing Channel Project

south and north lines. After heading north at the directional estuary to avoid the deep trough of the estuary, the shield tunnels of the south and north lines are separated: the south line tunnel passes through Qianzhou and Jiangxinzhou successively and goes to shore at the intersection of Yangtze River Avenue and Dinghuaimen Avenue, with a total length of 7.363 km (a total length of 4.135 km). The north line tunnel is 7.339 km long (3.537 km long) from the upper reaches of the Qinhuai River estuary after crossing the submarine. Nanjing Wei San Road crossing channel is designed for two-way double eight lanes, single pipe has independent capacity.

The site of Nanjing Wei San Road crossing the Yangtze River is mainly the Yangtze River alluvial plain, the Yangtze River waters and the river center. The Nanjing Yangtze River levee on both sides of the Yangtze River waters is a high-level levee, which is the key water conservancy facilities for its crossing. The settlement and stability control of shield construction are very important. The near-water sides of the embankments on both sides of the South and the North are mortar block stone feet and dry block stone slopes, and the annual average high water level of the Yangtze River is 8.37 m. Small structural stiffness, poor integrity, high water level, a small amount of settlement and horizontal deformation will endanger the safety of the embankment. When a large displacement occurs in the soil of the slope facing the water surface, the upper soil of the slope will move and even slide. Surface, triggering the overall sliding of the embankment slope. Moreover, the shallow overburden at the river bottom and the great variation of overburden depth at the bank slope increase the difficulty of shield tunneling under the bank. Therefore, it is necessary to analyze the influence of shield construction on the bank settlement and slope stability, and establish reasonable construction control indexes to ensure the smooth passage of large-diameter slurry shield.

This section takes the south line project crossing the north bank embankment as an example as an example for analysis. The corresponding interval mileage pile number SDK4+350–SDK4+850, the slope length of the embankment facing the water surface is about 180 m, and the slope height is 36 m. The stratigraphic distribution is shown in Fig. 4.5.

4.2.2 Calculation Model and Parameters

Figure 4.5 shows the downward slope of the tunnel when crossing the embankment. In order to simplify the calculation, it is considered that the tunnel axis level, that is, the slope model of the tunnel does not change. According to the actual working condition of Nanjing Wei San Road crossing the river channel, a semi-model is established based on the symmetry principle. The shield diameter is 15 m, the longitudinal dimension is 500 m, the transverse dimension is 100 m, and the vertical dimension is 74.5–110.5 m, with a total of 50,832 units and 55,349 nodes. The formation unit adopts the Mohr–Coulomb constitutive model and the isotropic fluid model considering the seepage effect. The initial pore pressure of the model is calculated automatically by the water level. The normal displacement is constrained by the

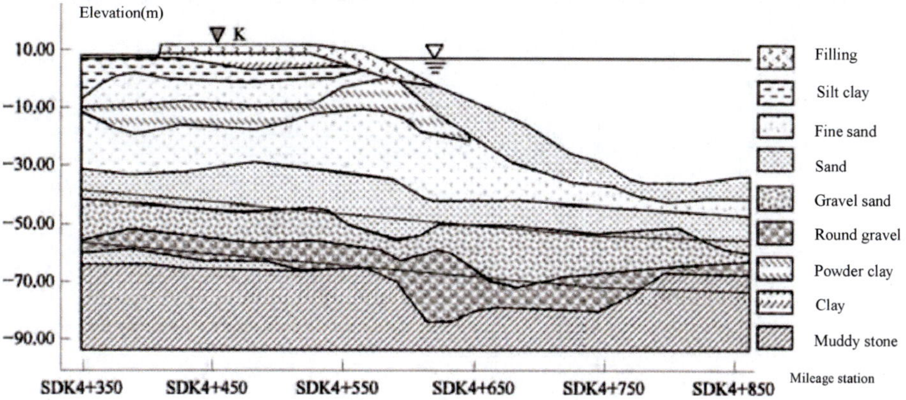

Fig. 4.5 Schematic diagram of the shield crossing the embankment stratum (sdk4+350~sdk4+850)

bottom boundary and the front, rear, left and right boundaries of the model. The bottom boundary and the left and right boundaries are permeable boundaries. The front and rear boundaries are impervious boundaries. The displacement and seepage of the top boundary of the model are free boundaries. The shield shell is simulated by shell element, and the lining segment is simulated by lining element. It is assumed that the constitutive relations are elastic, and the stiffness of segment ring is corrected according to the method. The correction coefficient is 0.75. The grouting behind the wall is simulated by "three stage consolidation cementation" elastic equivalent layer and shield tail clearance element, assuming that the filling rate of grouting body is 85%. The calculation model is shown in Fig. 4.6. According to the geological survey report and the experimental results of field soil, the relevant calculation parameters are shown in Table 4.1.

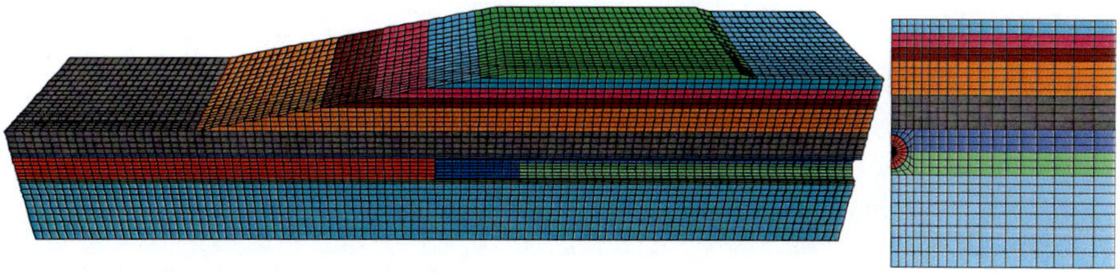

Fig. 4.6 Schematic diagram of the numerical calculation model

Table 4.1 Numerical calculation parameters (according to the geological survey report)

Number	Name	Thickness (m)	Density (kN/m^3)	Volume modulus K (MPa)	Shear modulus G (MPa)	Internal friction φ (°)	Sticky force C (kPa)	Pore ratio e	Saturation Sr (%)	Penetration factor K (× 10^{-7} cm/s)	Poisson by μ
①0	Filling	5.01	1.36	9.53	4.4	13.7	34	1.01	99	8.5	0.3
②1	Clay powder	7.06	1.37	9.83	4.53	13.3	20.5	0.97	97	5.15	0.3
②3	Silt	5.94	1.44	27.43	12.66	28.3	5	0.85	98	1.04	0.3
③1	Silt clay	7.5	1.39	9.63	4.44	15.18	24.1	0.957	92	1.85	0.3
④1	Silt	12.5	1.53	34.45	15.9	26.7	4	0.751	92	1.01	0.3
④3	Fine sand	7.5	1.55	33.28	15.38	33.9	1	0.736	89	2.82	0.3
⑤2	Gravel	10	1.6	40.58	18.73	26.72	0	0.577	93	11.96	0.3
⑥1	Pebble	10	1.7	62.5	28.85	22.24	0	0.512	95	80	0.3
⑦	Mudstone	>20.00	1.91	3107.36	3162.84	25.02	48.2	–	–	0.85	0.12
–	Shield shell	0.15	7.9	116,667	87,500	–	–	–	–	–	0.2
–	Lining	0.6	2.45	19,167	14,375	–	–	–	–	–	0.2

4.2.3 Calculation Results Analysis

4.2.3.1 Computational Rationality Verification

Numerical simulation requires reasonable calculation to have reference value. The usual verification method is to compare the calculated value with the actual monitoring data. The shield starts from the launching shaft in the north of the Yangtze River and enters the bottom of the Yangtze River through the north bank of the Yangtze River. In order to monitor the ground subsidence and optimize the control of shield tunneling parameters according to the ground subsidence, multiple ground monitoring sections are set up on the surface between the levee and the working well in Jiangbei for subsidence monitoring, numbered S1–S3, L1–Ln, M1–M2, arranged as shown in Fig. 4.7.

The S1–S3 monitoring section is located in the reinforcement area of the originating well, and the stratum has been reinforced. The settlement law cannot reflect the influence of shield tunneling on undisturbed soil, so it is not analyzed here. The surface of L1–Ln and M1–M2 sections is farmland and outside the reinforcement range of the originating well, there is no disturbance before construction. M1 and M2 are located in the above model, so the settlement values of these two monitoring sections are compared with the settlement values obtained by numerical simulation, and the results are shown in Fig. 4.8.

It can be seen from Fig. 4.8 that the measured settlement trough caused by shield construction is slightly wider than the calculation result. The measured settlement above the tunnel axis is slightly smaller than the calculation value, while the actual settlement on both sides is slightly larger, but the difference in settlement value is 2–3 mm. The settlement curve calculated by the model is generally consistent with the measured settlement curve, and the width and depth of the settlement trough are not significantly different, indicating that the selection of the above calculation model is reasonable.

Fig. 4.7 Schematic diagram of monitoring band distribution

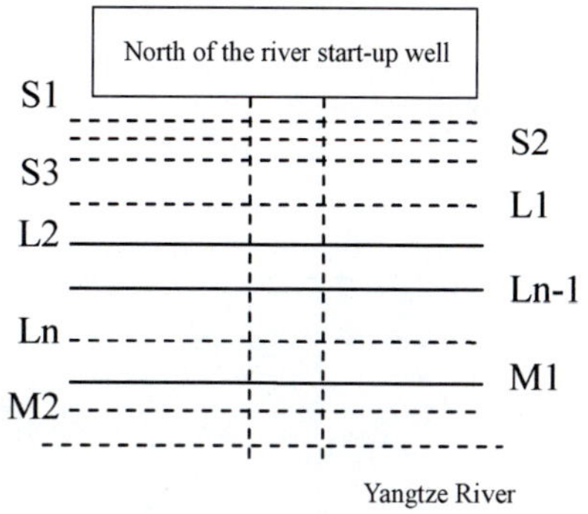

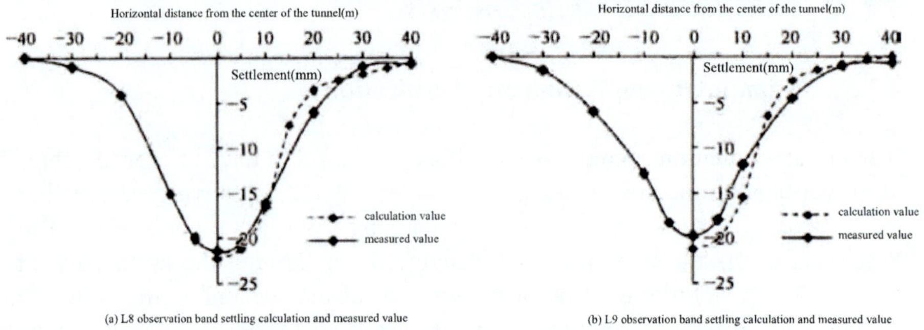

Fig. 4.8 Settlement calculation and measured curve of surface observation zone

4.2.3.2 Analysis of Surface Settlement and Horizontal Displacement

According to the excavation steps, the shield construction is simulated. After the tunnel runs through the model, the final vertical settlement contour and the final horizontal and vertical displacement contour (along the horizontal axis of the tunnel) are obtained, as shown in Fig. 4.9. In the figure, $X = 0$ corresponds to the surface above the tunnel axis, and $X = 100$ corresponds to the surface on the right boundary of the model in Fig. 4.5. The location of $Y = 0$ corresponds to the mileage pile number SDK4+350, which is the starting point of the tunnel excavation model. The location of $Y = 500$ corresponds to the mileage pile number SDK4+850, which is the location when the tunnel excavation is completed.

Figure 4.9a shows that shield construction causes ground settlement, the settlement above the tunnel axis is larger, and the ground settlement far from the tunnel

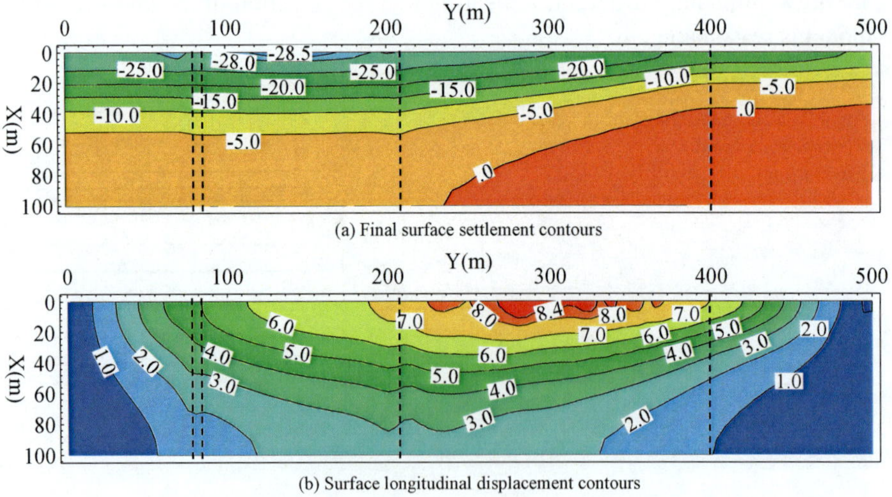

Fig. 4.9 Final settlement and longitudinal displacement contours of the surface

axis is smaller. At the central position at the top of the embankment ($Y = 150$ m), the settlement is the largest, reaching 28.5 mm, and the settlement trough is the widest. The width of the settlement trough at the bottom of the slope facing the water surface gradually decreases, and the settlement gradually decreases. Figure 4.9b shows that after the completion of the shield construction, the slope produces longitudinal displacement along the tunnel axis, and the slope facing the water shows the outward extrusion. For example, the horizontal displacement at the midpoint of the slope ($Y = 300$ m) reaches 8.4 mm. Therefore, when shield tunneling under embankments with thick overburden, it is necessary to ensure stable posture and set reasonable supporting stress, so as to prevent excessive surface settlement at the top of embankment and threaten the safety of embankment. When the shield passes through the water-facing slope section, the slurry pressure should be controlled to avoid the "extrusion" damage of the water-facing slope.

4.2.3.3 Coastal Slope Stability Analysis

For the embankment slope structure, besides the settlement and horizontal displacement, the stability of the slope is also crucial. It is necessary to extract each construction step file of the shield crossing the embankment, and use the strength reduction method to analyze the stability of the slope during the shield crossing the embankment. The shear strength of stratum soil is reduced step by step, and the calculation convergence criterion is selected as the judgment basis of slope instability combined with slope displacement nephogram and plastic zone. When the shear strength of soil is reduced to a certain extent, the calculation is no longer convergent, and the plastic zone is penetrated and the displacement of soil in the circular arc is significantly increased, the slope is judged as instability, and the initial strength of soil and the resistance at this time are judged. The ratio of shear strength is defined as safety factor. Figure 4.10 shows the variation curve of slope safety factor during shield construction.

It can be seen from Fig. 4.10 that before the shield construction, the safety factor of the initial slope of the embankment is about 1.63. When the shield excavation face advances to the $Y = 100$ m, it begins to affect the embankment slope, so that the safety factor continues to decrease, and remains unchanged after the shield advances to the $Y = 400$ m, maintaining at about 1.48. At the same time, it can be seen that the stability of the embankment slope is closely related to the position of the shield: when the shield is far from the slope, the disturbance influence area of the excavation surface has not yet reached the embankment slope, and the shield tunneling has little effect on the safety factor of the slope facing the water surface. When the shield excavation surface is gradually close to the slope section, the disturbance of shield construction on the water-facing slope is increasing, and the safety factor begins to decrease slowly. With the advance of the shield, the shield reaches the front half of the waterward slope. At this time, the disturbance of the cutterhead excavation, the shear friction of the shield shell on the soil, the deformation of the shield tail hole and the grouting body behind the wall, and the "extrusion" effect of the shield on the

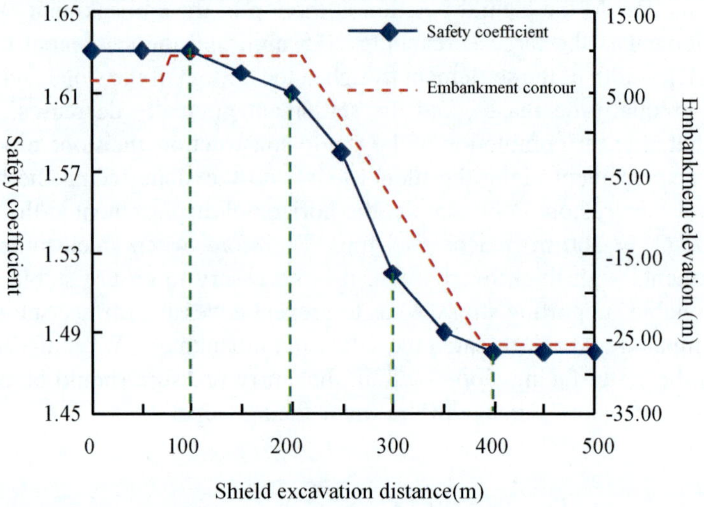

Fig. 4.10 Change of safety factor of slope during the process of shield crossing the bank

waterward side of the levee reach the maximum, and the safety factor of the slope is significantly reduced. When the shield enters the rear half of the slope, the influence area of the excavation face gradually moves out of the bank slope, and the change of the safety factor gradually tends to be gentle. As the shield continues to advance, the safety factor will not change when the excavation face is away from the slope.

4.2.3.4 Analysis of Different Operating Conditions

The above analysis is only for the specific working conditions of Nanjing Wei San Road Project, for different construction parameters, different bank slopes and the settlement of the stratum and the stability of the bank slope under burying depth need to be further studied. This section selects the three conditions of grouting effect, embankment slope and tunnel depth to analyze.

1) Influence of different grouting effects

The settlement when the shield tail is detached is the most important settlement during shield construction, accounting for 25%–40% of the total settlement deformation of shield construction, which is the main source of shield construction settlement. Therefore, this construction parameter is selected for analysis. This settlement is mainly due to the formation loss caused by the shield tail gap produced by the shield tail out of the segment. In the project, the back wall grouting method is used to fill the shield tail gap to reduce the formation settlement. In general, the post-wall grouting effect is determined by the filling rate of the slurry, the consolidation and cementation characteristics of the grouting body (elastic modulus is often considered

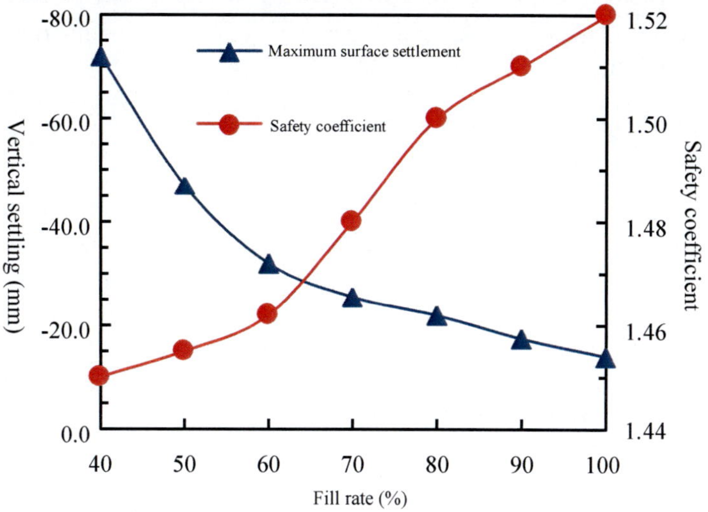

Fig. 4.11 Maximum settlement value and final safety factor of surface filling under different wall grouting filling rates

in the numerical analysis) and the grouting pressure. In order to simplify the calculation, the filling rate of slurry is used to approximately characterize the post-wall grouting effect. In order to consider the influence of different grouting effects, the post-wall grouting model which can consider the filling rate is adopted. It is assumed that the filling rate of the slurry is 40%, 50%, 60%, 70%, 80%, 90% and 100%, respectively. The slurry is considered as a "three-stage consolidation-cementation" elastic equivalent layer. The grouting pressure is 0.1 MPa higher than the formation pressure at the corresponding position. The final safety factor of the slope and the maximum surface settlement under different grouting effects are calculated as shown in Fig. 4.11.

It can be seen from Fig. 4.11 that with the increase of grouting filling rate, the maximum ground settlement gradually decreases, and the safety factor of embankment slope increases. When the filling rate is less than 60%, the vertical settlement increases obviously and the safety factor changes little. When the filling rate is greater than 60%, the trend of vertical settlement change slows down, and the safety factor increases significantly. Therefore, in practical engineering, grouting filling rate must be controlled more than 60%. Due to the complex grouting process, the maximum filling rate can only reach about 90%. Even if the grouting amount is large, it is difficult to achieve full filling of the shield tail gap. Excessive grouting amount or excessive grouting pressure is easy to cause surface uplift, threatening the safety of the embankment, and grouting at the bottom of the river. Based on this, it is suggested that the filling rate should be controlled at 70%–90% in the construction of the river-crossing channel project of Wei San Road in Nanjing. According to the engineering experience, the empirical over-excavation coefficient $\alpha = 1.21$ and the slurry loss coefficient $\delta = 0.42$ can be taken to obtain the corresponding grouting rate of 146.0%–187.8%, so the grouting amount of 16.9–21.65 m^3/ring can be used in the construction.

2) Influence of different bank slopes

Large-scale river crossing tunnels at home and abroad are frequently constructed, and crossing various complicated forms of dikes has occurred from time to time. Because it is different In the case of the bank form (especially the slope of the bank), the shield construction has different effects on the settlement and stability of the dike. In order to study this difference, on the basis of the above calculation model, change the slope coefficient of the bank's waterfront slope (6, 5.5, 5, 4.5, 4, 3.5, 3) without changing other parameters (after wall grouting) "Three-stage consolidation-cementation" equivalent model, assuming a fill rate (70%), the simulated shield crosses the dike different slopes.

From Fig. 4.12, it can be seen that with the increase of slope coefficient (i.e., the slope gradient decreases), the decrease of surface settlement at the top of the dam is not obvious, but the safety factor increases significantly. When the slope coefficient is 3.0, the water-facing slope is steep, and the safety factor is only 1.35. As the slope slows down (that is, the slope coefficient increases), the safety factor increases. When the slope coefficient is 5.0, the safety factor is 1.43. When the slope coefficient exceeds 5.0, the slope has become very flat and the safety factor increases more significantly.

3) Influence of different tunnel depths

In different river crossing tunnel projects, the buried depth of shield passing through the embankment is different. Under different buried depth, the influence of shield construction disturbance on the embankment settlement and bank slope stability is

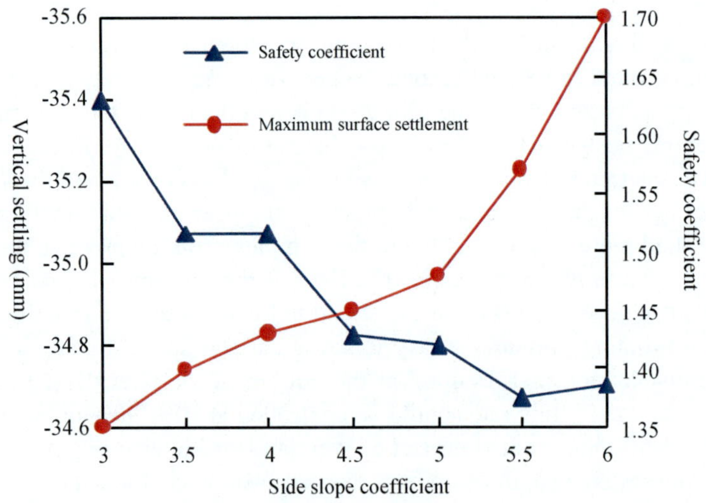

Fig. 4.12 Maximum settlement value and final safety factor of different surface slopes

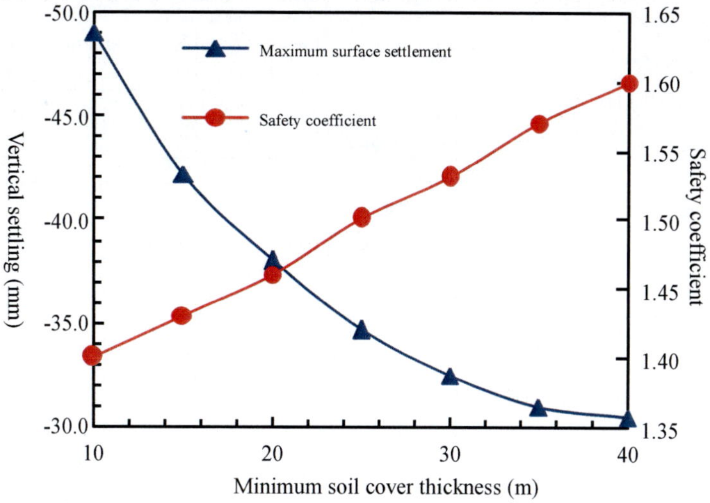

Fig. 4.13 Maximum settlement value and final safety factor of different surface thicknesses

different, which needs further research. In order to consider the different buried depth of the tunnel, the thinnest overburden thickness at the riverbed is assumed to be 40, 35, 30, 25, 20, 15 and 10 m without changing other parameters (the "three-stage consolidation-cementation" model is adopted for grouting behind the wall, and the filling rate is assumed to be 70%). The results are shown in Fig. 4.13.

It can be seen from Fig. 4.13 that with the increase of shield depth (i.e., the minimum overburden thickness in the figure increases), the vertical settlement value of the embankment top surface decreases. The safety stability coefficient of embankment slope increases approximately linearly, but the increase is small. The influence of shield depth on the surface settlement of embankment is more significant than that on the safety factor of embankment slope.

4) Sensitivity analysis of influencing factors

Figure 4.11 shows that when the filling rate is increased from 40 to 10%, the settlement value of the surface of the embankment is reduced from 72.0 to 13 mm, and the reduction value is 59 mm. Meanwhile, the safety factor of the levee increases from 1.45 to 1.52. Only 0.06 is increased. Figure 4.12 shows that when the slope coefficient of the surface of the water increased from 3 to 6, the settlement value of the surface of the embankment was reduced from about 35.4 to 34.7 mm, which was reduced by only about 0.7 mm, while the safety factor of the embankment increased from 1.35 to 1.70, increasing by 0.35. Figure 4.13 shows that when the minimum soil cover thickness is increased from 10 to 40 m, the sedimentation value of the surface of the embankment is reduced from 49 to 30.5 mm, with a reduction of 18.5 mm, while the safety factor is increased from 1.40 to 1.60, with an increase of 0.20. In order to visually represent the effects of the above three factors, the range of changes in the parameters is unified to 10%, and the corresponding settlement and

Table 4.2 Comparison of influencing factors

Parameters	Grouting effect (%)	Embankment slope (%)	Tunnel depth (%)
Parameter variation range	10	10	10
Settlement impact range	12	0.27	3.2
Safety factor influence range	0.42	3.20	0.92

safety coefficient changes caused by them are listed as percentages in the form of a percentage of Table 4.2.

Table 4.2 shows that the effects of grouting effect, bank slope and tunnel depth on bank settlement and slope stability are different: for stratigraphic settlement, grouting effect has the greatest impact, followed by tunnel burial depth. Impact of bank slope, it is almost negligible; for the safety factor of the bank's waterfront slope, the slope of the bank has the greatest impact, followed by the tunnel. The depth of the road is buried, and the effect of grouting is minimal.

So in different working conditions should be selected according to the actual situation of embankment settlement and embankment slope stability as the key control index: when the embankment slope is small, shallow tunnel depth, should be mainly to the embankment surface settlement as the main control target, to prevent the embankment structure due to uneven settlement and damage. When the embankment facing the water slope is large and the tunnel depth is deep, the control of settlement is no longer the key control target, and the slope stability control of the embankment becomes very important.

4.2.4 Analysis of the Value of Stress Field and Mud Water Pressure on Embankment

The selection of support pressure on the excavation face when the shield crosses the embankment is crucial to the settlement of the embankment and the stability of the embankment. The insufficient or excessive support pressure will lead to settlement or uplift, thus endangering the upper embankment structure. Before the shield reaches the embankment, the terrain is usually flat, and the support pressure of the excavation face can be calculated from the surface at the incision. As the excavation surface gradually approaches the embankment, the stress state of the stratum becomes complex, so the support pressure becomes difficult to determine. If the calculation method of the support pressure on the flat surface (i.e., from the surface at the notch) is still used, the difference between the obtained support pressure and the actual stress value of the stratum is large. It is a problem worthy of further study that how much influence the settlement and displacement of the embankment will be caused by the shield construction with such support pressure.

1) Analysis of the stress field of the embankment

During shield construction, with the excavation of the cutter head, the soil on the excavation face is unloaded and the initial stress balance of the stratum is destroyed. In order to ensure the stability of the excavation face, prevent the collapse of the soil and control the settlement of the stratum, a force must be applied to the excavation face to prevent the deformation of the stratum to the cement tank, that is, the support pressure, as shown in Fig. 4.14. The slurry shield balances the water and soil pressure at the excavation face through the slurry pressure: the slurry pressure balances the water pressure in the stratum in the form of similar water pressure, and the slurry continuously penetrates into the stratum ahead of the excavation, forming a micro permeable or impervious mud film on the excavation face, thus transforming part of the slurry pressure into an effective stress to balance the soil pressure at the excavation face. It can be seen that in order to determine the supporting pressure, the initial stress state of the stratum should be clarified, namely, the initial geostress and the initial pore water stress.

For the case where the ground is flat and the surface is not overloaded, it can be calculated from the self-heavy stress of the stratum and the conventional static pore water stress, which is why the calculation of the excavation face support pressure can be calculated from the surface at the incision. And for the embankment this special case, this method will no longer be reliable. Figure 4.15 shows the principal stress direction of the bank model of the Wei San Road Crossing River in Nanjing. The figure shows that the principal stress is deflected at the bank slope. The most obvious.

The stratum of the Wei San Road crossing channel in Nanjing is mostly sandy stratum. The mud water support pressure is usually set by the soil moisture in the project. According to the engineering routine, the mud support pressure is calculated

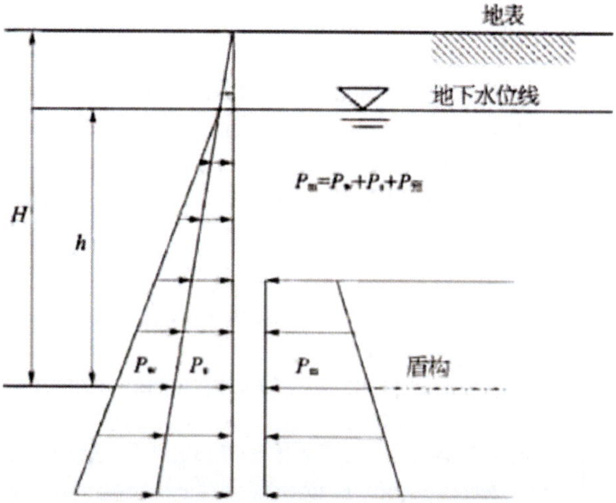

Fig. 4.14 Schematic diagram of shield support surface pressure

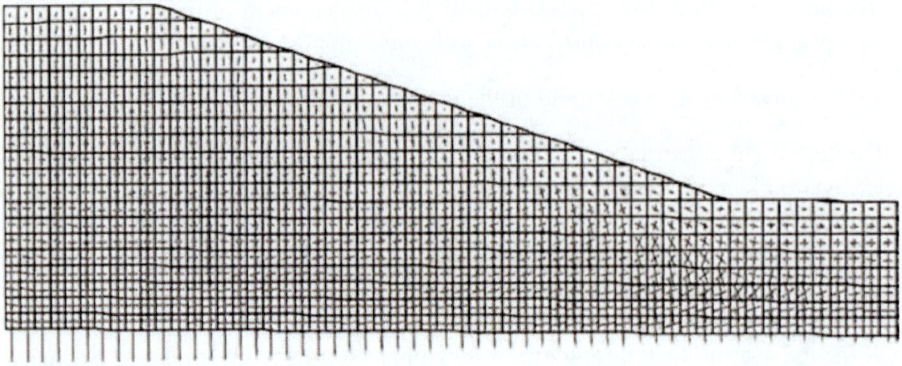

Fig. 4.15 Main stress field of the embankment slope of the Weisan Road crossing in Nanjing

and the initial stress value calculated by the model is used as the mud. The water pressure is set and the difference between the two is made, as shown in Fig. 4.16.

It can be seen from Fig. 4.16 that, at a distance away from the bank slope ($y = 0$), the mud support pressure values obtained by the two calculation methods have a small difference of about 2.3 kPa; gradually approaching the bank support pressure gradually increases, at the bank The top center reaches the maximum value. At this time, the support pressure value obtained by the conventional support pressure calculation method is greater than the numerical calculation value, and the difference between the two is 31.1 kPa; Slope, the support pressure is gradually reduced, and the support pressure value obtained by the conventional support pressure calculation method at the foot of the slope is less than the numerical calculation value, and the difference between the two reaches the maximum (26.6 kPa); near the bottom of the river bed, the two calculations The results are not much different. From this step analysis shows that if the special form of the bank is not taken into account, the engineering calculation method is adopted, that is, only from the surface or slope of the levee to calculate the muddy water pressure on the excavation surface may have the following consequences: ① When the shield gradually approaches the levee and passes through the embankment, the mud water pressure is set too high, which will cause a large disturbance to the excavation surface and the nearby stratum, and even lead Water surface slope "extrusion" damage; ② When the shield passes through the top of the embankment and enters the slope below the water surface, the mud water pressure setting is low, which will cause large settlement of the slope.

2) Analysis of support pressure value

The above two supporting stress setting methods are used for shield excavation. After the excavation is completed, the vertical displacement value of the surface above the tunnel axis, the horizontal calf displacement and the safety factor during the shield excavation are obtained as shown in Figs. 4.17, 4.18 and 4.19.

It can be seen from Fig. 4.17 that the difference in vertical displacement calculated by the two methods is not large. Near the dam crest, using a numerical meter. The

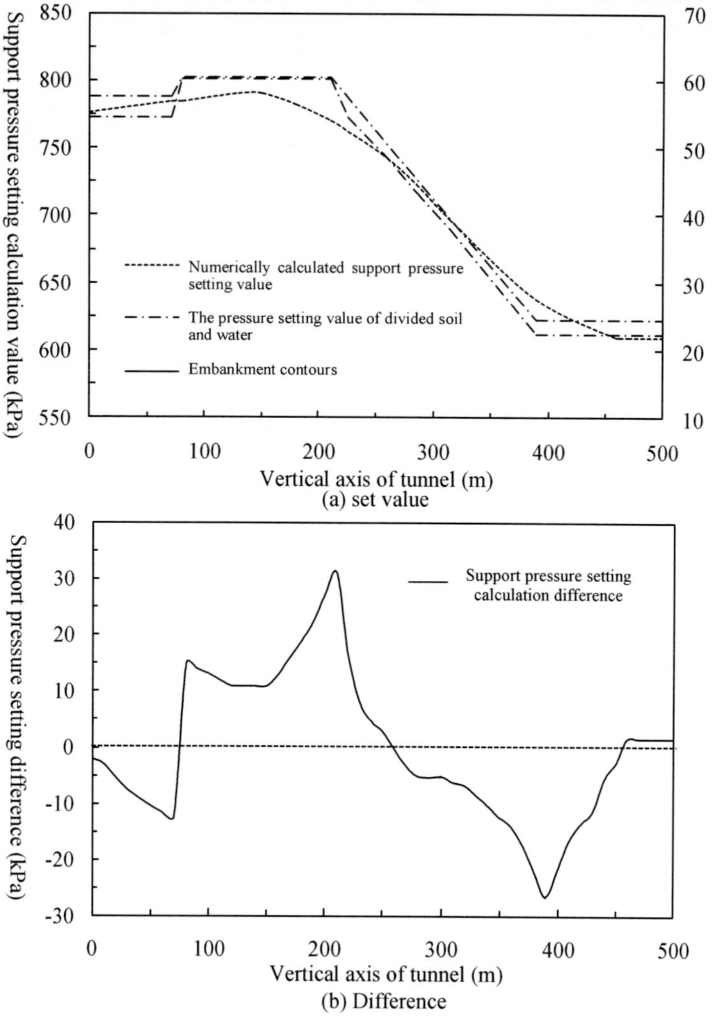

Fig. 4.16 Two kinds of support stress setting values and their difference

Fig. 4.17 Final settlement value of shield excavation ($X = 0$)

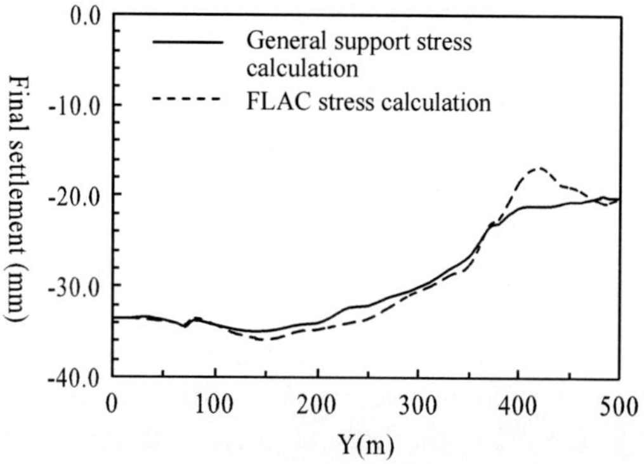

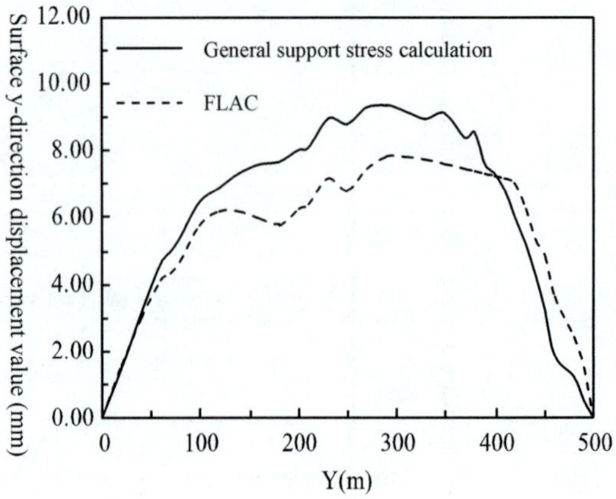

Fig. 4.18 Shield excavation final displacement value ($X = 0$)

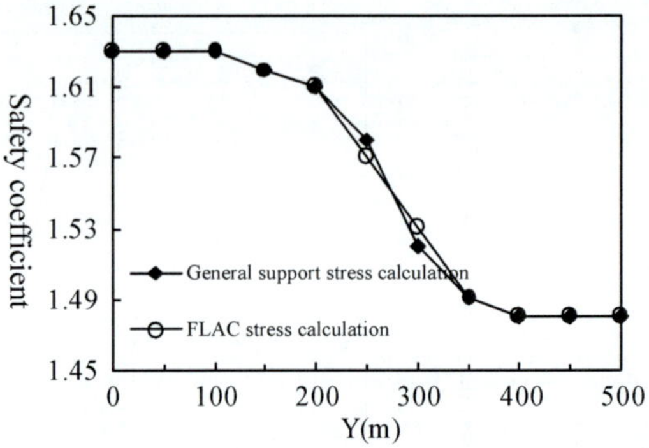

Fig. 4.19 Safety factor of bank slope during shield excavation

settlement of the in-situ stress value for excavation is slightly larger than the conventional support stress calculation, which is due to the position of the dam crest. The support stress obtained by the general calculation method is slightly larger than the initial stress in the formation; near the foot of the slope, the numerical calculation of the in-situ stress value for the excavation settlement is smaller than the conventional support stress calculation, which is due to the conventional calculation method. The stress value is small.

From Fig. 4.18, it can be seen that there is a big difference between the surface longitudinal displacement obtained by the two calculation methods. The surface longitudinal displacement obtained by the excavation of the support stress calculated by the initial stress is less than that calculated by the conventional support stress, which can effectively reduce the extrusion failure of the waterward slope.

It can be seen from Fig. 4.19 that the safety factors obtained by the two calculation methods have little difference.

From this point of view, for this case, the setting of slurry pressure according to conventional calculation is different from the initial stress state of stratum, but has little effect on the settlement and safety factor of embankment. This is due to the deep buried depth of the shield in this case, the additional stress caused by surface fluctuation diffuses in the soil, and the transfer to the shield position has been greatly weakened. On the other hand, due to the high water level in this case, a large part of the slurry support pressure is used to resist the water pressure. If the shield passes through such embankments with shallow buried depth or low groundwater level, the effect of special topography of the embankment should be fully considered. The slurry pressure near the top of the embankment can be slightly lower than the slurry pressure value calculated by the conventional engineering, so as to avoid excessive disturbance, surface uplift or "extrusion" of the water-facing slope caused by excessive support pressure. When the shield reaches the water-facing slope, the slurry pressure can be slightly larger than the slurry pressure value calculated by the conventional engineering, so as to avoid excessive settlement at the slope toe.

4.3 Shield Through Embankment Construction Control Technology

4.3.1 Control of Excavation Parameters

When crossing the embankment during the construction of the slurry shield, due to the complicated conditions of the stratum under the river, it is necessary to take corresponding measures strictly control the shield tunneling parameters to prevent deformation of the protection levees caused by shield tunneling and ensure construction safety.

1) Cutting water pressure control

The setting of the cut slurry pressure is the key to control the water-soil pressure balance of the excavation surface. Due to the shallow overburden at the top of the shield during the construction of the river bottom, it is more difficult to control the cut water pressure. If the incision water pressure fluctuates too much, it will increase the disturbance of the front soil, leading to the loss of the front soil, and the excavation face is difficult to maintain stability. Therefore, the fluctuation of incision water pressure should be reduced as much as possible. In technology, the operator is required to change from automatic control to manual control, adjust the construction parameters in real time, and control the fluctuation value of incision water pressure between ±20 kPa to ensure the stability of soil. In the actual construction process, the mud pressure value of the incision should be adjusted in time according to the monitoring information.

2) Shield attitude control

Good shield attitude control ensures uniform force on the segments and avoids uneven or uneven pressure on the segments damaged. At the same time, it can reduce over-excavation and construction disturbance, reduce settlement loss and disturb soil consolidation settlement. Poor attitude control of the shield will not only increase the building void and stratum disturbance, but also increase the displacement of the stratum, and will cause the tube to break due to the extrusion of the segment, reduce the assembly quality of the segment and increase the difficulty of assembling the segment. Therefore, the control of the shield attitude should be strengthened during the advancement process. Once the shield posture is found to be poor, measures should be taken immediately.

In order to avoid excessive deflection of the shield during construction, the cutter head torque can be appropriately reduced under the premise of ensuring smooth excavation of the excavation surface, and the cutter disc rotation should be alternated clockwise and counter clockwise. Once the shield is found to be deflected too large in one direction, the cutter head in this direction to correct the shield deflection. Once the lateral offset of the shield is found to be too large or the "head" and "head lift" are too large during construction, the adjustment jack should be used to correct the deviation, and the simultaneous grouting management should be strengthened to ensure the grouting effect, so that the looped tunnel remains stable and the correction is improved.

When the shield enters the influence range of the flood dike, the shield operation management should be strengthened to maintain the smooth advancement in the excavation process, reduce the deviation correction, reduce the shield deflection and lateral deviation, prevent the occurrence of snake, reduce the disturbance to the front soil, and maintain the stability of the stratum. The plane position is controlled within ± 50 mm, and the elevation is controlled around -50 mm when crossing the breakwater, considering the shallow overburden and the drift of shield. Shield tunneling speed is controlled at about 15 mm/min to prevent large impact on the front soil due to too fast tunneling speed.

3) Other major driving parameters

Excavation speed, cutter torque, mud discharge and cutter thrust have certain influence on construction. In order to reduce pressure fluctuation, low speed uniform propulsion is used in construction. On the premise of ensuring the smooth excavation of the excavation face, properly reducing the torque of the cutter head can reduce the disturbance of the cutter head rotation on the surrounding strata and the wear of the cutter head, and reduce the deflection of the shield, thereby reducing the circumferential friction between the shield shell and the surrounding soil layer. Strengthening mud management and monitoring of excavated volume can prevent formation disturbance caused by overbreak and underbreak. The thrust of the cutter head, i.e. the force of the cutter head supporting the excavation face, is one of the important factors to maintain the stability of the excavation face, and its value is still a difficult problem internationally. Usually in the construction process of the incision pressure setting is not taken into account, only as a safety reserve. In fact, in the process of tunneling, the supporting effect of cutterhead on the excavation surface cannot be ignored. It is of great significance to play the supporting role of cutterhead in maintaining the stability of excavation face to reduce the disturbance of soil. The shield tunneling parameters interact with each other, and each tunneling parameter should be coordinated, which not only reduces the formation loss, but also reduces the construction disturbance and reduces the long-term consolidation of disturbed soil.

4.3.2 Post-wall Pulp Management

With the progress of shield construction, stratum deformation is closely related to many factors such as stratum soil conditions, groundwater conditions, tunnel section, tunnel setting depth and construction technical parameters. When the assembled segment lining is removed from the shield tail, due to the original space occupied by the shield steel shell, the gap left for the assembly operation of the lining and the adhesion of some soil to the shield shell during the shield tunneling, there is a ring gap between the back of the lining ring and the actual excavation of the tunnel wall, so that the soil is in a state of no support, and this gap is the gap at the shield tail.

The size of the shield tail gap is determined by the thickness of the shield steel shell and the shield tail operation space, generally in 10–20 cm. Because the segment and surrounding rock soil are temporarily suspended without support when the segment is out of the shield tail, the soil moves to the gap of the shield tail to produce deformation or local collapse, and the loose range of the surrounding rock is expanded, which is an important reason for the deformation and settlement of the stratum caused by shield construction.

Shield tunnel is a kind of stable structure which is composed of segment lining and surrounding rock. Filling the gap of shield tail uniformly and compactly is the precondition to ensure the uniform action of earth pressure on segment lining. Although the gap can be controlled by post-wall grouting, the grouting material, grouting time, pressure, grouting amount and grouting position of post-wall grouting will affect the deformation of the stratum.

The purpose and function of post-wall grouting in shield construction mainly include preventing stratum deformation, improving tunnel impermeability and ensuring the early and late stability of segment lining. If the shield tail gap is not filled in time, it is bound to cause stratum loss, cause stratum deformation, and make adjacent surface buildings settlement or tunnel itself offset. Therefore, the most important purpose of post-wall grouting is to fill the gap of shield tail in time to reduce the stress release and formation deformation. After consolidation and hardening of grouting slurry behind the wall, it generally has certain impermeability, which can be used as the first waterproof line of the tunnel to improve the overall impermeability of the tunnel. The slurry with certain early strength is used to fill the gap of shield tail in time to ensure the early and late stability of segment lining.

4.3.2.1 Slip Casting Methods

1) Synchronous grouting

In the process of shield tunneling, grouting at the same time is the most commonly used grouting method to fill the gap of shield tail in time. Grouting is carried out through the grouting pipe installed at the tail of the shield (generally No. 3 shield body), and the built-in methods are generally convex and concave (Fig. 4.20c). What kind of method is from the shield design. The convex type can reduce the outer diameter of the shield to a certain extent, thereby reducing the thickness of the gap at the tail of the shield. However, the shield shell is not round, which is not conducive to the entry and exit of the shield, and is easy to wear when the soil strength is high. Although the concave type increases the outer diameter of the shield and the gap between the shield tail to a certain extent, it is not easy to wear and tear, and has better adaptability to the stratum.

2) Timely grouting

After the shield tunneling rings, when there is a large number of gaps in the shield tail, the gap in the shield tail is grouted through the grouting hole reserved on the segment (Fig. 4.20b). Because this grouting method cannot quickly fill the gap at the shield tail, it increases the disturbance to the soil and is not conducive to the control of land subsidence. Moreover, since the early segment is suspended after it is out of the shield tail, it is unfavorable to the stress of the segment and prone to dislocation. The grouting method is generally used only when the geological conditions are good and the surface settlement requirements are low.

3) Secondary grouting

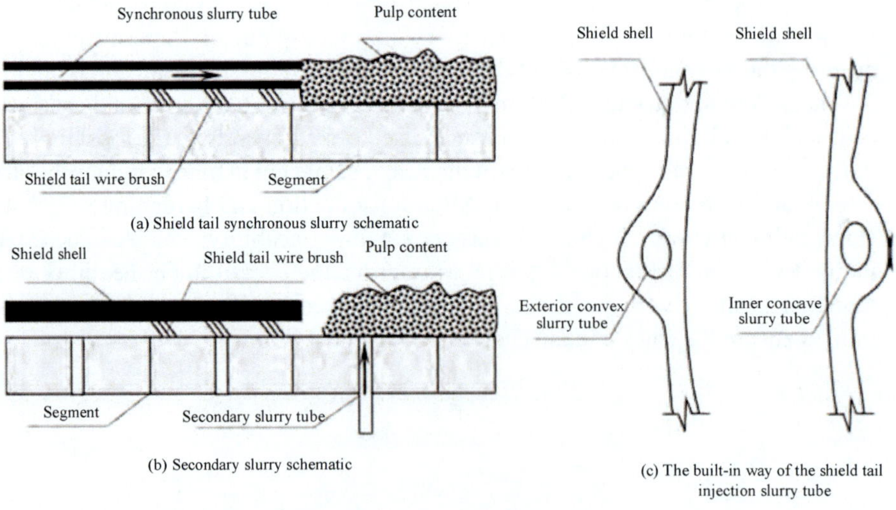

Fig. 4.20 Schematic diagram of shield grouting pipe arrangement

When the primary grouting effect is not ideal, it is necessary to supplement the early grouting through secondary grouting. This method is generally used in the abnormal settlement of the tunnel, the deviation of the tunnel or in some special sections (near the entrance and exit station of the shield and the contact channel), and can also be used for the reinforcement of the main body crossing buildings and underground pipelines. As shown in Fig. 4.20b, timely grouting or secondary grouting can be carried out through the reserved grouting holes on the segment.

In the actual shield construction, the choice of grouting method behind the wall is generally determined by the shield. In some special cases, it is conceivable not to use the grouting pipe provided by the shield for grouting, but to use another purchased grouting device to grout through the reserved grouting holes on the segment.

4.3.2.2 Pulp Engineering Characteristics

In order to achieve the purpose of post-wall grouting and meet the requirements of shield construction, in addition to the grout injection must be quickly and fully filled with shield tail gap, the slurry engineering characteristics of post-wall grouting must meet the following requirements:

(1) Good filling properties and is not easy to flow to areas other than the shield tail gap (not missing into the digging surface and surrounding rock soil).
(2) The slurry should have good fluidity and be less analytical.
(3) The slurry shall have an appropriate consistency and is not easily diluted by groundwater.
(4) The material is stable and not easy to layer, so that it can be pressed over long distances.
(5) After the wall slurry is filled, the restraint tube should be condensed early, and the early strength is best comparable to the strength of the original soil.
(6) The volume shrinkage and permeability coefficient of the slurry after hardening should be small.
(7) Pollution-free, cheap. Among them, the most important engineering characteristics are filling, fluidity, strength and shrinkage.

The post-wall grouting materials in shield construction can be divided into two types: single slurry and double slurry, and single slurry can also be divided into inert slurry and hard slurry (Fig. 4.21). Inert slurry is slurry which has no cementitious material such as cement in slurry preparation material, long setting time, low early strength and late strength. generally consists of lime, fly ash, sand, bentonite (clay), water, additives, etc. The hard slurry is the slurry mixed with cement and other cementitious materials, generally by cement, fly ash, sand, bentonite (clay), water, additives and so on in the mixer, with a certain early strength and late strength of the slurry. The setting time can be adjusted in a certain range according to the mixing ratio. The double liquid slurry is mixed by liquid A (single liquid slurry) and liquid B (generally water glass) at a certain volume ratio. According to the actual needs, the setting time can be adjusted from several seconds to tens of seconds. The early

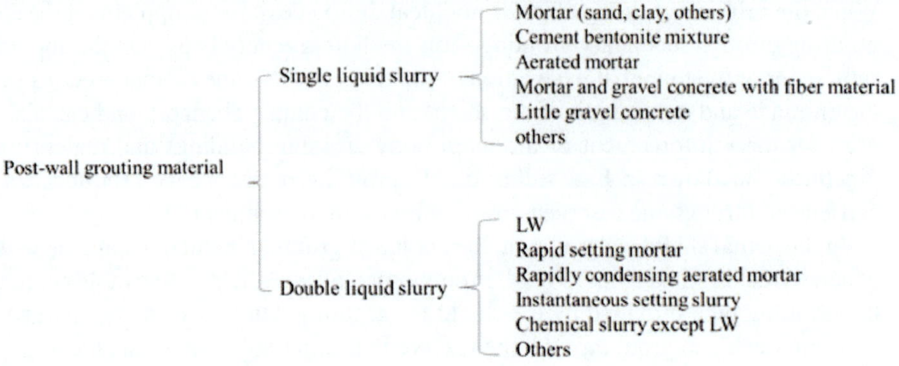

Fig. 4.21 Types of grouting material behind the wall

strength and late strength of this slurry are generally higher than those of single slurry.

The longer the setting time is, the more likely the slurry leaks to the excavation surface and loses in the soil. At the same time, it is also easy to be diluted by groundwater and mud channeling from the excavation surface, which is difficult to restrict the segment and control the formation displacement. However, when the shield tail gap is not fully filled, the setting time is too short to make the slurry lose liquidity, and it is easy to cause pipe blockage. When the construction is not smooth, this situation is more likely to occur, which requires higher construction control management.

Each kind of post-wall grouting has its own advantages and disadvantages due to different slurry types and injection methods, as shown in Table 4.3. The selection of slurry material types and injection methods is closely related to geological conditions, tunneling methods, construction conditions and cost control. Single liquid grouting and timely grouting can be considered in areas with low ground protection requirements, good geological conditions and certain self-stability of soil or rock. The soft strata and ground protection requirements are higher, especially in the areas with important buildings or underground pipelines, double slurry and synchronous grouting should be preferred. If the construction management level is high, single slurry can also be used, but synchronous grouting must be used, and the consistency of the slurry is large, water retention is good, which can restrict the segment and maintain the early stability of the tunnel. In the formation rich in groundwater or caves, double liquid slurry and synchronous grouting method with short setting time are preferred to prevent the dilution or loss of slurry by groundwater. In the construction of slurry pressurized shield, in order to prevent the influence of slurry on slurry or the dilution of slurry after slurry channeling, double slurry and synchronous grouting should be adopted.

From the perspective of cost control and ease of operation, single slurry is superior to double slurry, and inert slurry is superior to hard slurry. From the grouting effect, the double slurry is superior to the single slurry, and the hard slurry is superior to the inert slurry. In the current construction of large section cross-river slurry pressurized

Table 4.3 Comparison of advantages and disadvantages of various types of post-wall grouting

Grouting type	Injection site	Advantage	Disadvantage
Inert slurry	Shield tail/tube piece	Not easy to block the pipe; high efficiency of grouting; low cost	The setting time is very long; unfavorable control of settlement and stability of the sheet; Very low strength; poor water proofing
Hard pulp	Shield tail/tube piece	The strength is higher in the later stage; the cost is lower; the equipment is simple; the shield is easy to clean	The condensation time of the pipe is longer; the slurry is easy to be lost, high requirements for shield tail seals
Single slurry	Shield tail/tube piece	Low cost; easy to clean	Long setting time; easy to lose slurry; slower effect
Double slurry	Shield tail/tube piece	The setting time is fast and adjustable; the slurry is not easy to be lost; the strength is high in the early period	The early equipment is complicated, the cost is high; the pipe is easy to block, and it is not easy to clean; High construction management requirements
Shield trail grouting	Shield tail	Synchronous and uniform grouting can be realized; without injecting holes from the tube; reducing the possibility of water leakage	It is possible to increase the diameter of the shield it is more difficult to clean the slurry when plugging the pipe
Tube grouting	Segment	Flexible and convenient to operate; easy to clean; it's easy to inject	The grouting hole of the equipment may become a potential water leakage point, the surface of the tube piece is easy to be contamination; easy to cause uneven grouting

shield, inert slurry has also achieved good results. Therefore, the choice of slurry type and injection method should be determined according to the specific situation.

The unreasonable slurry engineering characteristics of post-wall grouting will lead to poor grouting effect, which is mainly manifested in poor slurry fluidity, short setting time, poor physical stability, serious segregation, excessive shrinkage after solidification, and low early strength after solidification. At the same time, it will also cause insufficient filling, unevenness, grouting pipe blockage of the shield tail gap during the implementation of post-wall grouting, as well as water conversion of the slurry at the top of the shield tail gap after the implementation of post-wall grouting, insufficient control of stratum loss, and the failure of segment lining to form a community with surrounding rock soil, which will cause tunnel uplift and axis deviation, thereby causing surface subsidence.

4.3.2.3 Pulp Control

Post-wall grouting is an important measure to prevent stratum settlement. The construction control parameters of post-wall grouting include grouting pressure, grouting amount, grouting location selection and grouting rate, among which grouting pressure and grouting amount are the main control parameters in the implementation of post-wall grouting. During tunneling, it is necessary to not only fully grout, but also control the grouting pressure to avoid the grouting pressure exceeding the high roof to break the overburden. There are two ways in construction: one is to control grouting by grouting pressure, and the other is to determine whether the grouting process ends by preset grouting amount. Since the grouting pressure is affected by many factors, at present, under the premise of ensuring a certain amount of grouting, the upper limit value of grouting pressure is usually set to determine whether the post-wall grouting is completed.

The setting of the grouting pressure is first determined by the engineering characteristics of the slurry. The slurry must overcome the groundwater pressure, earth pressure and pipe friction during the injection process to be pressed into the gap of the shield tail. The greater the grouting pressure, between the lining and the formation denser the void filling, the more stable the surrounding formation. However, the grouting pressure should not be too large, otherwise the splitting grouting will occur, the formation will be bulged, and the shield seal brush will be easily broken or the slurry will be sucked into the pressure chamber, and the force on the lining segments will not be very strong during the construction period. In particular, in the case of a single grouting hole, it is more likely to cause damage to the joint portion or to endanger the water proof performance. Therefore, the value of grouting pressure is quite critical, and the optimum value of grouting pressure should be determined on the basis of geological conditions, sheet strength, equipment performance, slurry characteristics, pressure chamber soil and water pressure. The formation of a uniform slurry layer around the tube sheet is a prerequisite for the final requirement of post-wall grouting. However, injecting excess slurry may cause the soil to cleft and tend to appear in soft clay formations. In general, when the grouting pressure is increased, the amount of injection will increase accordingly. When the thickness of the overburden increases, the grouting pressure will increase accordingly, and there is a certain positive correlation.

In the implementation process of post-wall grouting, only grouting amount is used as the control index, which will cause insufficient grouting amount or uneven filling in special sections (geological mutation, cutter head overexcavation, small curvature construction, etc.). If the grouting amount is high, it will lead to surface uplift. The grouting pressure setting is too large to cause the slurry to enter the pressure chamber along the periphery of the shield shell, resulting in surface uplift, breakdown of shield tail seal, segment cracking or dislocation. The unreasonable selection of grouting hole position and corresponding grouting pressure results in uneven or insufficient grouting filling, segment dislocation, tunnel floating and axis deviation. Therefore, the grouting pressure and grouting quantity should be strictly controlled when the shield passes through the embankment. During the construction, the construction gap is filled in time by synchronous grouting to reduce the deformation of soil in the construction process and ensure the stability of embankment buildings. In addition, according to the monitoring situation, when the synchronous grouting cannot meet the requirements, the double liquid slurry (cement slurry and sodium silicate) is poured through the secondary grouting hole reserved for the segment, so that the soil is consolidated and stabilized in a relatively short period of time and the embankment is reinforced.

4.3.2.4 Post-pulping Simulation of the End Wall of the Shield

The grouting process behind the shield wall is to use a fluid-like grouting material with a certain fluidity, using a grouting pump to grout. The pressure and grouting amount are pumped into the annular shield tail gap outside the segment, and the filling of the shield tail gap is ensured to be even and full, thereby reducing the stress release and deformation of the formation. After the wall grouting material gradually undergoes consolidation compression and cementation hardening, the grouting pressure gradually dissipates and returns to the state equal to the formation stress. The simulation of grouting process actually includes the simulation of grouting material and the simulation of grouting pressure. This section summarizes the grouting rate and filling rate by summarizing the research work related to the predecessors. The concept of a post-grouting simulation method that is more in line with the actual grouting process.

1) Discussion on the concept of "grouting rate" and "filling rate"

Through the investigation of relevant domestic literature, many scholars have mentioned "grouting rate" and "filling rate" in their studies. However, in the field of engineering, technology and scientific research, these two words are confused and often confused with each other. Figure 4.22 is a schematic diagram of the actual filling effect of the grouting behind the shield wall. Because the outer diameter of the shield cutter head and the shield shell is larger than that of the lining ring, there will be a shield tail gap between the planned (ideal) excavation boundary and the segment ring in the project. The volume V_{gap1} of this shield tail gap in the project is known as:

$$V_{\text{gap1}} = \pi(D^2_{\text{Design excavation}} - D^2_{\text{Pipe diameter}})/4 \tag{4.1}$$

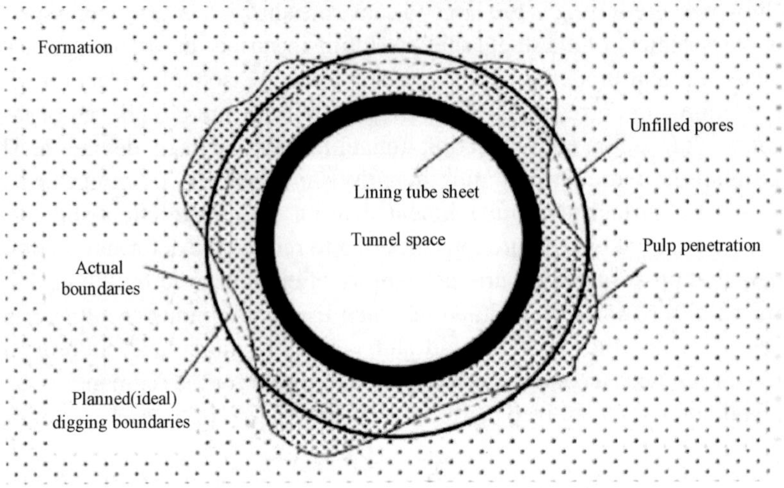

Fig. 4.22 Schematic diagram of actual filling effect after grouting of shield wall

At the time of construction, the actual excavation diameter of the shield is larger than planned due to shield rectification or other construction factors. The (ideal) excavation has a large diameter, so the actually produced shield tail gap volume is larger, and this book is defined as V_{gap2}. Due to the complexity of the shield construction control, it is difficult to obtain the actual shield gap volume V_{gap2} of a certain ring by direct calculation, which is often determined according to the management of the amount of slag combined with the construction experience. Therefore, define the empirical over-extraction coefficient α:

$$V_{\text{gap2}} = \alpha V_{\text{gap1}} \tag{4.2}$$

According to the construction situation of the Shield Tunnel of the Wei San Road in Nanjing and the Slender West Lake Tunnel in Yangzhou, the α is 1.1–1.3. Under the condition of good support pressure and attitude control, α takes a small value; the shield attitude is poorly controlled, α takes a larger value.

After the slurry is injected into the shield tail clearance, the diffusion process is very complex due to the influence of slurry properties, groundwater level, construction technology, grouting pressure and formation conditions. A large number of laboratory tests and field measurements show that the slurry is not evenly distributed in the gap, and is not only distributed in the shield tail gap. The diffusion mode of slurry in the pores and strata around the segment will experience the process of filling, infiltration, compaction and splitting, so there will be permeable slurry in the surrounding strata. And in some parts (especially the top of the lining) strata squeeze into the shield tail gap, making this part of the shield tail gap has not been filled with slurry. Therefore, the total grouting amount is the sum of the lost slurry penetrated into the stratum and the slurry filled in the shield tail gap.

Define the slurry loss coefficient δ, so that:

$$Q_{\text{fill}} = Q_{\text{inject}} - Q_{\text{permeable}} = (1 - \delta)\, Q_{\text{inject}} \tag{4.3}$$

In the formula, δ is related to slurry properties, formation gradation and grouting pressure, and the range is from 0.30 to 0.50.

Grouting rate and gap filling rate:

$$\beta_1 = Q_{\text{inject}}/V_{\text{gap1}} \tag{4.4}$$

$$\beta_2 = Q_{\text{fill}}/V_{\text{gap2}} = (1 - \delta) Q_{\text{inject}}/(\alpha V_{\text{gap1}}) = \beta_1 (1 - \delta)/\alpha \tag{4.5}$$

in this formula: β_1—grouting rate;

Q_{inject}—the total grouting volume of a ring can be controlled and accurately measured during construction;

β_2—gap fill rate;

Q_{fill}—The amount of slurry filled in the shield tail gap is not easy to be measured directly. The slurry loss coefficient δ can be obtained indirectly by indoor similar model test or theoretical calculation.

Thus, the concepts of "grouting rate" and "gap filling rate" are clarified, and the relationship between the two is established.

Taking the Wei San Road crossing channel in Nanjing as an example, according to the construction situation, $\alpha = 1.21$ and $\delta = 0.42$. The outer diameter of the segment is 14.5 m and the diameter of the excavation is 15 m. Then $V_{\text{gap1}} = 3.14 \times (152–14.52)/4 = 1.58\ \text{m}^3$, when the grouting volume is controlled to 18–22 m^3, the grouting rate β_1 is 155.4%–190%, and the corresponding gap filling rate β_2 is 74.5%–91.1%, which is corresponding to the relevant grouting parameters given by Wu Shiming, Zhu Caihui and Zhu Hehua.

2) "Equivalent generation" optimization

In order to reflect the influence of shield tail clearance, soil squeezing into the shield tail clearance and the distribution of grouting body behind the wall on the settlement and deformation of strata, Zhang Yun et al. put forward the concept of "equivalent layer": in numerical simulation, the soil, grouting body behind the wall and disturbed surrounding rock naturally squeezed into the shield tail clearance are generalized into an annular homogeneous elastic material layer with equal thickness after the segment is removed from the shield tail, which makes the settlement and deformation of the actual soil layer around the segment and the surface behind the grouting body remain unchanged. After putting forward the method of using "equivalent layer"' instead of wall grouting for numerical simulation, many scholars have carried out a lot of research. However, most of the literature did not consider the filling rate of grouting, and did not unify the grouting in numerical simulation with the grouting amount and filling rate in practical engineering. Moreover, the consolidation-cementation hardening effect of slurry with time and the dissipation of grouting pressure behind

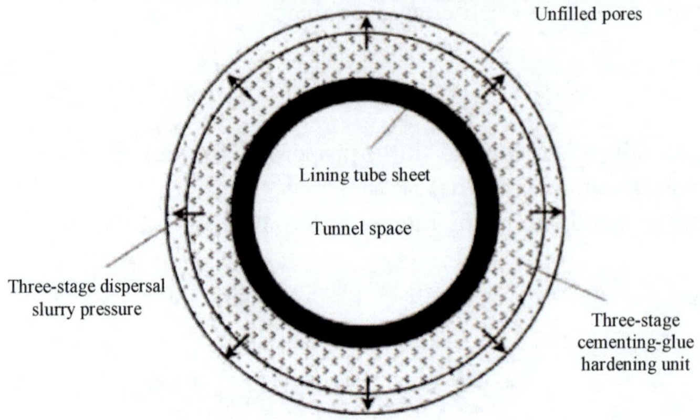

Fig. 4.23 Schematic diagram of the optimized generation layer

the wall were less considered. Therefore, based on the concept of equivalent layer proposed by Zhang Yun, a series of optimizations are carried out to make this abstract "equivalent layer" not only reflect the filling rate, slurry consolidation and cementation and grouting pressure dissipation, but also unify with the grouting rate and grouting effect in practical engineering.

The optimized equivalent layer is shown in Fig. 4.23. The grouting body adopts the "three-stage consolidation-cementation hardening" unit, and the three-stage dissipation pressure is applied to simulate the grouting pressure behind the wall. The low modulus elastic element is used to consider the outer circumference of the grouting as an unfilled void (the thickness is determined according to the filling rate).

3) Grouting unit and hardened slurry unit

When a shield tunneling ring is completed and grouting behind the wall of the ring is completed, the slurry is in a plastic flow state, and the segment lining and surrounding rock soil are mainly subjected to liquid-plastic slurry pressure. As the shield continues to move forward and time goes on, the slurry has the effect of consolidation, drainage and cementation hardening at the same time, and gradually becomes a solid state. A protective layer is formed around the segment, and the surrounding rock stress is acted on the lining. The strength and elastic modulus of grouting materials increase with the consolidation-cementation of grout. In order to simulate the mechanical properties of grouting materials at different stages and improve the efficiency of numerical model calculation, the "three-stage consolidation-cementation equivalent layer" method is adopted. After grouting, the plastic flow-like material composed of slurry and soil is formed outside the segment lining, which is approximately a homogeneous elastic material layer of equal thickness for simplified simulation. Considering the strength and elastic modulus of grouting material increases with time, the elastic modulus of three stages is used to calculate. According to the study of Han Yuewang et al., according to the actual consolidation-cementation process of grouting material behind the wall, the material is divided into the following three kinds:

(1) The initial slurry body, regarded as a flow-shaped slurry that has just been injected into the tail of the shield, with an elastic modulus of 1.0 MPa.
(2) Short-term hardened slurry, considered to be after 24 h hardening of the slurry, elastic modulus 5.0 MPa.
(3) Long-term hardened slurry, considered to be after 28 days hardening of the slurry, elastic modulus takes 300 MPa.

According to the excavation step material transformation to simulate the solidization and bonding process of the wall after the slurry body: 1 step after excavation using the initial slurry simulation injection, 1 excavation step after transformation into a short-term hardened slurry body, 4 excavation steps after the use of long-term hardening slurry simulation.

4) Grouting pressure simulation

When grouting behind shield tail wall, it is necessary to exert certain pressure on the slurry to distribute the slurry around the segment. If the grouting pressure is too large, it may lead to segment dislocation cracking, bolt shear, tunnel seepage accident. Grouting pressure is too small, it is easy to lead to insufficient grouting or uneven, is not conducive to the full filling of slurry, resulting in large settlement, so the selection of grouting pressure is very important. Usually, the grouting pressure is 0.1–0.2 MPa larger than the stratum resistance around the segment. In this paper, the initial grouting pressure difference of 0.2 MPa is selected according to the hydrogeological conditions and actual construction conditions of Nanjing Wei San Road crossing river channel and Yangzhou Shouxihu tunnel, and the surface force of a certain vertical gradient is acted on the lining segment and the surrounding rock of the tunnel as the grouting pressure behind the wall. When the three-stage grouting pressure is selected, the initial grouting pressure is 0.2 MPa larger than the initial formation stress at the corresponding position, and it decreases by 0.1 MPa after one excavation step, and it decreases to be equal to the initial formation stress at the corresponding position after four excavation steps.

5) Determination of the thickness of the equivalent layer

The joint type (4.1) and the formula (4.2) can be used to derive the actual shield gap thickness:

$$\lambda = \frac{1}{2}[\sqrt{\alpha(D^2_{excavation} - D^2_{diameter}) + D^2_{diameter}} - D_{excavation}] \tag{4.6}$$

Then the thickness of the grout after grouting:

$$\lambda_1 = \beta_2\lambda = \lambda\beta_1(1-\delta)/\alpha \tag{4.7}$$

The thickness of the void unit is:

$$\lambda_2 = \lambda - \lambda_1 = \lambda - \beta_2\lambda = \lambda[1 - \beta_1(1-\delta)/\alpha] \tag{4.8}$$

4.3.3 Mud Management

During shield tunneling, slurry must be injected into the stratum to form a dense mud film on the surface of the excavation face. The mud pressure acts on the mud film to balance the water and soil pressure in the stratum, so as to maintain the stability of the excavation face. Therefore, there are higher requirements for the quality of mud and mud film in the construction. The stratum of the middle section of the river crossing by slurry shield is generally complex. According to different geological conditions, the corresponding mud is used to ensure the safe and efficient tunneling of the shield. If there is a serious mud filtration in the construction, the excavation face is difficult to form a mud film and the instability of the accident, it will produce large ground disturbance, endanger the stability of surrounding embankments and other buildings, and increase the risk of river crossing. Therefore, the slurry management should be strengthened in the construction process, and the slurry properties should be adjusted according to the permeability of the surrounding strata in order to maintain the balance between the slurry tank pressure and the water and soil pressure of the excavation face.

1) Mud transport system work management

The mud conveying system consists of slurry pump, slurry pump, flow meter, control gate valve, slurry pipeline and ground mud separation system. The slurry transportation system mixes the excavated slag soil with slurry and then transports it to the ground through the slurry pump and slurry pipeline for slurry separation. The separated slurry is transported back to the slurry tank by the slurry inlet pump and slurry inlet pipeline for recycling. The configuration principle of slurry pump and pipeline is that the slurry flow rate in the pipeline is not less than 4 m/s, and the maximum flow rate is not greater than the flow rate when the pump efficiency is 70%. Reasonable distribution of mud tank slurry flow, close to 60% of the total flow of the mud circulation system is appropriate.

2) Work management of mud pressure regulation system

Mud pressure needs to balance groundwater pressure and soil pressure. The value of soil pressure has different calculation modes. When the thickness of the overburden above the tunnel (soft clay or sand and other soft strata) is less than the diameter of the tunnel excavation, the soil pressure is calculated by the formula of static earth pressure or active earth pressure. When the thickness of overburden layer (sand layer, gravel layer and hard clay layer) is larger than the diameter of tunnel excavation, the soil pressure can be calculated by the height of Tyzaghi loose soil due to the arching effect of strata. The automatic control system can automatically adjust the pumping pressure of mud pump according to the real-time monitoring data to balance the short-term fluctuation of mud pressure. However, the actual construction should be adjusted at any time according to the influence of surface buildings such as embankments above the tunnel.

3) Slurry treatment system work management

The index values of mud include density, sand content, viscosity, yield value, gel strength, filtration rate, pH value and so on. The index values of mud required for forming mud film on the palm surface of different formations are also different. The application range of each index calibration must be determined by targeted film forming test. In order to strengthen the supporting capacity of the front soil and prevent the ground slurry from falling, heavy slurry propulsion should be adopted. The specific gravity of slurry inlet was controlled at 1.15–1.20, and the viscosity was controlled at 2–25 s. The slurry was modulated by high-quality bentonite combined with different levels of macromolecular materials and new materials assembled by original plant fibers and inert minerals. As the excavated residual soil continuously enters the slurry tank, the slurry index value in the slurry tank will change greatly. In order to maintain the basic stability of the slurry properties, the residual soil entering the slurry system must be removed, and the removal of slag and mud is completed by the slurry treatment system. In order to ensure the quality of mud, the mud treatment personnel should increase the test frequency of mud and adjust the mud parameters in time to ensure the smooth progress of excavation.

4) Slurry quenching and tempering system work management

The remaining mud after treatment by the slurry separation device enters the mud treatment pool. The site laboratory samples and tests the slurry of the treatment pool, obtains the actual index value of the mud composition, and compares it with the calibration index value. If the test results show that the density is too high, a part of high density slurry must be discarded to the sedimentation tank for sludge removal. The high density mud can be diluted by adding fresh water or low density bentonite slurry to the mud of the treatment pond after some space is released. If other indicators of the slurry cannot meet the requirements, fine sand, carboxymethyl cellulose (CMC), polyacrylamide (PAA) and other additives can be selectively added to the slurry of the treatment tank to adjust and improve the slurry properties, and then the slurry is pumped back to the slurry tank and air cushion tank of the shield for recycling.

4.4 Embankment Deformation Monitoring and Reinforcement Technology

Although shield tunnel construction technology has little impact on the surrounding environment, it is affected by geological conditions and human factors. In the process of construction, it will inevitably disturb the soil, cause surface subsidence, and then leave some security risks to the construction. When shield passes through embankments, ground buildings, underground pipelines and other areas, the control requirements of surface subsidence and deformation caused by construction are very high. If the construction method is slightly wrong, it will cause immeasurable losses. Therefore, in order to control the ground settlement during the shield tunnel construction, the construction personnel often monitor the ground settlement during the construction period, and reinforce it if necessary. In order to ensure the safety of subway construction, it is necessary for relevant units to study the surface settlement monitoring during the construction of subway shield tunnels. When the embankment buildings are deformed, timely reinforcement is carried out, so as to better complete the tunnel construction.

Taking the construction of the flood control embankment of the north bank of the Yangtze River at RK3+73.7 in Nanjing Yangtze River Tunnel as an example to introduce the monitoring of embankment deformation with reinforcement technology.

The flood dike of the Yangtze River is an important flood control project, and the protection level is set to be level two. When the shield passes, it must be ensured that the flood dike is infallible. The position relationship between the Yangtze River flood dike and the shield tunnel is shown in Fig. 4.24. When the shield passes through the Yangtze River embankment twice, the stratum at Pukou Pass is muddy silty clay stratum. The construction time was selected in March 2008, belonging to the dry season of the Yangtze River. When the shield crosses the Yangtze River embankment, the shield tunneling face is prone to instability, resulting in stratum collapse, which leads to the collapse of the embankment, and the river water gushing endangers the life and property safety of the nearby masses. Shield tunneling may also be due to excessive mud pressure breakdown of the overburden, resulting in river water from

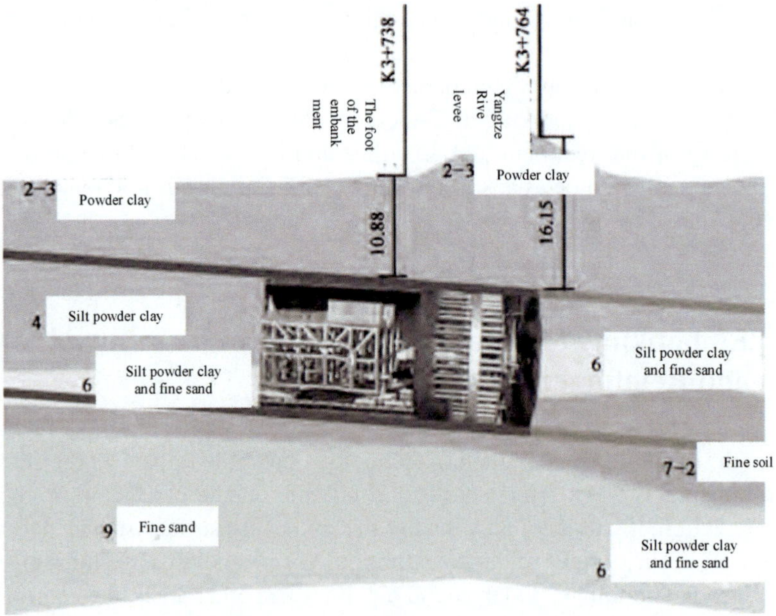

Fig. 4.24 Position relationship between shield tunnel and Yangtze flood control levee

the tail of the shield seal or segment waterproof weak position into the tunnel, pose a threat to construction personnel and equipment. Due to the high flood control level of the embankment, the surface settlement control requirements must be controlled at +10 to −30 mm. Moreover, due to the obvious change in the thickness of the ground cover, it is difficult to control the technical parameters of shield tunneling construction.

1. The initial crossing of the Yangtze River embankment section mainly consists of 4 layers of silty silty clay and 6 layers of silty clay with silt, 7–2 layer of silt.
2. The maximum risk of crossing the Yangtze River Dike is k3+738~k3+78. The cover soil in this area is increased rapidly and shallow, and the shield tail is still in shallow soil. The difference between the incision and the pressure of the shield tail water and soil is large, which can easily lead to the settlement of the embankment at the excavation surface or the roof of the shield tail, endangering the safety of the flood control embankment.

4.4.1 Monitoring of Embankment Deformation

The monitoring work in the construction process can accurately predict and predict the deformation effect. The monitoring data are compared with the predicted value to determine whether the early construction technology and construction parameters meet the expected requirements, so as to determine and optimize the next construction parameters, and do a good job in information construction. The results of surface subsidence analysis and research are applied to the information feedback to optimize the design, and the relationship between stratum characteristics, groundwater conditions and ground subsidence is studied. As the basis for improving the design, the design is more quality, safe, economic and reasonable, and the construction is more rapid. Real-time monitoring can understand the interaction force between stratum and tunnel structure and provide basis for establishing and adjusting shield construction parameters, such as determining shield tunneling speed, controlling stratum disturbance, reducing deformation, ensuring the safety of structure itself and surrounding environment during construction, and accumulating experience and providing guidance for similar projects in the future.

At present, there are many techniques for monitoring modern shield tunneling engineering, which can be divided into two types: deformation monitoring and nondestructive monitoring. Among them, the deformation monitoring has the characteristics of relatively lining installation tube and absolute deformation, and nondestructive supervision. The test can not be used to set up destructive signs, and can be used in monitoring work where the work space is narrow and the monitoring personnel have limited time. As far as the actual situation is concerned, the deformation monitoring technology has been more widely used.

1) Principles for the development of monitoring plans

In the development of monitoring projects, certain principles should be followed in order to ensure the effectiveness of monitoring. First of all, following the reliability principle to develop the monitoring plan, that is, to ensure the reliability of the monitoring program, you need to use reliable instruments and protect the measuring points during the monitoring. Secondly, it is necessary to follow the principle of hierarchy: Specifically, displacement is the main monitoring object, and other monitoring items are taken into consideration; at the same time, instrument monitoring is the main method, and the inspection method is used for coordination; when monitoring instrument selection is made, It is necessary to use machine-based instruments as the mainstay, supplemented by electric measuring instruments. In addition, separate points should be placed on the surface and underground pipelines in order to make the monitoring network have a certain measuring point coverage. It is also necessary to follow the principle of key monitoring of key areas: Specifically, for different geological conditions and different stability standards, it is important to monitor the areas with poor stability. In addition, the principle of convenient use should be followed: In order to prevent the monitoring work from interfering with the construction, the monitoring plan should be made more practical and convenient. Finally, in order to reduce the cost of monitoring, it is necessary to select practical instruments without the need to pursue advanced instrumentation.

2) Selection of monitoring sections

In the selection of monitoring section, geological conditions, engineering requirements and construction conditions should be comprehensively considered. In the layout of the monitoring section, it is necessary to consider the spatial–temporal relationship and adopt the combination of focus and general, local and overall, so as to systematically control the key parts of the project. At present, the monitoring section can be divided into main monitoring section and auxiliary monitoring section. Among them, the main section can be embedded with various instruments in order to monitor various monitoring projects.

3) Arrangement of monitoring points

In the arrangement of monitoring points, the type and number of monitoring points should be determined first according to the engineering properties, geological conditions and construction requirements. On this basis, the design data of monitoring points should be verified in the most unfavorable position and section in the design. At the same time, for the measuring points set to guide the construction, it is necessary to place the measuring points in the first construction site under the same working conditions for testing. Secondly, when determining the position of the surface deformation point, it is necessary to enable the measuring point to reflect the deformation characteristics of the monitored object well, and this position is conducive to the use of instruments and the protection of measuring points. The arrangement of measuring points cannot affect the normal force of stratum structure, nor can it weaken the structural deformation stiffness and strength. In addition, the arrangement of various

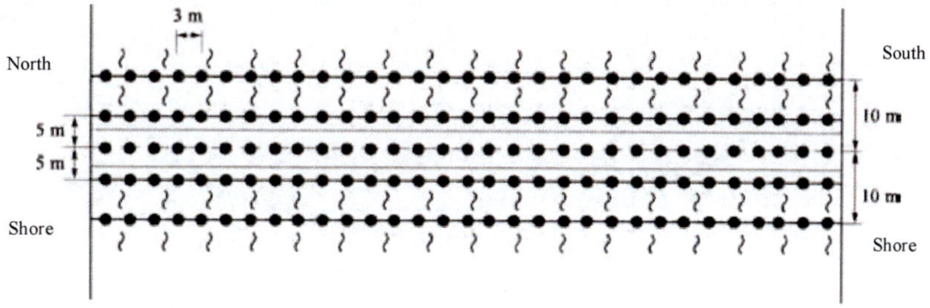

Fig. 4.25 Schematic diagram of sonar monitoring point layout

monitoring points should be organically combined in time and space in order to reflect the change relations and laws of different physical quantities.

4) Nanjing Yangtze River Tunnel Monitoring Example

The Nanjing Yangtze River Tunnel Project needs to cross the Yangtze River, and the shield adopts sonar and pressure sensor methods when crossing under the Yangtze River. The two methods simultaneously monitor the settlement of the river bottom when the shield passes over the Yangtze River.

(1) Sonar method, mainly through the use of ultrasonic velocity method to measure the depth of water, automated mapping, drawing flat water depth map and section. The method of measuring point is as follows: along the direction of the tunnel axis in the tunnel's central axis line is laid a central axis line, and then the left and right sides each 5 m apart a measuring line, a total of 4 parallel lines parallel to the central axis; shown in Fig. 4.25.
(2) The pressure sensor method, mainly by measuring the change of water pressure at the bottom of the river to reflect the size of the river bottom subsidence. The method of measuring point is: along the tunnel line every 10 m arrangement of a sensor point, and the use of a special signal receiver can automatically real-time access to the underwater pressure of each measuring point, by setting the water level benchmark to correct the impact of the tide rise and fall, and finally converted to each point of the elevation, as the basic parameters of data processing, At the same time, the corresponding software is developed for graphical processing, so that the water pressure and subsidence of each measuring point under water can be visually described.

Before the shield reaches the bank of the Yangtze River, a sensor monitoring point is placed every 10 m along the axis of the tunnel, about 54 in total; and a background measurement is performed by the sonar method before the shield crosses the river to understand the topography of the river bottom, as shown in Fig. 4.26 is shown. Simultaneous monitoring of river bottom settlement is carried out by sonar method and pressure sensor method. The monitoring frequency of the sonar method is once a day, and the monitoring frequency of the pressure sensor method is twice a day.

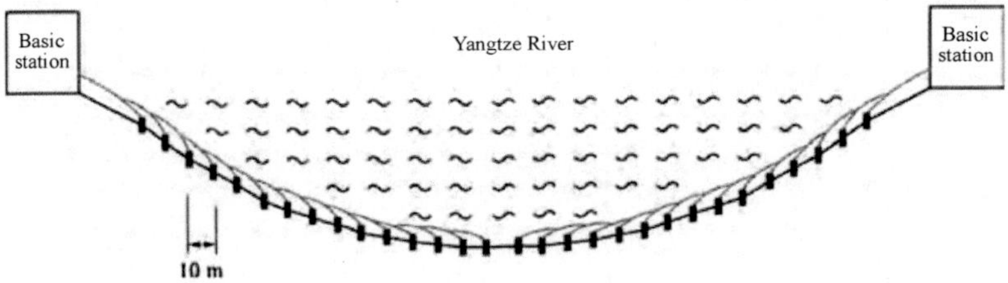

Fig. 4.26 Schematic diagram of the monitoring of the pressure sensor method

4.4.2 Embankment Reinforcement Technology

There are potential safety hazards such as land subsidence, cracks, collapse and water seepage in flood season in the embankment. The hazards are mainly in two aspects. First, the loose soil in the embankment reduces the safety of seepage resistance. In particular, the gaps and cavities above the shield will form seepage channels, which are prone to piping. Second, the physical and mechanical indexes of soil are reduced due to soil loosening, which affects the anti-sliding stability of embankment slope and further affects the embankment safety. Therefore, the use of corresponding embankment reinforcement means is the key to safe construction.

1) Foundation reinforcement

In the process of shield tunneling, the foundation of flood dike is reinforced according to the monitoring situation. The commonly used foundation reinforcement methods include stirring method and grouting reinforcement method. The stirring method uses special mixing machinery to mix the foundation soil and cement slurry in situ into cement soil. After 2–3 months of curing, the strength of cement soil can generally reach 0.5–1 MPa. The foundation strength after cement-soil reinforcement will be greatly improved, which is an effective foundation reinforcement method. The reinforcement technology of this method is mature and reliable without earthwork, but a certain amount of cement is needed. The grouting reinforcement method uses special machinery to force the cement slurry into the parts that need reinforcement at a certain pressure. The cement and the reinforcement parts jointly form a composite foundation, which greatly improves the foundation strength. The advantages and disadvantages of this method are the same as stirring method.

The Nanjing Yangtze River Tunnel was drilled and reinforced with sag grouting in the construction of the flood control embankment on the north bank of the Yangtze River at RK3+73. The scheme is to pre-bury the PVC grouting pipe at the angle of the water surface of the levee within 25 m around the shield axis. The grouting pipe is at an angle of 30° to the vertical plane, and is 3 m away from the top of the tunnel with a spacing of 1 m. If the displacement of the levee is found to be abrupt or the crack width is increased, the construction must be stopped. At the same time, the foundation is reinforced with spiral bored piles during construction, the pile diameter is 40–60 cm, and C30 concrete is used, which is arranged in plum blossom shape. It can be designed and constructed according to the specifications, as shown

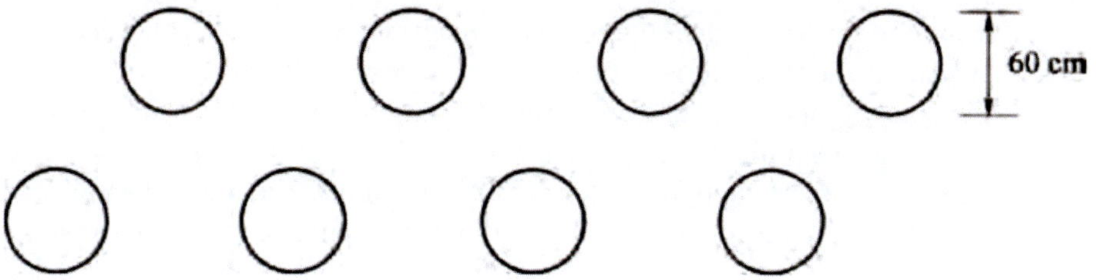

Fig. 4.27 Schematic diagram of spiral pouring pile

in Fig. 4.27. C30 concrete should be used No. 425 slag cement or ordinary cement, plus medium or coarse sand (with less than 5% mud), mixed stones (pebble or gravel, large diameter is 5–32 mm, with less than 2% mud).

The construction steps are as follows.

(1) The drill is in place: The drill must remain stable when in place. In order to accurately control the depth of drilling, a control scale should be made on the rack or on the tube for observation and recording during construction.
(2) Drilling: straightening the frame bout, on the good pile bit, start the machine drilling, unearthing, to reach the control depth after stopping drilling, drilling.
(3) Check hole quality, including drilling depth measurement and aperture control. The depth of hole and the thickness of virtual soil were measured by sounding rope (hammer) or portable lamp. The thickness of virtual soil equals the difference of drilling depth. In general, the thickness of virtual soil should not exceed 10 cm, so as to prevent the borehole diameter expansion caused by drill pipe sloshing, resulting in disturbance soil attached to the hole wall and increased fallback soil at the bottom of the hole.
(4) Pore bottom soil cleaning. After drilling to the predetermined depth, the soil must be cleared at the bottom of the hole, and then stop rotation. The drill pipe shall not be twisted. The soil scattered on the ground during drilling must be removed at any time.
(5) Move the rig to the next bit. After the hole inspection, the construction record of the pile hole should be filled. Then cover the hole cover and prevent driving or walking on the cover. Finally, remove the rig to the next level.

2) Dike crack reinforcement

In the study of embankment reinforcement, the reinforcement technology of weak parts such as cracks is described in detail, and the commonly used methods are excavation backfill and filling grouting. These methods are relatively mature in technology, and have been applied in many dam flood control and emergency projects at home and abroad. The effect is good, and the crack treatment is relatively mature in technology.

(1) Excavation backfill. Through analysis and research, it is concluded that the surface and embankment cracks caused by shield tunneling are developed

from the surface to the bottom, and the development depth is not large. This development characteristic of cracks provides conditions for the excavation and backfill of cracks. In engineering, the large-width cracks can be treated by excavation backfill. The implementation method is as follows: a trench is excavated downward along the crack, and the depth is 0.3–0.5 m below the crack. In order to facilitate the combination of backfill soil and embankment soil, the slope of groove slope should meet the relevant requirements, and the shape of groove slope can be made into steps. When backfill is carried out, the filler with good compactness and close combination with the original embankment is used, and the water content of backfill soil meets the design requirements. Stratification backfill ramming, backfill height should be slightly higher than the crest elevation, prevent rainwater infiltration, affect backfill effect.

(2) The cross wall is separated. The cross-wall cut-off method is suitable for the treatment of cracks crossing the embankment. The treatment method is to excavate trenches perpendicular to the cracks along the direction of the embankment in the center of the embankment. The excavation requirements of the trenches are the same as those before. The layered backfill compaction is carried out with soil materials with good water insulation performance and close combination with the original embankment. It should be noted that if the crack has communicated with the river water, it may lead to a large number of river water escaping along the groove during excavation. The closure should be made in the embankment first, and the filling should be completed. After the embankment is stable, the embankment is allowed to contact with the river water.

(3) Grouting treatment. Grouting technology is a method to use pressure to inject grout that can be consolidated into the embankment through drilling, filling weak surfaces such as cracks and pores of the embankment, and realizing the treatment of embankment cracks. In the treatment of embankment crack grouting, the commonly used grouting methods include curtain grouting, high pressure jet grouting, cone penetration grouting, etc. The cracks on the embankment are generally not deep into the foundation, and the water retaining height is not high. Clay slurry can be used for filling and filling, and cone penetration grouting is generally used. The cone penetration grouting belongs to the filling grouting, which is a method of injecting the slurry into the hidden danger area by the weight of the slurry to block the hidden danger. The grouting range of the affected embankment extends 50 m outward. The arrangement of grouting holes, row spacing, hole spacing, grouting depth, grouting materials and so on need to be determined according to the damage to the embankment caused by shield tunneling.

As a sharp weapon for urban rapid transportation across rivers, super-large diameter slurry shield will inevitably cross the river embankment. Whether it can smoothly cross the embankment not only affects the construction progress, but also relates to the life safety of the people. Through the successful experience of Nanjing Yangtze River Tunnel Project, it is proved that through detailed construction investigation, reasonable tunneling parameters (slurry pressure, slurry index, grouting pressure, grouting amount, mortar index, etc.), fine operation management (shield tail protection, segment assembly, etc.) and complete emergency measures, the ultra-large diameter slurry shield can completely complete the crossing of Yangtze River embankment or similar embankment without damage, which provides a reference for other large diameter shield underwater crossing embankment construction.

Chapter 5
Technology of Slurry Shield Crossing High Infiltration Induced Gravel Composite Stratum in River

The stratum in the middle section of the river crossing by slurry shield is generally complex. Therefore, according to different geological conditions, the corresponding construction technology is adopted to ensure the safe and efficient tunneling of shield. Due to its high permeability coefficient and low clay content, the high permeability gravel composite stratum is prone to serious mud filtration in construction, and the excavation face is difficult to form a mud film and lose stability. For example, during the construction of the middle section of the Yangtze River Tunnel in Nanjing, the mixed stratum of fine sand and gravel layer, and the round gravel layer with a length of about 370 m and a thickness of 0–3.9 m and the strongly weathered calcareous mudstone layer are encountered. Due to the soft and hard lithology of the stratum, the nature of the stratum is obviously different, and the pebble stratum is easy to cause the blockage of the excavation chamber and the damage of the cutter. The formation permeability coefficient is as high as 10^{-2} cm/s, and the coarse particles larger than 2 mm account for about 40% of the whole formation content, and the clay content is very small. This situation is not conducive to the formation of mud film on the excavation surface of slurry shield, and it is prone to the problem that the pressure balance on the excavation surface cannot be established due to the large amount of mud filtration. At the same time, the maximum working pressure of about 0.65 MPa of Nanjing Yangtze River Tunnel shield also increases the risk of crossing the river, which is also a test for the shield itself and the construction processed the risk of crossing the river, which is also a test for the shield itself and the construction process.

This chapter introduces the technology of slurry shield crossing the middle Yangtze River composite stratum from three aspects: high permeability and water-rich gravel composite stratum tunneling, slurry preparation and film forming technology, slurry circulation and slag carrying capacity calculation and analysis, which provides reference for future shield tunnel crossing high permeability composite stratum.

J. Chen et al., *Construction Technology of Large Diameter Underwater Shield Tunnel*, Key Technologies for Tunnel Construction under Complex Geological and Environmental Conditions, https://doi.org/10.1007/978-981-16-5896-9_5

5.1 High Permeability Water-Rich Gravel Composite Stratum Excavation Problem

5.1.1 Technical Problem

In recent years, there have been many large-scale cross-sea cross-river tunnel projects at home and abroad. The common feature of these tunnels is that they are large in diameter and often pass through coarse-grained strata such as flowing sand, gravel and sandy silt rich in groundwater. In the harsh working conditions, especially in a large number of water-bearing gravel layers, non-cohesive, extremely unstable soil layers and shallow overburden tunnel projects, as well as the construction of ultra-large diameter shields and areas with special high requirements for ground deformation, slurry shield construction is mostly used.

Slurry shield is a kind of sealed mechanical shield. In the construction process, the slurry needs to meet the following three requirements. ① A reliable mud film is formed on the excavation to stabilize the excavation surface. ② Slag soil was successfully transported to avoid sediment. ③ Slurry separation was completed by screening equipment. The stability of excavation face is maintained by the following three factors: ① Mud pressure balances soil pressure and water pressure. ② The impervious mud film is formed on the excavation surface, so that the mud pressure can effectively play a role. ③ The slurry penetrates into a certain stratum from the excavation surface, and the stratum on the excavation surface increases the cohesion. With the continuous excavation of the excavation face, the slurry pressure is always required to be one. The form of surface acts on the excavation surface. In this process, the formation of impervious cement film on the excavation surface is the core factor.

In low permeability clay and silty clay strata, the formation pore is small, and the mud is easy to form mud film on the excavation surface. However, in the high permeability formation such as fine sand and gravel sand, mud easily penetrates the formation and directly seeps out, and cannot form mud film on the excavation surface, which will lead to a large number of mud filtration, mud pressure is difficult to maintain, thus inducing excavation surface instability and other accidents, the consequences will be very serious. At the same time, large size particles in the formation are unfavorable for carrying slag, but the separation of mud and water is relatively easy. Therefore, how to form a good mud film on the surface of high permeability stratum is an urgent problem to be solved in the cross-river tunnel project.

5.1.2 Research Status

5.1.2.1 Research Status of Slurry Shield Mud Preparation and Mud Film Formation

Many scholars in China and abroad have done a lot of research work on the influence of slurry properties on film formation. In Japan, where slurry shield technology is relatively developed, Takeshi Watanabe et al. found through the self-developed test device that with the increase of slurry density, the slurry filtration gradually decreases. In high permeability formation, it is difficult to form dense mud film only by increasing the density and viscosity of mud. The sand content of mud plays a vital role in reducing mud filtration and forming dense mud film. Combined with specific projects in early Japan, a suitable mud formula was proposed. Similar to this study, the Swiss scholar Fritz et al. used the limit supporting pressure that the mud film could withstand to evaluate the quality of the mud film. By adding silt, sand, vermiculite, polymer and other additives to the mud to adjust the properties of the mud, a reasonable mud ratio was provided for the Zimmerberg tunnel (using a slurry pressure shield with a diameter of 12.35 m) to cross the coarse gravel sand stratum with the permeability coefficient of (1–4) $\times$ 10^{-3} m/s, which ensured the safe tunneling of the shield in the gravel sand stratum with a diameter of 780 m. German scholar Heinz studied the influence of additives on the rheological properties and penetration of bentonite mud. Fillers such as silt, sand and other particles can effectively reduce the penetration distance of mud in the formation.

Chinese scholars have also carried out research on mud preparation and film formation in specific projects in combination with the construction of Nanjing Yangtze River Tunnel, the Yellow River Tunnel of the Middle Route of South-to-North Water Diversion Project and some underwater tunnels in Shanghai. In 2001, Cheng Zhanlin et al. of the Yangtze River Academy of Sciences, took the tunnel of the Middle Route of South-to-North Water Diversion Project crossing the Yellow River as the background, carried out experimental research on the stability model test system of the shield excavation face, put forward the calculation formula of the critical slurry pressure of the slurry balance shield in the medium and coarse sand stratum, and obtained the correlation between the critical slurry pressure and the strength of the foundation stratum, the pore water pressure and the initial stress state at the excavation site. In 2007, Wei Liangwen and Zhang Qinghe of Tongji University took the Shanghai Fuxing East Road cross-river tunnel project as the background, through the indoor slurry characteristics test and microscopic analysis, as well as the engineering field test and monitoring data analysis, found that the PMS slurry system can quickly form high quality mud film, which is very beneficial to the stability of large slurry shield excavation face.

It can be seen that there has been a certain understanding of the formation law of mud film on the excavation surface of slurry pressurized shield, but these studies are mostly based on solving the problems of mud preparation and film formation in concrete slurry shield engineering. The influence of mud density, viscosity and particle composition on the formation of mud film is discussed, which promotes the development of slurry shield engineering. At present, there is still a lack of discussion on the mechanism of mud film formation in high permeability sand-gravel composite strata.

5.1.2.2 The Research of Water Shield Excavation Surface

There are many research results on the excavation face of slurry shield, but due to the importance of this issue, the work is still in-depth. At present, the research methods on the stability of slurry shield excavation face mainly include the simplified limit support pressure determination model which assumes the sliding shape of the soil ahead of the excavation, the complex centrifugal model test and the numerical analysis model which is convenient for a large number of research work and low cost. In 1994, Anagnostou et al. put forward a wedge "penetration model" considering slurry penetration, and proposed that in sandy soil, the support force of excavation face required for shield tunneling can be calculated by "film model", and the slurry penetration will greatly affect the stability of excavation face under the static state of shield stopping excavation. In addition, due to the complexity of the actual geological conditions, when the shield is advancing, if special geological conditions are encountered, such as lens or local sand layer, if the mud parameters are not adjusted in time and no high-quality mud film is formed, the mud penetration is inevitable. At this time, the "penetration model" should be used to analyze the stability of this excavation face. At the same time, the study also found that in the sand with good particle size distribution, increasing the slurry pressure can effectively improve the safety factor of excavation face stability, while in the larger soil layer such as coarse sand and gravel layer, only increasing the slurry pressure has little effect on improving the safety factor of excavation face stability. In this case, increasing bentonite content to improve the dynamic shear strength of slurry can effectively ensure the stability of excavation face. However, increasing the content of bentonite will greatly increase the construction cost, so it is unreasonable to increase the viscosity of slurry to stabilize the cost of shield excavation face in gravel stratum. In 1995, according to the two-dimensional test and the collapse situation during construction, the two-dimensional Murayama formula for the stability calculation of excavation face was proposed by Murayama scholars in Japan. In 2007, Li Yun and Zhang Zixin further applied the mud "permeability model", combined with the Murayama formula, considering the geometric relationship between the seepage area of the mud and the sliding surface, the effective mud pressure was reduced, and the safety factor of the excavation face stability of the mud permeability model was obtained.

Akira Mori et al. studied the increase of excess pore water pressure caused by slurry infiltration and the influence of excess pore water pressure on the stability of sand excavation face by using slurry filtration test device and shield simulation construction device. In 2002, Broere et al. from the University of Delft in the Netherlands established the groundwater flow model based on the wedge-cabin model, and analyzed the influence of the excess pore water pressure in the stratum on the stability of the excavation face. The study showed that the formation of excess pore water pressure in the stratum would reduce the effective mud pressure, the effective stress on the soil and the friction of the soil. It was proposed that the excess pore water pressure in front of the excavation was proportional to the penetration distance and the shear strength of the bentonite mud, and inversely proportional to the characteristic particle size d_{10} of the stratum. In 2009, Li et al. used the upper bound solution analysis method and FLAC^{3D} three-dimensional numerical simulation method to study the influence of the gradient of slurry pressure on the excavation face and the soil characteristics changing with depth on the stability of the excavation face for the stability of large-diameter slurry shield excavation face in soft soil layer. The study shows that the gradient of mud pressure will significantly improve the local stability of the excavation face, and the formation characteristics changing with depth will affect the overall failure mode. At the same time, the upper bound solution that can estimate the possible failure mode of the excavation face under given geological conditions is obtained. In 2010, American scholars Seung Hankim and Fulvio Tonon assumed that an ideal mud film was formed on the excavation surface. The three-dimensional finite element method was used to analyze the influence of tunnel diameter, the ratio of overburden thickness to tunnel diameter, soil lateral pressure coefficient and soil strength parameters on the stability of tunnel excavation surface. It was proposed that the minimum mud pressure should be determined by the relationship between mud pressure and expected displacement, and the influence of tunnel diameter on supporting pressure was small in non-viscous soil.

It can be seen that the research on the stability of excavation face has made great progress in the theoretical limit support pressure and numerical simulation, and it is believed that the penetration of slurry into the excavation face has an important influence on the stability of the excavation face. At present, there is a lack of research on the stability of excavation face in sandy gravel composite stratum, and the mechanism of mud film in the stability of excavation face is not revealed from the perspective of force transformation.

5.1.2.3 Research Status of Excavation Face Stability Control for Slurry Shield with Pressure Opening

In the domestic and foreign construction of shield with pressure cabin maintenance, there are many examples of successful cabin opening under pressure conditions. Typical examples of cabin opening with pressure in foreign countries are: the fourth tunnel of Yibei River in Germany in the bottom of the sand drift and other strata, in 0.4–0.45 MPa pressure cabin opening repair. The slurry shield of Weser tunnel in Germany is maintained in iced gravel stratum under pressure above 0.45 MPa. When the red line subway of St. Petersburg in Russia is constructed by slurry shield in silty sand, the repairable open cabin under pressure of 0.55 MPa is carried out. The Westerschelde Tunnel in the Netherlands conducts active cabin opening maintenance in clay stratum. The cabin opening is carried out after the mud pressure is replaced by the pneumatic method, and the air pressure is 0.45 MPa. The most representative example of pressure cabin opening in China is the pressure cabin opening of Nanjing Yangtze River Tunnel Project in gravel stratum at about 60 m below the river bottom. In this project, the slurry film with good air-tightness is formed on the excavation surface through the two-step slurry scheme of first penetrating and then forming the film, and then the slurry level within 3.0 m of the shield upper arch is reduced, and the compressed air is replaced by 0.6 MPa. However, in these cases, more attention is paid to the process of cabin opening and the health problems of cabin entry personnel in the process of decompression, and there are few analysis and experimental studies on the stability of excavation face.

At present, although the cabin opening with pressure has attracted enough attention in the engineering field, the research on its system is rare, especially the research on the air tightness of mud film. The research on water and gas movement in porous media such as clay and clay can be used for reference. Richards found that the air pressure was applied on the side of the saturated porous ceramic plate. When the air pressure reached a certain value, the ceramic plate began to drain, and the air entered but could not penetrate the ceramic plate. The concept of intake value and foaming point was proposed. Ye et al. found that the gas in saturated clay exists initial pressure value, only when the pressure reaches a certain value, will appear obvious permeability. According to the research of Jieying Revolution, the bulk density of soil will change during the dehydration process, and there is no inflection point in the soil water characteristic curve due to the abrupt change of slope caused by the change of intake value. These studies have important reference significance for studying the air-closure law and mechanism of the mud film.

It can be seen that although there have been many successful cases of pressurized cabin opening, there are few studies on the failure mode and action mechanism of mud film under the condition of high permeability gravel composite stratum.

5.2 Slurry Preparation and Film Forming Technology of High Permeability Sand—Gravel Composite Formation

5.2.1 Mud Material Composition and Basic Properties

5.2.1.1 Mud Material Composition

1) Bentonite

Bentonite is a layered aluminosilicate clay mineral with montmorillonite as the main component. Its cell is composed of two layers of silica tetrahedron and a layer of aluminum oxide octahedron. The oxygen atoms shared between the tetrahedron and the octahedron form a highly ordered quasi-two-dimensional layer, which has a large specific surface area. The silicon and aluminum ions can be replaced by low-valent ions such as magnesium, iron and lithium, resulting in unsaturated interlayer electricity price of bentonite and permanent negative charge between layers. This negative charge is usually balanced by interlayer exchangeable hydrated cations. Therefore, bentonite has a series of valuable characteristics such as good ion exchange, water absorption, expansion, adhesion and adsorption. Using its good ion exchange performance, the natural Ca-bentonite is sodiumized, and its hydration performance is optimized, which can be used as the material of drilling mud and slurry shield mud. The evaluation items and test methods of bentonite are as follows.

(1) Montmorillonite content: Montmorillonite is a kind of silicate-aluminate clay mineral. Montmorillonite is a unit crystal layer composed of two layers of silica tetrahedron layer and one layer of aluminum hydroxide octahedron layer. Tetrahedron and octahedron are connected by shared oxygen atoms. Its molecular formula is $Al_2(Si_4O_{10})(OH)_2H_2O$, which is expressed as $4SiO_2{\cdot}Al_2O_3{\cdot}H_2O$ with oxides. The corresponding theoretical contents are 66.9%, 28.3% and 5.0%, respectively. Bentonite with high montmorillonite content is easy to disperse, large cation exchange capacity, high pulping rate and good stability due to its mineral structure and lattice substitution characteristics. Therefore, montmorillonite content is the most important index for evaluating pulping bentonite. At present, the commonly used method for the determination of montmorillonite content is the blue absorption method, as detailed in "Benton" (GB/T 20973—2007). It uses methylene blue to form monovalent organic cations in aqueous solution, which can react with bentonite by cation exchange to form organic bentonite complex and dye it.

(2) Expansion index: The expansion index of bentonite refers to the volume of 2 g bentonite after 24 h expansion in water. The test method is detailed in GB/T 20973—2007. The greater the expansion index of bentonite, the better its dispersion, the higher the pulping rate. In common pulping bentonite, the expansion index of sodium bentonite is higher than that of calcium bentonite and acid bentonite. This is because the water absorption and expansion process of bentonite is the continuation of the interlayer cation hydration process. The interlayer hydration process is carried out step by step, and roughly forms 1–4 layers of water molecules. The types of interlayer cations are different, and the hydration energy is also different. Ca^{2+} hydration energy is larger, Ca^{2+} hydration film is thin and irregular, and Na^+ hydration film is thick and arranged regularly, which inevitably leads to the expansion of sodium bentonite. For the same type of bentonite, the more montmorillonite content is, the higher expansion index is. Expansion index is one of the technical indexes for evaluating bentonite quality.

(3) Cation exchange capacity: Cation exchange capacity of bentonite CCEC refers to the total amount of cations (Na^+, K^+, H^+, Ca^{2+}, Mg^{2+}, etc.) adsorbed by bentonite at $pH = 7$ (meq/100 g soil), including exchangeable hydrogen and exchangeable base. The size of CCEC is related to the hydration, expansion and electrification of bentonite, which is not only an important indicator for identifying the mineral types of bentonite, but also an important basis for evaluating

the quality of pulp bentonite. Bentonite can be divided into sodium bentonite, calcium expansive soil and alkali treated calcium expansive soil according to the type and quantity of adsorbed ions. The exchangeable cations adsorbed were mainly sodium bentonite, and calcium expansive soil, etc. The exchangeable cations adsorbed by bentonite are different, and the pulping performance will be greatly different. In terms of pulping rate, only sodium bentonite can have high pulping rate, and other calcium bentonite can have high pulping rate only after pure alkali treatment.

(4) Particle gradation: In soil mechanics, the soil particle content of a certain particle group in the soil is defined as the ratio of soil particle mass to total dry soil mass in the particle group, expressed by percentage. The relative content of soil particles in grain groups of various sizes in soil is called soil gradation. In soil mechanics, the particles with particle size greater than 75 μm are defined as sand, the particles with particle size of 5–75 μm are silt, the particles with particle size less than 5 μm are clay, and 2 μm is also used as the upper limit particle size of clay in foreign countries. In colloidal chemistry, particles with particle size less than 0.1 μm are called colloidal particles. In fact, particles with particle size of 0.1–5 μm have some properties of colloidal particles. In addition, the smaller the particle size is, the larger the specific surface area is. For clay soil, due to the small particle size of clay particles, it has a large specific surface area. When clay particles meet with water, it will produce complex physical and chemical changes and have some characteristics of colloidal dispersion system. The higher the clay content in the soil, the greater the adsorption and hydration ability, the stronger the dispersion ability, the thicker the particles, and the worse the pulping performance. Therefore, the content of impurities such as sand can be determined by measuring the particle size distribution of bentonite, and the content of clay beneficial to pulping can be measured. Therefore, the gradation curve of bentonite can be used as a basis for evaluating the quality of bentonite.

2) CMC

As an important part of mud materials, carboxymethyl cellulose (CMC) plays a major role in regulating the viscosity of mud. CMC is the largest, most widely used and most convenient product of cellulose ethers, commonly known as "industrial monosodium glutamate". CMC is an important cellulose ether and an anionic cellulose compound with good water solubility obtained from chemical modification of natural fibers, which is easily soluble in hot and cold water. CMC is a macromolecular chemical substance that can absorb water and expand. When swelling in water, it can form a transparent viscous glue, which is neutral in pH. Solid CMC is stable to light and room temperature, and can be stored for a long time in dry environment. CMC is white or yellow powder, granular or fibrous solid, odourless, tasteless, non-toxic. The important characteristics of CMC are the formation of colloids and solutions with high viscosity, which have the characteristics of adhesion, thickening, flow, emulsification and dispersion, shaping, water retention, protective colloids, film forming, acid resistance, salt resistance and turbidity, and are physiologically harmless.

3) Mud filling materials such as clay and silt

Clay, silt and fine sand with different particle size ranges are used as mud filling materials. The main function is to adjust the particle size of solid particles in the mud and the density of the mud. The filling materials with different particle size ranges are added to adjust the mud gradation, and different amounts are added to adjust the density of the mud.

4) Other additives

Sometimes in order to increase the stability of bentonite mud, will also be added to the mud sodium carbonate Na_2CO_3, play a scattered, uniform role, the principle can be called salt immersion principle. Ingredients such as calcium silica and magnesium in the monoshide can react with alkaline substances, and the alkali is designed to destroy the lattice structure of the monosacis, making the main components of the monomorphic state of polysetic sodium hydrochloric acid and sodium polysilicate and partial al-aluminic acid. Because the monosacis contains the gender substance Al_2O_3, generally in high temperature conditions, will react with NaOH or Na_2CO_3 under the melting state, thus destroying the cell of the montiline, when the Al_2O_3 dissolution in the montiline reaches 75–85%, the lattice structure is often destroyed. Coupled with the solubility of other ions, the amorphous silicon in the lattice structure is converted into active silicon, thus achieving the function of activation. Sodium carbonate, an additive, can solve a certain amount of Na^+ and polymerized anions. Ca^{2+} on the surface of Na^+ and clay is an ion exchange, due to the intervention of the hydration of sodium ions, increased the thickness of the diffusion layer, increased the potential ξ of the clay particles, in addition to the polymerion of anion can be partially adsorbed by the positively charged clay particles end surface, so that the end of the charge modification, breaking the bond structure formed by the side of the clay particles, Release part of the free water, so can play a certain dilution, dispersion role.

At present, some special pulping agents have been developed in the engineering field. NSHS-1, NSHS-2, NSHS-3, NSHS-4, NSHS-5 and other five kinds of pulping agents provided by North China Oilfield are composed of several products, and their structures are complex. Due to the use of a variety of polymer, and the same polymer has different molecular weight, substitution degree and polymerization degree, the actual effect of each pulping agent is still uncertain. Through analysis and research, the main functions of five kinds of mud are determined. The main components and functions of different pulping agents are shown in Table 5.1.

5.2.1.2 The Basic Nature of the Mud

1) Density

Mud density characterizes the content of solid particles in mud, which has an important impact on the penetration of mud in the formation. It is one of the most important

Table 5.1 Pulping parameters

The name of the pulper	Main ingredient	Main effect
NSHS-1	It consists of large and medium molecular weight polyacrylates, cellulose salts, anticalcium agents, modified soils, synergistic additives, etc.	Tackifier
NSHS-2	It consists of different grades of mica, vermiculite, nut shell, asbestos grain, rubber powder, single seal and other raw plant fibers and inert minerals	Plugging agent
NSHS-3	It consists of different large and medium molecular weight polyacrylates, cellulose salts, anti-calcium agents, vinyl multi-monomer copolymers, plugging microporous expansion agents, modified soils, synergistic additives, etc.	Tackifier
NSHS-4	Composition of polyacrylate and polyacrylate salt additive additives with a molecular weight of more than 1.5 million	Flocculant
NSHS-5	It consists of large, medium and small molecular polyacrylates, cellulose salts, polyacrylonitrile salts, inhibitors, modified soils and synergistic additives with different adsorption groups and hydration groups	Flocculant

properties of mud and an important indicator of mud management in slurry shield construction. The mud density is small, the mud pump will not overload operation, but the density is small, the content of soil particles in the mud is small, the speed of mud penetrating into the formation will be slow, which is unfavorable to the stability of the excavation face. The density of mud is relatively large, that is, there are many solid particles in the mud. Solid particles are easy to deposit at the bottom of the mud drainage pipeline, which not only causes great wear and tear of the pipeline, but also increases the driving power of the equipment, and makes the mud pump overload operation. So in determining the density of mud, not only to consider the stability of excavation face, but also to consider the bearing capacity of equipment.

The commonly used instrument for measuring mud density is a simple and easy-to-operate mud specific gravity scale. The ratio of mud specific gravity to 4 °C pure water is tested, and the reading on the scale is mud specific gravity. The specific gravity scale used is 1002 type slurry specific gravity scale, as shown in Fig. 5.1. It is composed of slurry cup, beam, moving weight and support, and there are weight regulating tubes and horizontal bubbles on the beam.

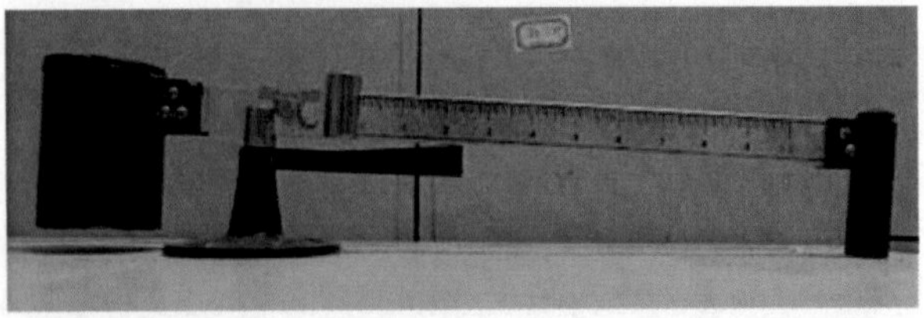

Fig. 5.1 Type 1002 gravity scale

2) Viscosity

Mud viscosity represents the viscosity of mud flow and reflects the fluid characteristics of mud. When the liquid flows, the measurement of the resistance to shear deformation rate is called the viscosity of the liquid (referred to as viscosity). Viscosity is the nature of the liquid itself, and does not change with the shear rate. Viscosity is expressed by viscosity. The greater the viscosity of the mud, the better the physical stability of the mud, and the mud is not easy to segregation. It is not easy for particles to settle when transporting slag. However, the viscosity of the slurry is too large, and there are not only many viscosifiers needed in pulping, but also high economic costs. When the slurry is pumped, the pumping capacity of the slurry pump is also limited. The mud viscosity is small, the physical stability of the mud is poor, and the mud is prone to segregation. The combination of particles and water in the mud is weak, and the carrying capacity of the mud is poor. Although the particles in the mud plug the formation pores, the water in the mud is easily pressed out under the action of mud pressure, leaving the particles in the mud plug on the surface of the formation, which makes the formed mud film thick and loose, and is not conducive to the stability of the excavation face.

Mud viscosity can be divided into two kinds: one is more intuitive and simple funnel viscosity, which characterizes the macroscopic viscosity of the mud. The funnel viscometer is used to measure the viscosity of the mud. The commonly used funnel viscometers are Sue funnel viscometer (Fig. 5.2) and Mae funnel viscometer (Fig. 5.3). The other is the plastic viscosity that reflects the properties of slurry fluid, representing the ability of slurry to hinder the flow. It is one of the basic parameters of hydrodynamic properties. The rotational viscometer is used to measure the commonly used rotational viscometers, including six-speed rotational viscometer (Fig. 5.4) and fifteen-speed rotational viscometer (Fig. 5.5). However, in order to simplify the measurement in site construction, only the funnel viscosity related to yield is usually measured (in a certain range, the apparent viscosity of mud can be

Fig. 5.2 Su-type field funnel viscometer

Fig. 5.3 Horse field funnel viscometer

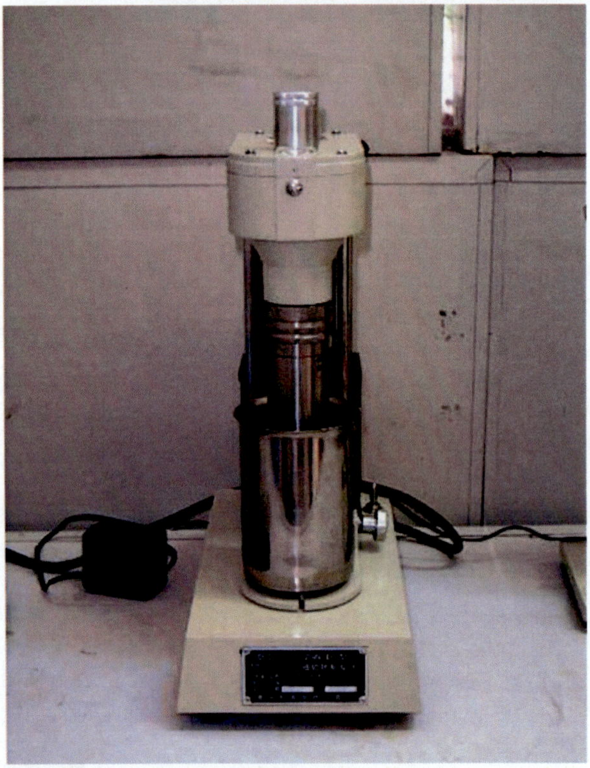

Fig. 5.4 ZNN-D6 electric six-speed rotary viscometer

characterized by the funnel viscosity of mud, and finally the viscosity and fluidity of mud can be reflected).

(1) Funnel viscosity. The slurry viscosity commonly used in engineering is basically funnel viscosity, because the funnel viscometer is cheap, compact,

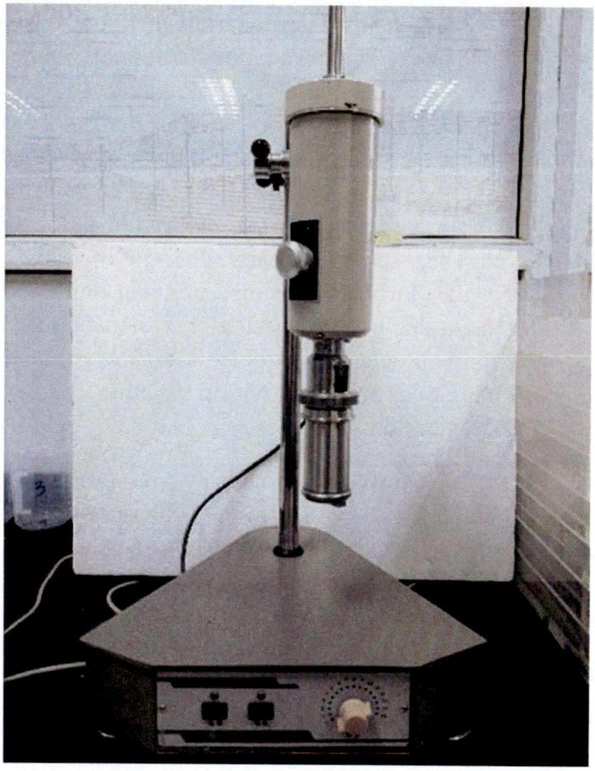

Fig. 5.5 NXS-11A type fifteen speed rotary viscometer

portable, and easy to operate in the field. It measures the viscosity of slurry at low flow rate. At present, the mud viscosity used in engineering is mainly Sue funnel viscosity and Mae funnel viscosity, and the conversion relationship between them is shown in Table 5.2. If there is no special description of the funnel viscosity of mud in this book, it is measured by the funnel viscometer commonly used in the construction site (the Soviet field funnel viscometer).

Table 5.2 Specifications and comparison of commonly used funnel viscometers

Name	Total funnel volume (mL)	volume of liquid flowing out during measurement (mL)	Funnel viscosity value of calibrated water (s)	Note
Markov funnel viscometer	15,400	946	26 ± 0.5	U.S. api standard
Japanese funnel viscometer	500	500	17 ± 0.5	Japanese data provided
Susie funnel viscometer	700	500	15 ± 0.5	Common funnel viscosity in China

Method for determination of funnel viscosity (taking the soxhlet funnel viscometer as an example): the funnel is vertical, gripped by hand and blocked by index finger, and then poured 200 and 500 mL mud into the funnel at both ends of the measuring cylinder. Put the measuring cylinder 500 mL upward under the funnel, open the index finger, and start the stopwatch to record the time needed to fill the 500 mL mud, that is, the viscosity of the measured mud.

(2) Plastic viscosity (PV). Under laminar flow conditions, the slope value when the shear stress is linearly related to the shear rate is called the plastic viscosity of the slurry, and the measurement unit is "mPa s". Plastic viscosity reflects the strength of internal friction between suspended solid particles, between solid particles and liquid phase, and within continuous liquid phase when the damage and recovery of grid structure in slurry is in dynamic equilibrium under laminar flow.

Plastic viscosity is measured by rotary viscometer, which consists of four parts: dynamic part, variable speed part, measurement part and support part. The power part provides power for rotary viscometer by motor. Variable speed part variable six speed, respectively, 3 r/min, 6 r/min, 100 r/min, 200 r/min, 300 r/min, 600 r/min. The measurement part is composed of a torsion spring, a dial and an inner and outer cylinder. The inner cylinder is matched with the taper of the shaft, and the outer cylinder is connected with the chuck. The bracket part adopts the tray to lift the tested vessel.

The working principle of the rotary viscometer is as follows. The liquid is placed in the annular space of two concentric cylinders. The motor drives the outer cylinder to rotate at constant speed through the transmission device. With the help of the viscosity of the measured liquid, it acts on a certain torque of the inner cylinder to drive the inner cylinder connected with the torque spring to rotate. The size of the rotation angle is proportional to the viscosity of the liquid, so the viscosity measurement of the liquid is converted to the measurement of the inner cylinder rotation angle.

The use of rotary viscometer (taking NXS-11A rotary viscometer shown in Fig. 5.5 as an example): The appropriate measurement system is selected according to the estimated viscosity value of the material. The selection basis is that in the 15 speed measurement, the scale reading cannot be all less than 25 or all greater than 75, and then the corresponding inner cylinder is inserted into the coupling of the measuring head to tighten the screw fixation. Appropriate amount of measured material is added to the outer cylinder, and the outer cylinder is installed on the measuring head, and fixed with a screw sleeve, so that the material in the outer cylinder is just immersed in the inner cylinder. Level the measuring head with leveling bubble, turn on the power supply, adjust 15 speed in turn, record the corresponding scale of each speed. The rheological characteristic curve was drawn according to the corresponding rotational speed and the shear force value converted by the viscometer reading.

3) Particle grading

The mud gradation reflects the size and distribution range of solid particles in the mud, which is an important parameter affecting the permeability of the mud in the formation. Studies have shown that for high permeability formation, the mud contains a certain amount of sand, vermiculite and other coarse particles can change the permeability form of mud. Therefore, the slurry gradation is another important parameter reflecting the slurry properties. Generally, for strata with permeability coefficient exceeding 10^{-2} cm/s, such as coarse sand, gravel, pebbles or strong weathered strata with cracks, mud is easy to escape and mud pressure cannot be maintained, which is easy to cause the collapse of excavation face to lose stability. At present, the practice and theoretical research in Nanjing Yangtze River Tunnel and Nanjing Wei San Road Crossing Channel have proved that the requirements of slurry shield construction on slurry performance in this stratum need to consider the gradation of particles in the slurry to ensure that this slurry can form a stable and dense mud film on the excavation surface. The traditional soil sieve is mainly suitable for testing the gradation of sand (>75 μm), and the gradation of particles with particle size less than 75 μm cannot be measured. Since there are many fine particles in the slurry, the commonly used test instrument for slurry gradation is laser particle size analyzer. The particle size distribution curve of mud is often measured by MS2000 laser particle size analyzer (Fig. 5.6).

4) Sand content

The sand content of bentonite refers to the percentage of sand volume that cannot pass through the No. 200 sieve (equivalent to the diameter greater than 0.075 mm) in the bentonite slurry, which reflects the amount of coarse particles in the slurry. For the bentonite slurry, the sand content can also be called the impurity content, reflecting the advantages and disadvantages of the grinding process of a bentonite ore. For high permeability formation, a certain sand content of mud is conducive

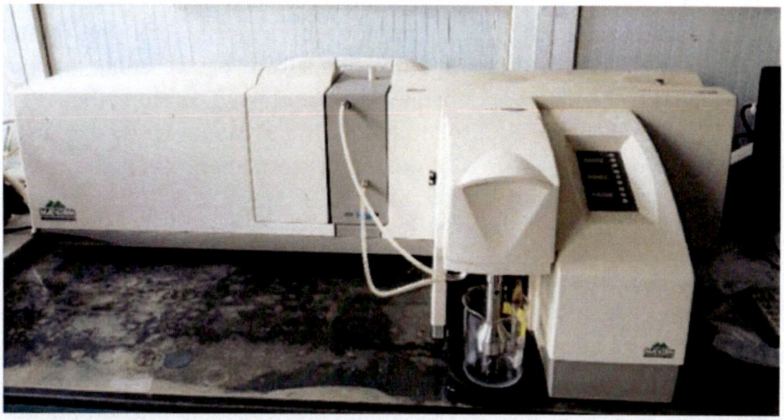

Fig. 5.6 MS2000 laser particle size analyzer

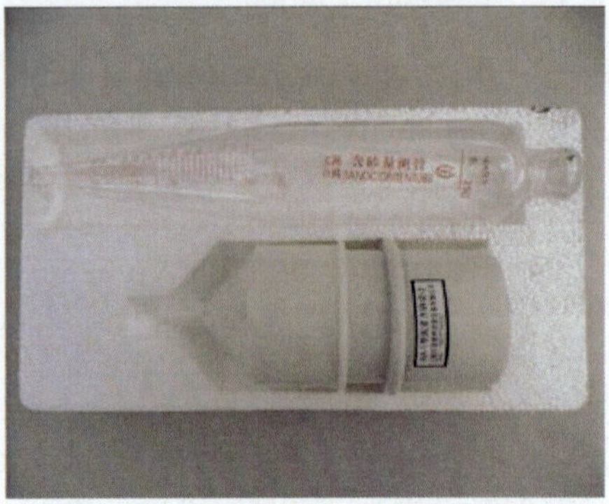

Fig. 5.7 LNH type mud sand content measuring device

to plugging the pores of the formation. The determination of sand content in this experiment mainly adopts LNH mud sand content tester, as shown in Fig. 5.7. The instrument is mainly composed of filter tube, funnel and glass measuring tube.

5) Water loss of mud

In the field of oil drilling, the research on drilling mud filtration is relatively mature, and there are currently test standards. Since the temperature and pressure difference have a great influence on the filtration of drilling mud, the filtration of drilling mud can be divided into low temperature and low pressure filtration (or API filtration) and high temperature and high pressure filtration. At present, API filtration is usually used to evaluate the filtration of drilling mud in China and abroad, that is, under the specified pressure difference (0.69 MPa), a certain volume (240 mL) of mud is measured by the filtration of a certain area of infiltration medium within 30 min (mL/30 min). Commonly used instruments are ZNS 1 mud dehydrator, as shown in Fig. 5.8. The special filter paper produced by American Baroid Company is used in the filtration medium, the filter paper area is $45.6 \pm 0.5\ cm^2$, and the pore size is 0.5–51 μm. Because the API filter loss is measured under the same pressure and the same permeability medium, and the formation is changeable and the mud pressure is different, the same mud has different permeability types in different formations and forms different forms of particle accumulation. However, according to the API filter loss test standard, one mud has only one API filter loss. Therefore, the API filter loss of mud can only reflect the quality of mud itself and cannot be used as a standard to evaluate the permeability process of mud in the formation.

6) Physical stability and colloidal rate of mud

Physical stability refers to the ability of solid particles in mud to maintain the physical state of floating suspension, which is usually described by the interface height. After a certain amount of slurry is placed in the measuring tube for a period of time, some soil particles will lose the suspension characteristics and precipitate. Clear water appears on the surface of the slurry, soil particles appear at the bottom, and mud is still in the middle. Clear water and mud appear at the interface. The precipitation degree of soil particles in the slurry can be identified by observing the variation of the

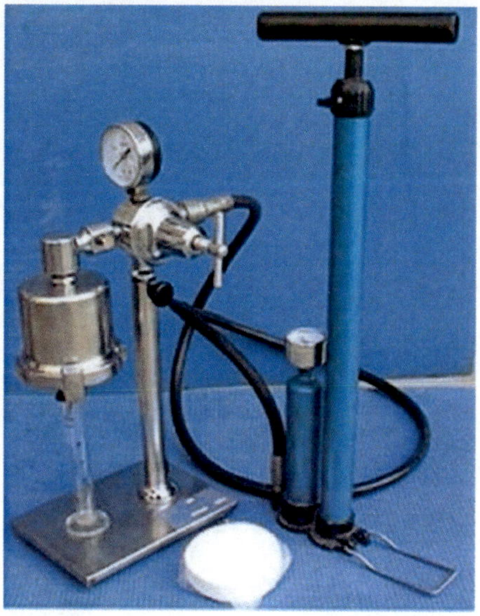

Fig. 5.8 ZNS-1 mud water loss meter

boundary height between clear water and slurry over time. The slurry colloidal ratio is commonly used to measure the physical stability of slurry in engineering. Colloid ratio represents the degree of dispersion and hydration of clay particles in bentonite, which is a rough indicator of the degree of dispersion and hydration of clay particles in mud. It characterizes the hydration ability of mineral components in mud. It is mainly the process of water absorption of clay particles, colloidal particles and other particles, the stretching of molecular bonds, and the thickening of the binding water film around clay particles. Macroscopically, it shows that the particles absorb water and disperse into uniform colloids. The higher the colloid ratio is, the more stable the physical properties of the mud are, and the mud is not easy to segregation and water precipitation, so the formed mud is relatively uniform. Therefore, the colloid rate is the basic index to characterize the physical stability of mud. The test of mud colloid rate is shown in Fig. 5.9.

Test method: 1000 mL mud was poured into 1000 mL measuring cylinder, and glass was placed on the cover for 2 h or 24 h. The volume of the upper clarification liquid and the sedimentation volume of the lower sand particles were observed. If the upper clarification rate was 3% and the lower sedimentation rate was 2%, the slurry colloid rate was 95%.

7) The pH of the mud

The pH value is mainly used to reflect the acidity and alkalinity of bentonite mud. The main measuring tool is pH meter (Fig. 5.10). The pH value of the test mud indirectly reflects the chemical stability of the mud. Chemical stability refers to the chemical deterioration of the mud film-forming function when positive impurities (including Ca^{2+}, Na^{+}, Mg^{2+}, etc.) are mixed into the mud. The reason is that clay particles are negatively charged. When positive ions are encountered, clay particles change from

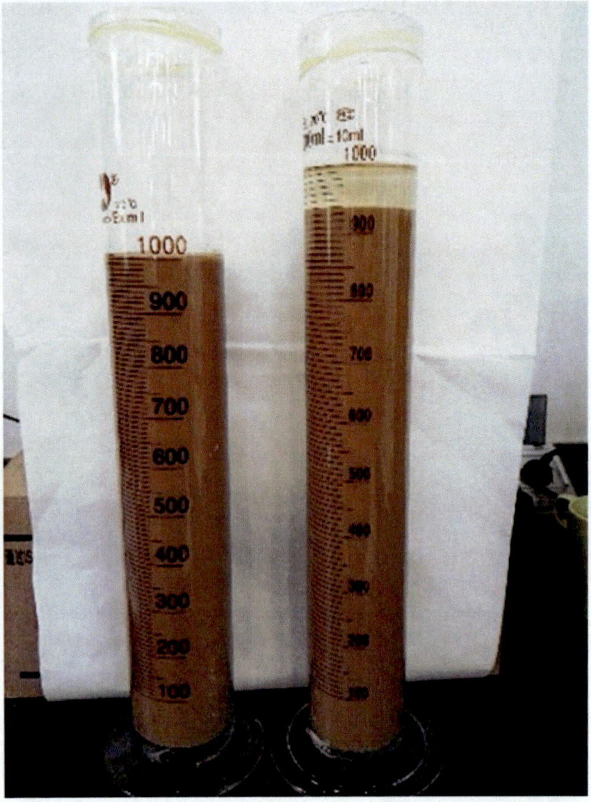

Fig. 5.9 Mud colloid rate test

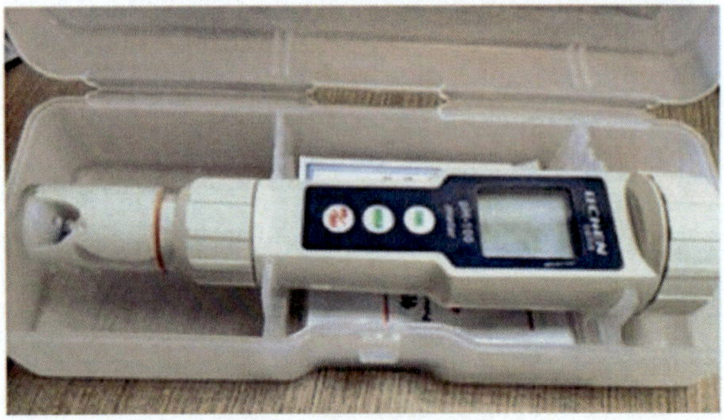

Fig. 5.10 pH machine

suspended state to condensed state, and the number of suspended clay particles in mud is sharp. Less, resulting in difficulty in forming mud film. Studies have shown that the pH range of the slurry without positive ion pollution is 7–10, showing weak alkaline. When the sludge is polluted by positive ion impurities, the pH value is far more than 10. Therefore, the degree of deterioration caused by positive ions can be determined by measuring pH value, so as to identify the chemical stability of sludge. The mud with pH 8–10 has good chemical stability.

5.2.2 Mud Preparation Technology

Taking Nanjing Yangtze River Tunnel as an example, this section introduces the mud preparation technology for the typical high permeability fine sand and gravel composite strata crossing by slurry shield.

5.2.2.1 Normal Excavation Mud Preparation

1) Mud ratio and basic properties

According to the typical fine sand stratum, gravel stratum and round gravel stratum, two groups of mud ratio were designed and film-forming tests were carried out. Generally speaking, the preparation of mud is to mix the basic slurry with bentonite, and add clay and silty clay to adjust the particle size and density of solid particles in the mud. The addition of CMC increases the viscosity of mud and anhydrous Na_2CO_3 improves the physical stability of mud. Therefore, different contents of clay (silty clay), CMC and anhydrous Na_2CO_3 were added to the bentonite slurry during construction to prepare slurry with different specific gravity, viscosity and gradation. Among them, the slurry in the silty sand stratum is prepared with the old slurry and five slurry preparation agents provided by North China Oilfield. The old slurry is taken from the slurry used after the slurry circulation of the slurry shield in Nanjing Yangtze River Tunnel. However, due to the small content of fine particles in the gravel stratum, the slurry is prepared by base slurry, clay, fine sand and polymer. The slurry ratio is shown in Table 5.3.

It can be concluded from the basic properties that in the silty sand formation, the density of the first six groups of mud has little change, and the funnel viscosity and plastic viscosity have been improved after adding the slurry preparation agent, among which the physical stability of No. 3 and No. 5 mud is better. The physical stability of No. 7 mud is the worst, and water absorption seriously produces flocculation. There is little difference in mud density between each group in gravelly sand and round gravel formation. The funnel viscosity and plastic viscosity of mud are greatly improved after adding fine sand and polymer, and the more the polymer is added, the greater the mud funnel viscosity.

2) Mud film formation test

(1) Test device. The test adopts self-made mud film forming test device, which is composed of organic glass column, flange plate, beaker, electronic scale, electronic scale data acquisition device, air compressor, pressure gauge and pressure regulating device, as shown in Fig. 5.11. The organic glass column simulates the working chamber excavation surface stratum system in the construction, and the upper end is sealed with seal rings, bolts and flange plates. The beaker receives the water discharged from the test, namely the seepage flow. The water is weighed by an electronic scale, and the data acquisition device automatically records the change of seepage flow at different times. The data

Table 5.3 Test mud ratio

Formation	Number	The proportion of mud	Density (g/cm^3)	Viscosity (s)	$\varphi600$ (mPa s)	$\varphi300$ (mPa s)	Plastic viscosity (mPa s)	Water-destructing rate of 24 h (%)
Fine sand formation	1	Old pulp	1.15	17.3	10	5	5	41
	2	Old pulp + 1‰1#	1.15	21.4	16	8	8	40
	3	Old pulp + 2‰1#	1.14	22.8	20	10	10	20
	4	Old pulp + 1‰3#	1.15	22.5	17	8	9	39
	5	Old pulp + 2‰3#	1.14	24.4	23	11	12	14
	6	Old pulp + 1‰5#	1.15	21	14	7	7	40
	7	Old pulp + 2‰5#	The mud produces flocculation, the sediment is like sludge, and the mud is seriously water					
Gravel sand and round gravel formation	8	Basic slurry + <0.075 mm clay + 0.3‰ high polymer	1.1	35	31	18	13	–
	9	Basic slurry + <0.075 mm clay + 0.6‰ high polymer	1.2	63	56	34	21	–
	10	Basic slurry + <0.075 mm clay + 0.9‰ high polymer	1.3	136	66	43	23	–
	11	Basic slurry + <0.01 mm fine sand + 0.6‰ high polymer	1.1	59	40	27	13	–
	12	Basic slurry + <0.01 mm fine sand + 0.9‰ high polymer	1.2	95	54	34	21	–

(continued)

Table 5.3 (continued)

Formation	Number	The proportion of mud	Density (g/cm^3)	Viscosity (s)	φ600 (mPa s)	φ300 (mPa s)	Plastic viscosity (mPa s)	Water-destructing rate of 24 h (%)
	13	Basic slurry + <0.01 mm fine sand + 0.3‰ high polymer	1.3	56	49	31	18	–
	14	Basic slurry + <0.25 mm sand + 0.9‰ high polymer	1.1	94	50	33	17	–
	15	Basic slurry + <0.25 mm sand + 0.3‰ high polymer	1.2	57	52	32	20	–
	16	Basic slurry + <0.25 mm sand + 0.6‰ high polymer	1.3	67	62	36	26	–

acquisition time interval can be set by itself. The air compressor provides air pressure to the organic glass column. The air pressure gauge shows the internal pressure of the device, and the pressure regulating device can regulate the air pressure. The organic glass column used in this experiment is 60 cm high, 8.4 cm inner diameter, 0.5 MPa maximum withstand pressure, 0.01 MPa pressure control accuracy of pressure regulator, 0.01 g electronic scale accuracy.

(2) Test method. A certain amount of water is added to the organic glass column to saturated stratum. Firstly, the filter layer with particle size of 2–5 mm is added, and then the test stratum is added. Open the drain valve so that the liquid level is aligned with the surface of the formation, and then slowly inject mud into the device along the guide rod to avoid disturbing the formation. The sealing device was used to adjust the pressure regulating device to the test osmotic pressure, and the film forming test was carried out for a certain time under a stable air pressure. The change of filter water was recorded by data acquisition. After the infiltration film is formed, the remaining mud in the organic glass column is poured out, the mud film is taken out, and the thickness of the mud

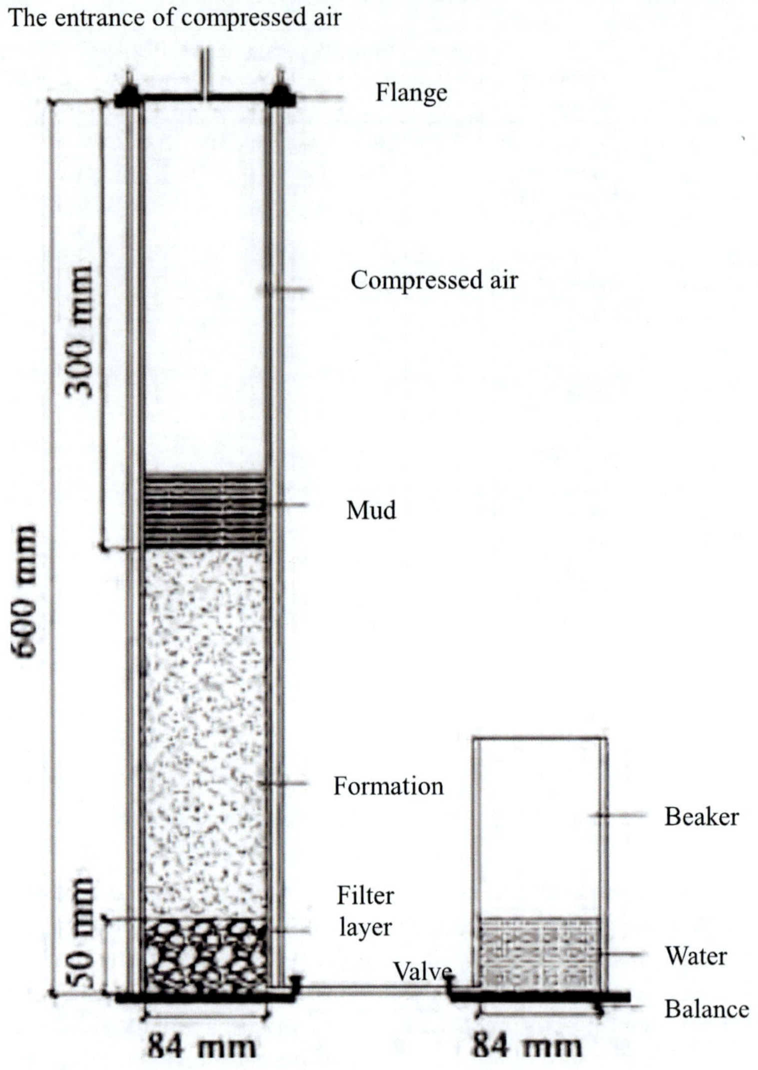

Fig. 5.11 Mud film forming test instrument

film and the thickness of the infiltration zone are measured. Finally, sample the mud film into the oven to measure moisture content.

(3) The mud film form is shown in Fig. 5.12.

Through the test results, the film formation effect of the mud was analyzed, and the mud plan suitable for different formations was finally determined, as shown in Table 5.4.

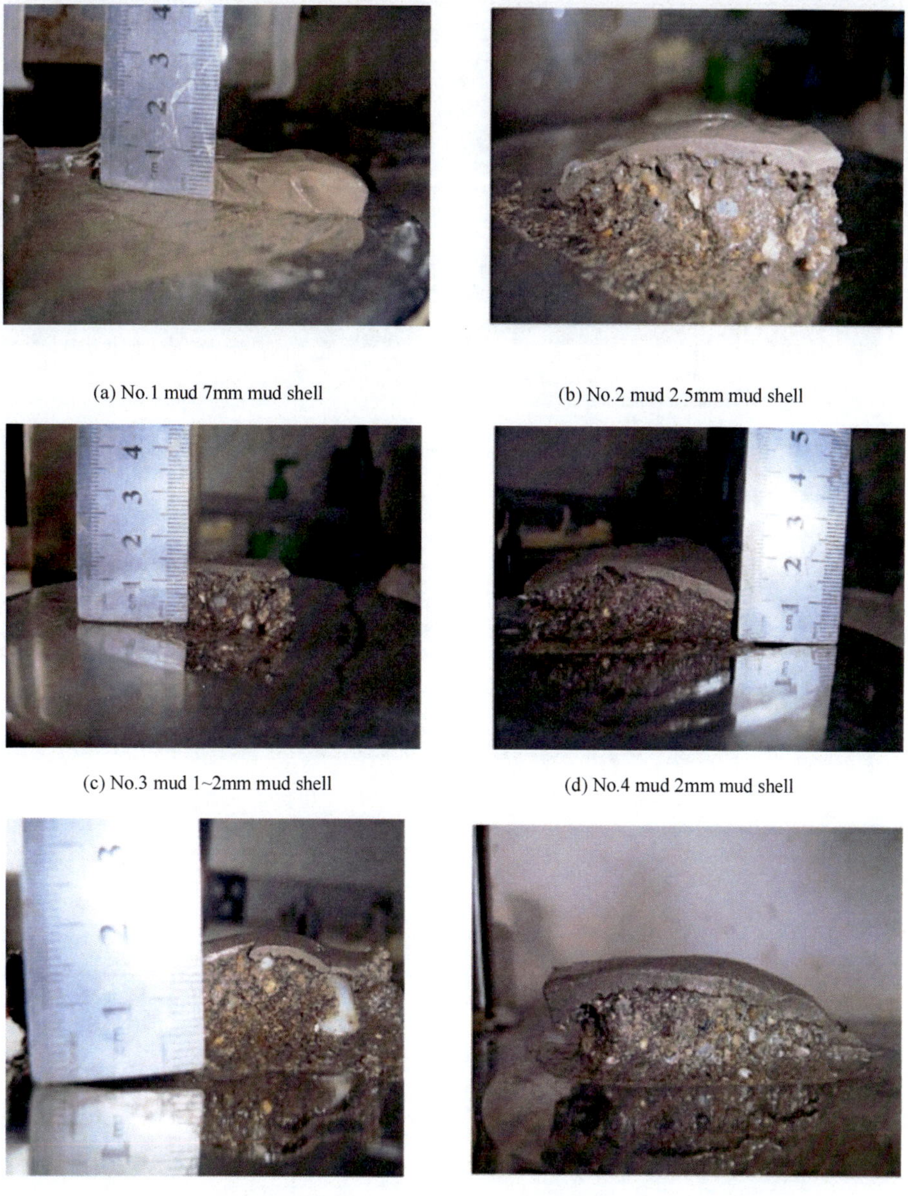

(a) No.1 mud 7mm mud shell

(b) No.2 mud 2.5mm mud shell

(c) No.3 mud 1~2mm mud shell

(d) No.4 mud 2mm mud shell

(e) No.5 mud 1~2mm mud shell

(f) No.6 mud 5mm mud shell

Fig. 5.12 Mud film formation

5.2.2.2 Preparation of Shutdown Opening Mud

At present, the general tunneling distance of river crossing and subsea tunnels constructed at home and abroad is relatively long, and due to the complex geological and hydrological conditions, shield opening inspection is sometimes carried out. In

(g) No.13 mud + penetration belt

(h) No.14 mud shell

(i) No.15 mud forming film

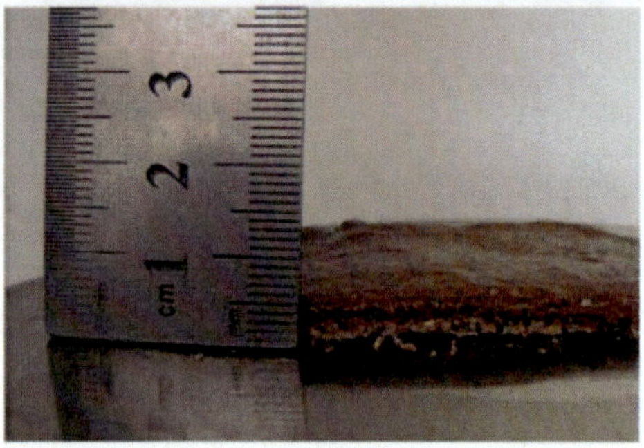

(j) No.16 mud forming film

Fig. 5.12 (continued)

Table 5.4 Different formation mud schemes

Formation	The feature of the formation	Slurry plan
Fine sand formation	A high permeability coefficient, and the particle size of about 70% of the stratum is concentrated in 0.075–0.25 mm. The clay particles below 45 μm account for 11.5%, with low fine particle content and difficult yin to form a membrane	The old pulp + slurry agent of the slurry pool, (NSHS-1 and NSHS-3)
Fine sand gravel mixed formation	High permeability coefficient, more coarse particles, fewer fine particles, uneven distribution, mud difficult to form a membrane	The old pulp + poor pulp + slurry agent of the slurry pool (NSHS-1 and NSHS-3)

general, the shield is mostly under the condition of high water pressure at the bottom of the river when opening the cabin. At the same time, in order to ensure the safety of the personnel entering the cabin, the setting value of the slurry pressure will be reduced. Therefore, it is very important to study how to effectively use the slurry film to ensure the stability of the excavation face when opening the cabin. In the construction process before the opening of the cabin, the density and viscosity of the slurry used are relatively large, and only mud-skin mud film can be formed on the excavation surface where the shield stops, and no permeable zone can be formed. This mud film is relatively weak, the opening risk is relatively large. In order to ensure the safety of cabin opening operation, it is necessary to adjust the slurry ratio so that the slurry can not only form the mud crust, but also form the permeable zone. The formula of new mud is based on bentonite mud and waste mud, and the reasonable ratio is determined through experiments.

1) Film formation test of pure bentonite slurry

Four groups of mud were used in the experiment, and sodium bentonite was used for expansion. The ratio was 3%, 5%, 8% and 12%, and the expansion time was 24 h. In this experiment, the most unfavorable gravel sand layer for film formation is selected as the test stratum, and the coarse particles larger than 2 mm account for about 40% of the whole stratum content, and the permeability coefficient of the stratum in the test is about 3.95×10^{-2} cm/s, which is basically consistent with the permeability coefficient of the gravel sand stratum in the actual project. For each formulation and film forming test results, see Table 5.5.

The bentonite used in the test had a general expansion effect. After 48 h, there were still particles not completely expanded in the first three groups of mud. It can be seen from Table 5.5 that the bentonite slurry with the concentration of 3%–5% has low viscosity and the density is less than 1.05 g/cm^3, which can form a permeable zone-type mud film in the gravel stratum. However, the mud loss is relatively large and the film-forming time is long. The viscosity of bentonite mud with the concentration of 8%–12% is more than 20 s. After the mud penetrates into the formation for a certain distance, it quickly stabilizes, and the mud loss is less, forming a mud film with mud

Table 5.5 Four groups of mud film formation test

Bentonite mud concentration (%)	Density	Funnel viscosity	Whether to form mud	Whether to form an infiltration zone	Mud film formation time
3	1.03	16.14	No	10 cm penetration zone	Hard to identify
5	1.04	17.52	1 mm mud skin	8 cm penetration zone	About 3 s
8	1.06	21.05	2 mm mud skin	3 cm penetration zone	About 2 s
12	1.08	22.33	1 mm mud skin	Thinner osmosis belt	About 2 s

skin permeability zone, and the film forming time is very short. The results of slurry film-forming test in each group are as follows.

The first group of 3% bentonite mud can be infiltrated in the gravel stratum, forming a permeable zone of about 10 cm thickness. The filtration water is large, and almost no mud skin is formed, the test condition is shown in Fig. 5.13.

The second group of 5% bentonite mud can be infiltrated in the gravel sand stratum, forming a muddy skin of about 1 mm thick and a permeation zone of 8 cm thick. The test is shown in Fig. 5.14.

The third group of 8% bentonite mud can be infiltrated in the gravel sand stratum, forming mm thick mud and 3 cm thick infiltration zone. The test condition is shown in Fig. 5.15.

The fourth group of 12% bentonite mud can also be infiltrated in the gravel sand stratum, forming a 2 mm thick mud, a thinner permeate zone, and the mud film is formed in a short time, the test situation is shown in Fig. 5.16.

The test results show that the pure bentonite slurry with smaller concentration, density and viscosity is more likely to penetrate into the formation, forming a mud film with permeability zone, and the depth of mud infiltration, that is, the thickness of permeability zone increases with the decrease of concentration, density and viscosity.

Fig. 5.13 3% bentonite slurry film formation test

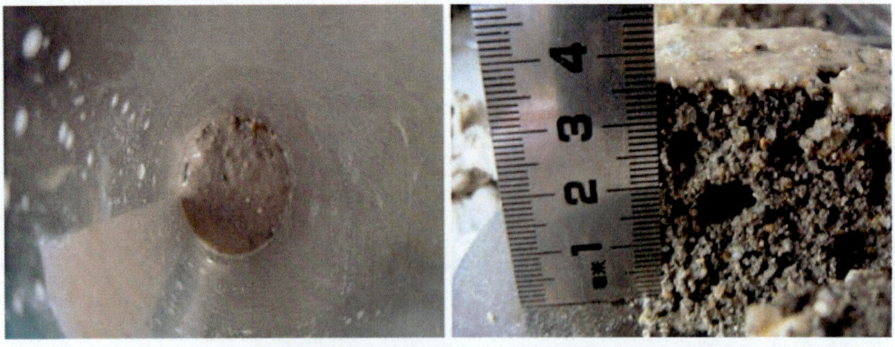

Fig. 5.14 5% bentonite slurry film formation test

Fig. 5.15 8% bentonite slurry film formation test

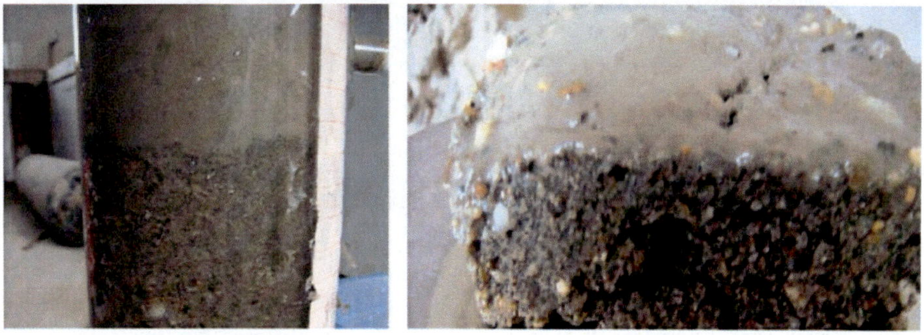

Fig. 5.16 The mud film test situation

2) Mixed mud preparation and film formation test

The volume of pure bentonite will increase after expansion, so the density of pure bentonite mud is small. When the density of pure bentonite mud is 1.15–1.20 g/cm^3, the viscosity is very high and difficult to pump. Therefore, considering the mixing of low specific gravity bentonite slurry with high viscosity and natural clay slurry, waste slurry stored during shield tunneling through silty clay stratum can be added to obtain better mixed slurry. In the actual mixing process, the bentonite slurry in the slurry tank is directly mixed with waste slurry, and the density meets the requirements, but the viscosity cannot reach the standard. Therefore, the dry bentonite powder should be added to the slurry tank according to the mass ratio of 1:20, and then the waste slurry and the slurry of the slurry tank should be mixed according to the volume ratio of 1:2. Finally, the density of the slurry is 1.16 g/cm^3 and the viscosity is 30.27 s, which meets the requirements. The water separation rate test and mud film test were carried out by using the mixed mud with this ratio.

(1) The physical stability of the mud, expressed at the water-analysis rate of 24 h mud. The smaller the water analysis rate of mud, the better its physical stability. The water-analysis rate test shows Fig. 5.17, the mud is loaded into a capacity

Fig. 5.17 Water precipitation rate test

of 100 mL, after 24 h the water-analysis volume of 1 mL, that is, 24 h water-smomography rate of 1%, indicating that the physical stability of the mud is very good.

(2) Mud film forming test. This test is carried out in gravel sand stratum, the test method and steps are the same as the above mud film test, but this test is grading pressure, drawing the change curve of mud filter water. The mud-skinned mud film formed by the mud is shown in Fig. 5.18, and the relationship between the filtration rate and time and the mud pressure is shown in Fig. 5.19.

Fig. 5.18 Mud-type mud film

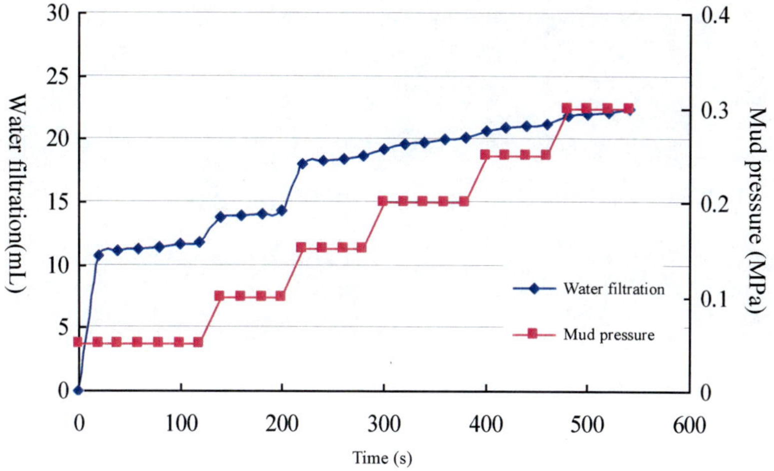

Fig. 5.19 Water filtration curve

It can be seen from Fig. 5.19 that under the pressure of each stage, the amount of water filtered by the mud is very small, and the amount of filtered water per minute is about 0.6 mL. The filtered water volume change curve is converted to the actual excavation surface and the water loss is about every 2 h 3 m^3 (the amount of water loss is satisfied: 2 h 10–20 m^3).

The above test results show that the bentonite slurry with a concentration of 8–12% can quickly form a mud film containing the osmotic zone on the excavation surface; the mixed mud of bentonite and natural clay with specific gravity of 1.15–1.20 and viscosity of 25 s or more is good in physical properties. The stability is high, and a micro-permeable mud-type mud film can be quickly formed on the excavation surface.

3) Evaluation of mud film formation after opening

After adjusting the mud by using the above method in the opening of the Nanjing Yangtze River tunnel, the mud of the original mud tank is replaced. The mud level of the slurry tank is lowered when opening, and only the air pressure is used to support the excavation surface. From the scene situation, a thick mud film is formed on the excavation surface, and the mud film is well sealed, and there is no gas leakage phenomenon, so that the air pressure can effectively act on the excavation surface and maintain open. The surface of the excavation is stable. The specific situation is shown in Fig. 5.20.

5.2.2.3 Waste Mud Reuse

When shield tunneling passes through the silt silty clay stratum, a large number of abandoned clay will be produced. This stratum has strong self-made slurry ability,

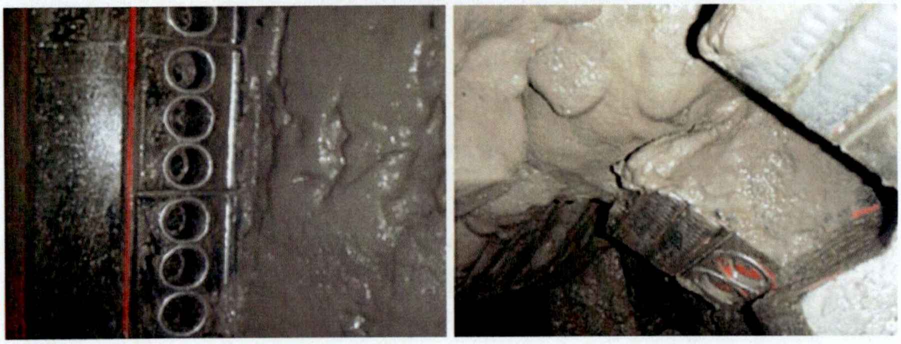

Fig. 5.20 The surface of the excavation

and it is difficult to separate mud and water. The cost of these abandoned soil treated by conventional methods is high, and the waste will cause environmental pollution. When the tunnel passes through the gravel and round gravel stratum, the coarse grain content is very high, and the fine grained soil with self-pulping function is very small. In this formation, in order to ensure the quality of mud, the need to add a large number of bentonite and pulping agent, high cost. If the two can be combined, the waste slurry containing a large number of clay particles originally produced in silty clay stratum can be used to supplement the slurry of gravel sand and round gravel stratum, which not only avoids the difficulty of fine particle separation, but also solves the raw material supply demand of slurry mixing in gravel sand stratum. In this section, through calculation and test, the reasonable proportion of recycled mud is determined, and the film forming quality of recycled mud is evaluated.

1) Disposal slurry reuse ratio test

The waste slurry used in the test was taken from the construction site of the Nanjing Yangtze River tunnel. When the slurry was mixed, the waste mud was first mixed with the expanded bentonite slurry. The mud density ranged from 1.15–1.2 g/cm^3 and the viscosity requirement was greater than 25 s. If the viscosity does not meet the requirements, adjust with 1# and 3# pulp.

By calculating and repeating the comparison test of the slurry, the ratio of the mixed mud with a density of 1.15–1.2 g/cm^3 is finally obtained. The waste slurry and 8% bentonite slurry can be mixed at a volume ratio of 0.4–1.1.

2) Mud and mud film quality evaluation

In order to evaluate the various indexes of the reclaimed mud, the mud and silt with high clay content stored on site were mixed in different proportions, and five groups of muds with different clay content and the same density were mixed to simulate the waste slurry with different quality. Then in CMC solution (Compared to the pulping agent, only change the mud viscosity, do not change the mud grading) to adjust the viscosity of the mud to the same. The gradation curve and basic properties of the mud are shown in Fig. 5.21 and Table 5.6.

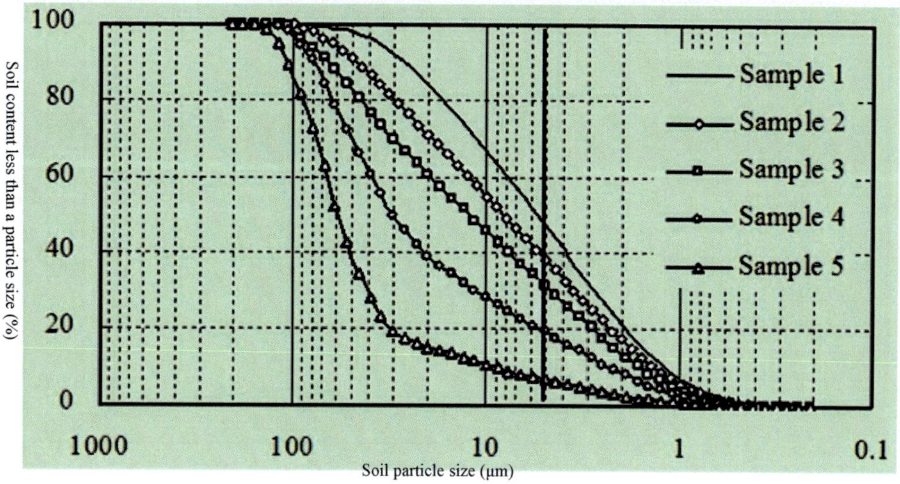

Fig. 5.21 Soil particle size distribution

Table 5.6 Basic properties of mud

The number of the mud	Clay's content (%)	Funnel viscosity (s)	Density (g/cm^3)
Sample 1	47.32	23	1.18
Sample 2	38.43	22.4	1.18
Sample 3	31.75	22.5	1.18
Sample 4	19.2	22.9	1.18
Sample 5	6.83	22.7	1.18

Note The classification criteria for soils in China define that the clay is less than 5 μm

(1) The effect of the mucous grain content on the physical stability of the mud.

Good physical stability is one of the necessary characteristics of mud, the physical stability of mud refers to the solid particles in the mud to maintain the suspension and dispersion state of the ability, mainly reflects the mud to prevent particle subsidence and transport, slag, is one of the important indicators of mud. Usually with a certain amount of mud placed in the barrel in a period of time to express the water analysis rate, generally selected 2 h and 24 h water-reducing rate, to reflect the shield excavation process in the assembly of pipe pieces of the short-term shutdown and a longer period of downtime. The smaller the water analysis rate, the better the physical stability of the mud, the less likely the mud to occur separation, precipitation, the larger the water analysis rate, the poor physical stability of the mud, mud is prone to separation, precipitation, the stability of the excavation surface, mud circulation and slag discharge are not conducive. The results of the water analysis rate test are shown in Fig. 5.22.

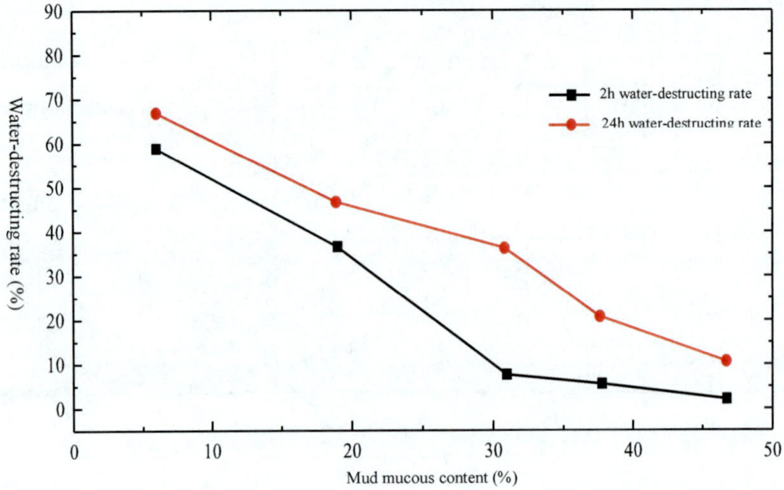

Fig. 5.22 Water test rate test curve

When the density and viscosity are the same, with the increase of clay content in the mud, the water release rate of the mud decreases significantly, that is, the physical stability of the mud is getting better and better. Slurry with clay content of more than 30% has little 2 h precipitation rate, less than 10%. When the clay content is less than 20%, the 2 h water separation rate is relatively large, more than 30%, which will affect the quality of mud film and the transport of slag. The physical stability of the mud is good when the clay content is high, because the clay particles are small and the settlement is very slow. At the same time, it has large specific surface area, negative charge and good dispersion.

(2) The effect of the clay content on the amount of additives.

The CMC (carboxymethyl cellulose sodium) selected for the mud tackifier can stabilize the mud. The action between CMC content and clay content was analyzed by experiment, as shown in Fig. 5.23. During the test, CMC is added in the form of a solution. In the mud, it can increase the viscosity of the mud without affecting the mud grading.

Under the premise of reaching the same viscosity, the higher the clay content of natural mud, the less CMC content is needed. The CMC content of mud with clay content less than 20% is more than 2 times higher than that of mud with clay content more than 30%. This is because the tackifier plays the role of tackifying, mainly relying on the bridging effect between polymer and soil particles. Due to the large specific surface area and negative charge, clay particles are the most active particles in the mud. The mud with high clay content is conducive to the tackifying effect of tackifier. Therefore, when preparing the slurry, the slurry with high clay content (generally above 20%) should be selected as the basic material of the slurry. The less the tackifier is needed for preparing the slurry, the lower the cost is.

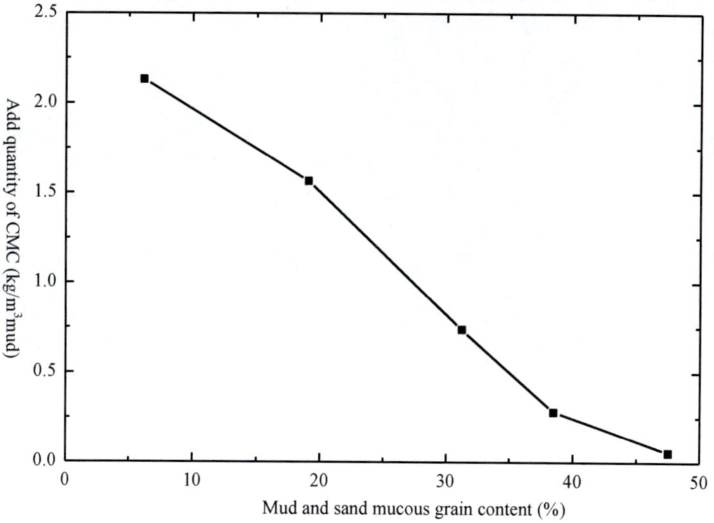

Fig. 5.23 Curve of CMC content and mud clay content

(3) The effect of mud clay content on the film quality of slurry shield.

In the scenario of high permeability and high pressure in the gravel sand stratum crossing the simulated mud water shield, five layers of clay with different clay content were used for the osmosis film formation test in the self-made mud osmosis film forming test device. Figure 5.24 is five sets of mud, the curve of water filtration as a function of time.

It can be seen that when No. 1–5 mud is filmed, the filter water increases in turn. The filtration capacity of No. 1–4 mud is small and stable in a short time, while the

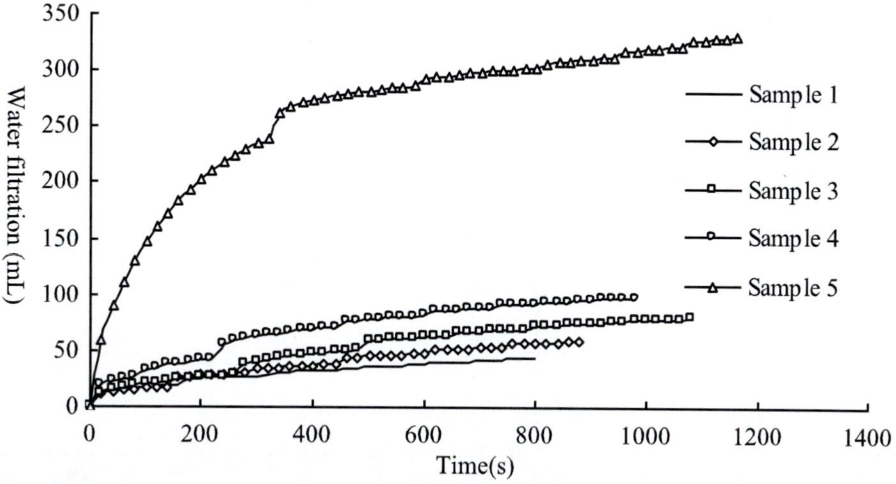

Fig. 5.24 Curve of mud filtration with time

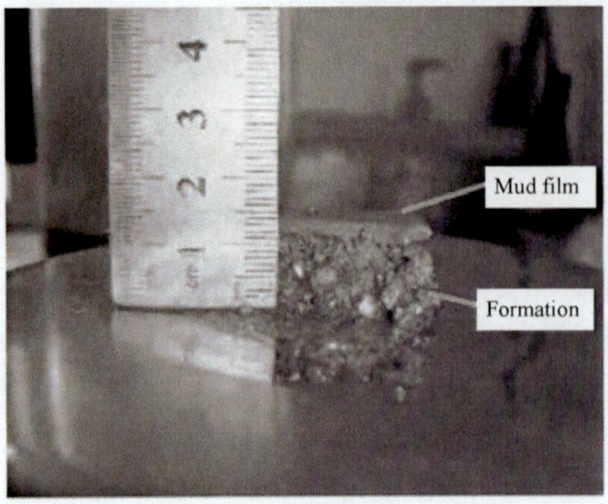

(a) Sample 1 formed

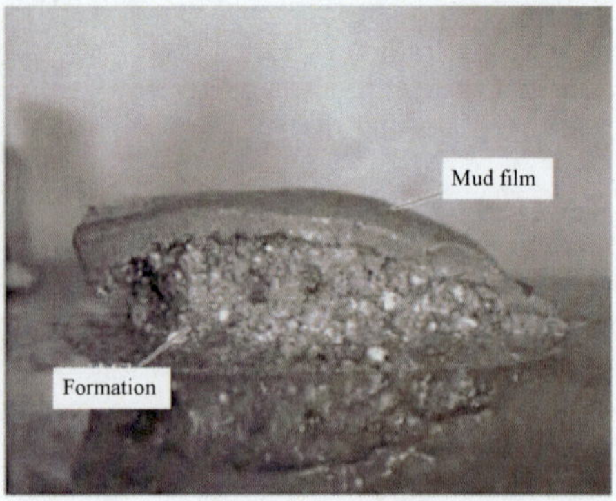

(b) Sample 5 formed

Fig. 5.25 Mud film formed in the test

filtration capacity of No. 5 mud is much larger than the previous four groups of mud, and the mud permeability tends to be stable after a long time. Figure 5.25 is the form of mud film formed after the test of Sample1 and Sample 5 mud. It can be seen that the mud film formed by Sample1 mud is only 1–2 mm thick and relatively dense, while the mud film formed by Sample5 mud is about 5 mm thick and slightly loose.

Figure 5.26 is the curve of permeability coefficient of mud film formed by five groups of mud with mud clay content. It can be seen that with the increase of mud clay content, the permeability coefficient of mud film formed by mud decreases gradually. The permeability coefficient of the mud film formed by Sample 5 mud (clay content of 6%) is about 5.36×10^{-5} m/s, which is close to the permeability coefficient of common silty sand. The permeability coefficient of the mud film formed by Sample 1 mud (clay content of 47%) is about 2.86×10^{-7} m/s, and the permeability coefficient of the mud film is reduced by nearly two orders of magnitude. Therefore, the larger the clay content of the mud, the smaller the filter water, and the shorter the time

Fig. 5.26 Relationship between mud membrane permeability coefficient and mud clay content

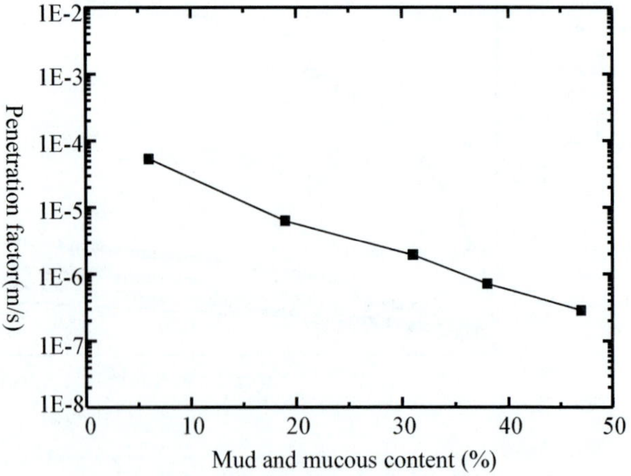

when the mud reaches the seepage stability, the denser the mud film is formed and the better the quality is. When the clay content is less than 20%, the filtration amount of the mud film is significantly increased, and the time to achieve stable permeability is also prolonged, so the quality of the formed mud film is also poor.

This large-scale use of slurry with high clay content to adjust the properties of slurry not only improves the film-forming quality and physical stability of slurry, reduces the use of pulping agent, reduces the economic cost, but also reduces the adverse environmental impacts such as large area of waste slurry, which has good engineering effect.

5.2.3 Film Formation Law of High Permeability Formation Mud

Due to the complex formation conditions of shield tunneling, the slurry ratio will be adjusted according to the actual situation in the actual construction. Although the different proportion leads to the difference of the basic properties of the mud, the law of the penetration film is consistent. When bentonite is used as the base slurry, clay minerals generally form flocs in the mud. When this floc is larger than the formation pore, the flocs are blocked on the surface of the formation under mud pressure. The subsequent pressure makes the floc dehydration and produces a similar effect as consolidation. The water in the floc is discharged and infiltrated into the formation, and the solid particles gather on the surface of the formation to form a dense mud. When the formation pore is large (when the average pore diameter is about 60 μm and k is greater than 1.0×10^{-1} cm/s), the floc can also be squeezed into the pore by pressure. However, these flocs may be intercepted and blocked in the small part of the pore diameter. If there are more (amount) of such flocs in the mud, the pores will be blocked first, and then the flocs will gather on the surface to form a mud film. Therefore, the film-forming type of mud is related to the particle size and formation pore of mud. In order to obtain the universal film-forming law of mud in sand stratum, the orthogonal experimental design is adopted to conduct the film-forming experiment of mud, and the film-forming law is summarized by factor analysis.

5.2.3.1 Mud Parameters

The mud ratio is designed according to three levels and four factors orthogonal table, there are nine groups. Orthogonal arrangements are shown in Tables 5.7 and 5.8.

Table 5.7 Factor table

Horizontal	Main material of mud	Mud density (g/cm³)	The addition of CMC (%)	Na_2CO_3 (%)
	(A)	(B)	(C)	(D)
1	Basic slurry + < 0.075 mm particles	1.1	0.03	0.30
2	Basic slurry + < 0.1 mm particles	1.2	0.06	0.40
3	Basic slurry + < 0.25 mm particles	1.3	0.09	0.50

Table 5.8 Orthogonal test table

The number of test	A	B	C	D	Density (g/cm³)	Fennel viscosity (s)	Plastic viscosity (MPa· s)	Effective particle size (μm)
1	A1	B1	C1	D1	1.1	35	13	61
2	A1	B2	C2	D2	1.2	63	21	71
3	A1	B3	C3	D3	1.3	136	23	76
4	A2	B1	C2	D3	1.1	59	13	110
5	A2	B2	C3	D1	1.2	95	21	116
6	A2	B3	C1	D2	1.3	56	18	118
7	A3	B1	C3	D2	1.1	94	17	155
8	A3	B2	C1	D3	1.2	57	20	177
9	A3	B3	C2	D1	1.3	67	26	187

5.2.3.2 Strati Graphic Parameters

The high-permeability sand is used in the stratum, and the sand used in the test is sand with a single particle size range, which is obtained by natural sand screening. The sand was stratified into the test vessel in strict accordance with the controlled dry density of 1.5 g/cm³, and the permeation test of the normal head was carried out. The formation permeability coefficient is shown in Table 5.9.

Table 5.9 Stratigraphic parameters table

The number of formation	1	2	3	4	5
Particle size range (mm)	0.1–0.25	0.25–0.5	0.5–1	1–2	2–5
Formation permeability coefficient (cm/s)	1.4×10^{-2}	7.7×10^{-2}	3×10^{-1}	1.2	5.2

5.2.3.3 Test Analysis

Nine kinds of mud in the orthogonal table were used to test in five formations, a total of 45 groups. The changes of filter water with time, the type of mud film and the thickness of mud film were recorded.

1) No. 1 formation

Nine kinds of mud in No. 1 formation seepage test process and results are very similar, that is, after each stage load is applied, the filter water is very small, and the filter water is stable quickly, and the total filter water under the sixth stage load is also very small. Finally, nine kinds of mud are formed on the surface of the formation. The curve of filter water per unit area versus time in No. 1 formation is shown in Fig. 5.27.

In this kind of mud can only form the formation of mud film, as long as the mud has the following two conditions can form high quality mud film.

(1) There are enough powder-stick particles in the mud that can silt the pores on the surface of the formation, and at the same time have a lower sand content.
(2) The mud has high viscosity and good physical stability, which makes the solid particles and water in the mud have strong binding force, and the mud is not easy to be separated and dehydrated under pressure.

2) No. 2 formation

The curve of filtration water per unit area with time in No. 2 formation is shown in Fig. 5.28. When No. 2 mud is used for mud filtration test, a large amount of filtration water is produced when the primary load is added, and some mud penetrates into the formation. With the increase of mud pressure, the production of filtration water is less and the stability speed is faster. Finally, some mud particles penetrate into the formation, and some mud particles are blocked on the surface of the formation,

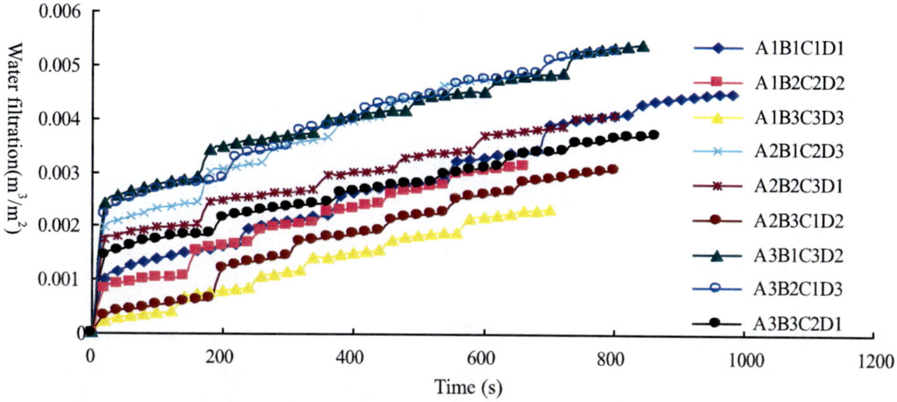

Fig. 5.27 Curve of water filtration per unit area of No. 1 formation with time

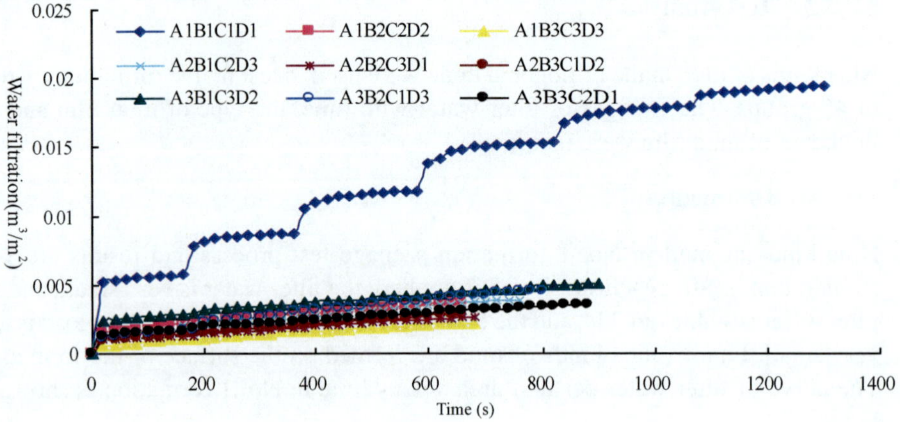

Fig. 5.28 Curve of water filtration per unit area of No. 2 formation with time

forming a thick mud film. The other eight kinds of mud form solid particle thin layers on the formation surface.

3) No. 3 formation

The curve of filter water per unit area of No. 3 formation with time is shown in Fig. 5.29. No. 3 mud after the sixth stage load, completely lost, filtered water turbidity, mud failed to form any form of mud film in the formation. During the test of No. 6 and No. 9 mud, a thin layer of solid particles was formed on the formation surface. Some particles in the other mud could penetrate into the formation, while some particles could be silted up on the formation surface to form a thin mud film.

The permeability coefficient of No. 3 formation is large, and if the mud does not match the formation, the mud will appear excessive filtration and even unable to film.

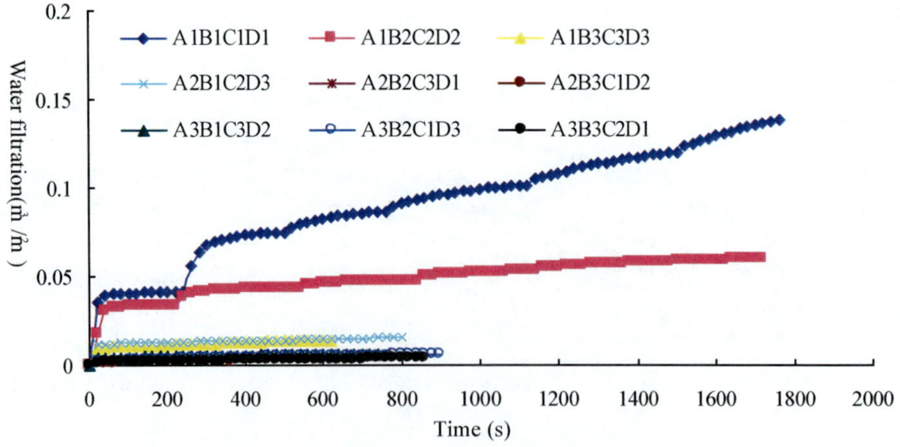

Fig. 5.29 Curve of water filtration per unit area of No. 3 formation with time

In order to avoid this situation and ensure that the mud can form a good quality mud film on the excavation surface, the mud must meet the following requirements:

(1) The mud must contain a certain amount of powder or silt particles to ensure that the effective particle size of the mud is large enough to block the formation pores.
(2) Due to the large effective particle size of the mud and the large particle size of the solid particles in the mud, it is necessary to make the mud have a certain viscosity to ensure that the larger solid particles in the mud can be uniformly dispersed in the mud for a long time without precipitation.

4) No. 4 formation

No. 1–No. 5 muds were completely filtrated under the second stage load, and no mud film was formed. Some particles of mud No. 6–No. 9 penetrated into the formation, and some particles were silted up on the surface of the formation, forming a mud film with zone (Fig. 5.30).

The permeability coefficient of No. 4 stratum is large, and the mud is not easy to form film in this stratum. In order to ensure that the mud can form a good mud film on the excavation surface, the mud must meet the following requirements:

(1) The mud must contain a certain amount of particles with a large particle size as a leak-blocking particle to block the largepores in the formation.
(2) The mud must contain a certain amount of powdered particles to further silt the formation pores that have been smaller than the leak particles.
(3) Because the mud contains a large particle size of large leakage particles, and whether the leak particles can be spread evenly in the mud for a long time to determine whether the mud in the formation can be filmed, so it is necessary to ensure that the mud has a large degree of viscosity, so that the particle

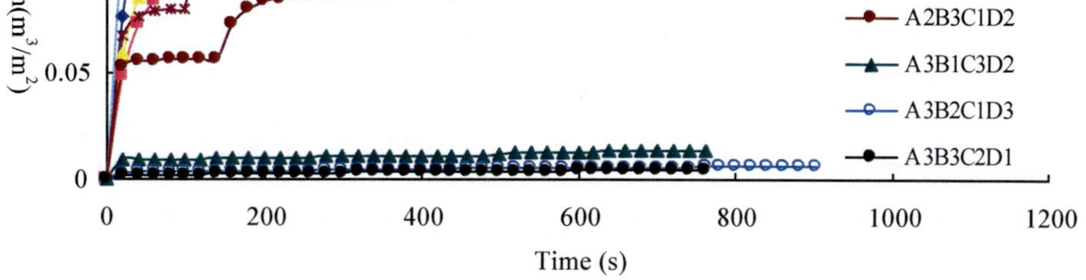

Fig. 5.30 Curve of water filtration per unit area of No. 4 formation

size of large leakage particles for a long time uniform lysing in the mud, no precipitation phenomenon.

5) No. 5 Formation

No. 1, No. 2, No. 3, No. 4 and No. 6 mud were completely filtered under the first load, No. 5 and No. 7 mud were completely filtered under the second load, No. 1–No. 7 mud failed to form a mud film, No. 8 and No. 9 mud formed a mud film with a permeable zone, and the mud film was thin (Fig. 5.31).

The permeability coefficient of No. 5 stratum is large, and the mud is very difficult to form a film in the stratum. Only the mud is guaranteed to meet the following requirements. Mud may permit the formation of a mud film on the excavation surface.

(1) The mud must contain a large number of particles large enough to block the leak particles, in order to block the formation of large pores, in this type of osmosis coefficient of the formation, the effective particle size of the mud must be at least more than 180 m.
(2) The mud must contain a large number of powdered particles to further silt the formation pores that have been smaller than the leak particles.
(3) Since the mud contains plugging particles with large particle size, and whether the plugging particles can be uniformly dispersed in the mud for a long time determines whether the mud can form a film in the formation, it is necessary to ensure that the mud has a large viscosity so that the plugging particles with large particle size are uniformly suspended in the mud for a long time without precipitation.

Five groups of experiments show that the density, viscosity and grade of mud have an effect on the formation of mud film. In different formations, the main mud parameters affecting the formation quality of mud film are different.

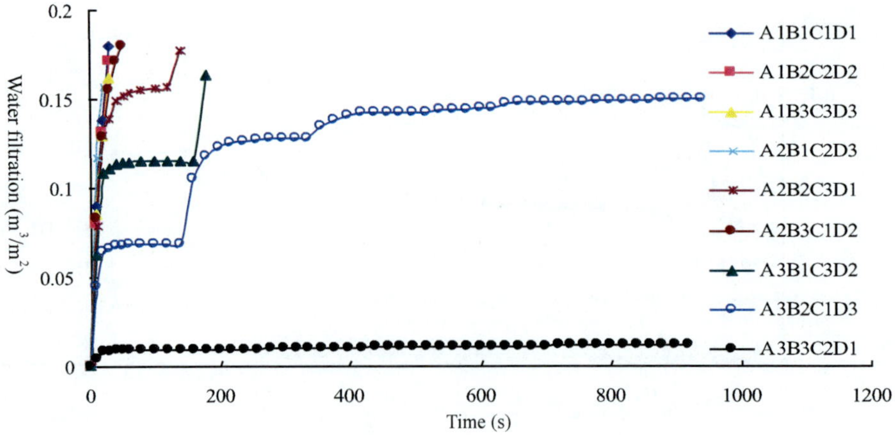

Fig. 5.31 Curve of water filtration per unit area of No. 5 formation with time

(1) When the permeability coefficient of the formation is less than 1×10^{-2} cm/s, the solid particles in the mud are difficult to penetrate into the formation due to the small particle size of the formation, and can only be blocked on the surface of the formation to form a mud film. In this kind of formation, the formation of the mud film is easy and the formation speed is the fastest. The main factor affecting the formation quality of the mud film is the proportion of the mud, followed by the particle size of the solid particles in the mud and the viscosity of the mud.
(2) When the permeability coefficient of the formation is between 1×10^{-2} cm/s and 1×10^{-1} cm/s, the particle size of the formation increases slightly. If the particle size of the solid particles in the mud is small and the viscosity of the mud is small, a large amount of mud will penetrate into the formation when the film is formed, resulting in a large amount of filtration. However, once the specific gravity, viscosity of the mud or any index in the particle size of the solid particles is increased, the formation of the mud film will become easy, and the filtration of the mud will be greatly reduced. In this kind of formation, the proportion of mud, viscosity and particle size of solid particles have great influence on the quality of mud film.
(3) When the formation permeability coefficient is between 1×10^{-1} cm/s and 1 cm/s, the particle size of the formation increases further, and the dependence of the mud film forming quality on the particle size of the solid particles in the mud is prominent. The increase of the particle size of the solid particles in the mud (that is, the increase of the effective particle size of the mud) can effectively reduce the filtration of the mud, improve the formation speed and the quality of the mud film. In this kind of formation, the biggest influence on the formation type and quality of mud film is the particle size of solid particles in mud, followed by the proportion of mud, and finally the viscosity of mud (Fig. 5.32).
(4) When the permeability coefficient of the formation is greater than 1 cm/s, in this kind of formation, the general mud will form a large amount of filtration (or even complete filtration), and cannot effectively plug the formation pores or plug the formation surface to form a mud film. The formation of the mud film depends entirely on whether there are enough solid particles in the mud to plug the formation pores or plug the formation surface to form a mud film. In this kind of formation, the size and amount of solid particle size in the mud determine whether the mud can form a film and the quality of the film. The specific gravity and viscosity of the mud play an auxiliary role in the formation of the mud film. The viscosity of the mud plays a very important role, that is, in order to ensure that large solid particles in the mud can be uniformly dispersed in the mud without precipitation, segregation and dehydration.

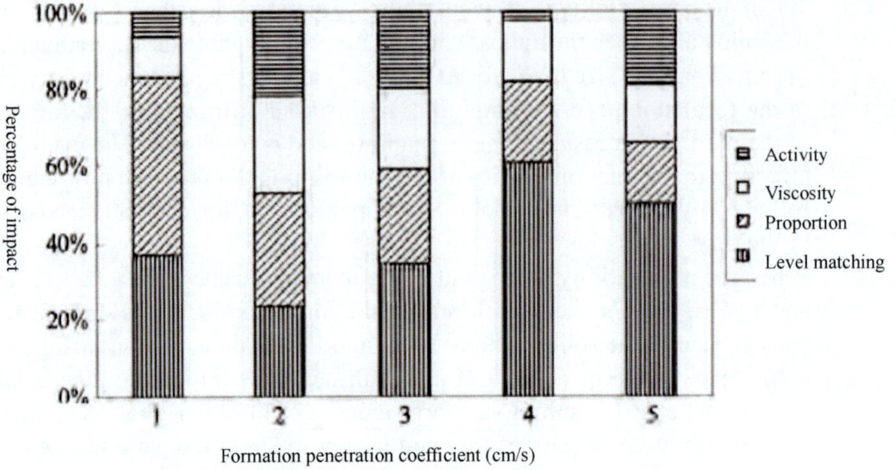

Fig. 5.32 Influence of mud parameters on film quality of mud

5.2.3.4 Film Formation of Mud in Various Types of Strata

The formation law of mud film is to reveal the law of what kind of mud can form what kind of mud film in what formation. This section aims to obtain the law of mud film formation in various formations from the perspective of mud and formation matching through the analysis of mud film formation test results.

1) Formation pore diameter

For sand and gravel layers, the formation pore diameter $L = 0.2D_{15}$, where D_{15} is the particle size corresponding to 15% of the cumulative curve of formation particle size. All the strata selected in this experiment are sand layers, so the pore diameter of each stratum can be calculated by this method, as shown in Table 5.10.

2) Effective particle size of mud

The effective particle size of the mud is the particle size corresponding to the 85% cumulative curve of the solid particle size in the mud. The effective particle size of the nine mud used in this test is shown in Table 5.8.

3) Film formation regularity of various types of formation mud

Table 5.10 Pore diameter of five formations

The number of formation	1	2	3	4	5
Particle size range (mm)	0.1–0.25	0.25–0.5	0.5–1	1–2	2–5
Formation permeability coefficient (cm/s)	1.4×10^{-2}	7.7×10^{-2}	3×10^{-1}	1.2	5.2
Formation pore diameter (μm)	24.5	57.5	115	230	490

Note The dry density of the formation is controlled to 1.5 g/cm^3

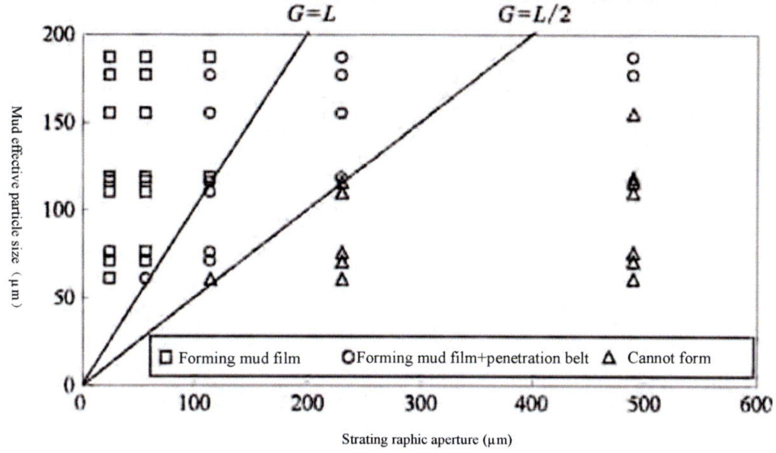

Fig. 5.33 Film formation test result of different muds in different formation

In this study, forty-five experiments were conducted on five formations of nine kinds of mud shown in Table 5.8 and Table 5.9, respectively. The experimental results (filtration loss of mud during the formation of mud film, formation type of mud film and permeability coefficient of mud film) are shown in Sect. 5.2.3.3.

Now taking the formation pore size as abscissa and the effective particle size of mud as ordinate, forty-five film-forming tests are described as shown in Fig. 5.33. Each point (a, b) in the figure represents a test, that is, the film-forming test of mud with effective particle size of a in the formation with pore size of b. The square point represents the formation of mud film as mud skin type, the circular point represents the formation of mud film as mud skin + permeability zone type, and the triangle point represents the formation of mud film.

Figure 5.33 shows that the distribution regularity of the test results is strong, the square points are all distributed in the left side of the line $G = L$, the triangle points are almost all distributed in the right side of the line $G = L/2$, and the circular points are mostly distributed between the line $G = L$ and the line $G = L/2$, which means:

When $G > L$, the mud forms a muddy mud film in the formation;

When $L/2 \leq G \leq L$, the mud forms a mud film + permeable belt type mud film in the formation;

When $G < L/2$, the mud cannot form a mud film in the formation.

Among them: G is the effective particle size of the mud, that is, the particle size corresponding to the 85% cumulative curve of the solid particle size in the mud; L is the pore diameter of the formation, and for the sand and gravel layer, the pore diameter of the formation is $L = 0.2D_{15}$, where in D_{15} is the particle size corresponding to the formation curve of the formation particle size of 15%.

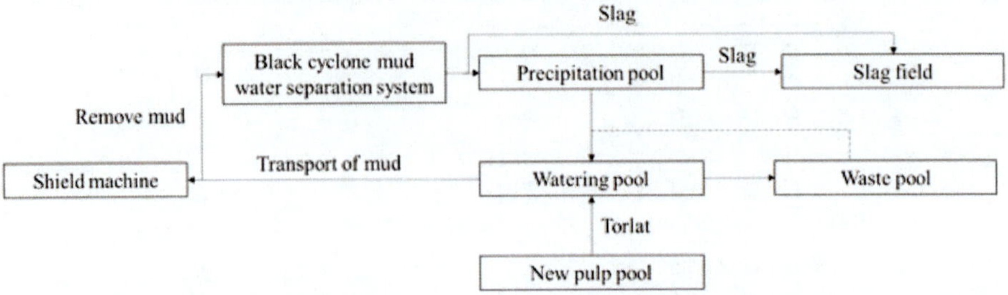

Fig. 5.34 Mud separation flow chart

5.3 Calculation and Analysis of the Mud Circulation and Slag Carrying Capacity of Gravel Formation

With the construction of a large number of river-crossing and sea-crossing tunnels in China, slurry pressurized shield has been more and more widely used. The slurry shield carries the excavated residue soil to the ground slurry separation system by inputting mud, and forms a set of circulation system.

The mud circulation system is composed of inlet pump, discharge pump, flowmeter, control gate valve, inlet and discharge pipeline, and ground mud separation system. The soil sand excavated by the shield enters the slurry tank, and mixes with the slurry to form high-density slurry, which is pumped by the slurry pump (slurry pump) to the slurry separation system for processing. The slurry treatment system of the separated slurry shield consists of a black cyclone screening system and a sedimentation tank. The slurry separation process is shown in Fig. 5.34.

On the basis of existing equipment, it is one of the key problems to be solved before construction whether the sand and residue carried in the slurry can be effectively separated to prevent the deposition of sand and stone in the slurry tank. By calculating the slag carrying capacity of the discharge pipe under normal construction conditions, the mud sampling at different separation stages is tested and analyzed to demonstrate whether the composition of the screening system can meet the separation requirements.

5.3.1 *Material Balance Calculation and Screening Effect Analysis of Two Typical Sand Strata*

Taking the fine sand and gravel sand of Nanjing Yangtze River Tunnel as an example, the balance calculation is considered according to the material quality of the shield tunneling ring, and the characteristics of the two typical sand strata are calculated and analyzed. The basic physical properties of the two formations are shown in Table 5.11. In the calculation, favorable and unfavorable conditions are considered respectively, as shown in Table 5.12.

Table 5.11 Basic physical properties of the stratum

Formation	Bulk weight (kN/m^3)	Moisture content (%)	Pore ration
Fine sand formation	18.9	32.22	0.736
Gravel sand formation	19.5	27.8	0.608

Table 5.12 Calculating the formation conditions

Working condition	Coarse mud sand filtered by black cyclone	Sedimentation tank	Pulping pool
Working condition 1	>75 μm	75–45 μm	<45 μm
Working condition 2	>45 μm	45–30 μm	<30 μm

1) Calculation of mud circulation system in fine sand formation.

The main component of the stratum is fine sand, and the proportion of particles in each stratum is shown in Table 5.13.

The particle size of <0.075 mm is only about 20% of the whole formation particle, and about 70% of the formation particle size is in the range of 0.075–0.25 mm. The clay particle content of this formation powder is small, and the separation of mud and water is not difficult. At the same time, since the particle size of about 70% of the particles is between 0.075 mm and 0.25 mm, and the pores between the formation particles are small, it is not difficult to form impervious or slightly permeable mud film on the excavation surface. The difficulty is how to smoothly discharge the sand excavated from the excavation surface through the circulation system without forming two flow velocities of sand and mud.

It can be seen from Fig. 5.35 that the black cyclone separation system has a good screening effect on the mud in the silty sand stratum. After the secondary screening, only 16% of the mud particles above 75 μm are left, which is far less than the proportion of the particles above 75 μm in the silty sand stratum. The sand in the stratum can be basically screened out. However, the classification screening effect of the black cyclone is in series and parallel, and the separation effect is gradually reflected.

2) Calculation of mud circulation system in gravel sand formation

The stratum is a non-full-section gravel layer, the upper fine sand stratum, and the lower part is a gravel sand stratum. The proportion of particles in each layer in the formation is shown in Table 5.14.

Table 5.13 List of soil particle content of each group of fine sand stratum

Fine sand formation	>2 mm	2 mm–75 μm	75–45 μm	<45 μm
Percentage	0	81.37%	7.17%	11.45%

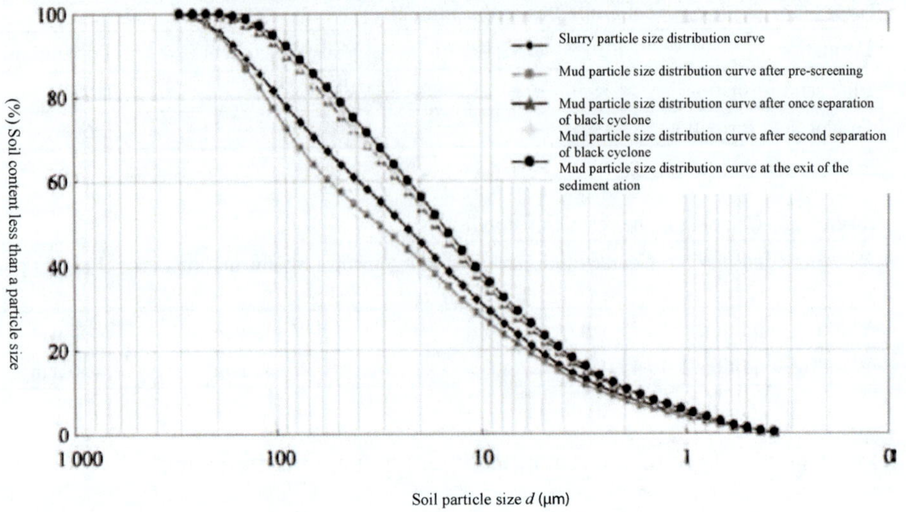

Fig. 5.35 Mud particle fraction curve at various locations in the fine sand formation circulation system

Table 5.14 Summary of soil particle content of each group of gravel sand stratum

Gravel sand formation	>2 mm	2–75 μm	75–45 μm	<45 μm
Percentage	39.60%	47.50%	6.63%	6.27%

The local layer has a large amount of large particles and a small amount of small particles. The existing screening equipment can smoothly complete the separation of mud water. How to smoothly discharge the sand excavated from the excavation surface and avoid blocking the pipe is one of the main problems to be solved.

5.3.2 Calculation and Analysis of Carrying Capacity of Drain Pipe

1) Calculation of carrying capacity

The sand carrying capacity of mud is closely related to the flow rate of mud. The Durand formula is used to calculate the boundary velocity of soil particles transported by slurry to ensure that there is no precipitation in the pipe. The calculation results are shown in Table 5.15.

The calculation results show that the current construction technology can basically ensure that no large amount of sand is precipitated in the pipe. In actual shield construction, there have been no accidents such as pebble precipitation, unable to discharge, pipe blockage and so on.

Table 5.15 Calculation table of limit flow rate of mud sand particles

Tube diameter (m)	Amount of drain mud flow	Drain mud flow rate	Density of drain mud	Coefficient	Boundary flow rate of sand	Note
0.5	2400	3.395	1.3	1.33	4.24	Partial coarse sand particle settlement
			1.4		3.93	Partial coarse sand particle settlement
			1.5		3.65	Partial coarse sand particle settlement

2) Pebble starting speed calculation

The particle size of pebbles is relatively large, and the mud must have a certain flow rate to have enough kinetic energy to carry pebbles. The starting velocity of pebbles is calculated by the Shamoff formula, and the calculation results are shown in Fig. 5.36.

It can be seen from Fig. 5.36 that the size of pebble particles that the slurry can carry increases with the increase of slurry flow rate. When the slurry flow rate exceeds 1.6 m/s, a small amount of pebbles with a diameter of 18 cm can be carried. At present, the slurry flow rate of slurry discharge pipe in shield construction is much higher than 1.6 m/s, which has strong carrying capacity. Studies have shown that under normal construction conditions, the discharge pipe of slurry shield in Nanjing

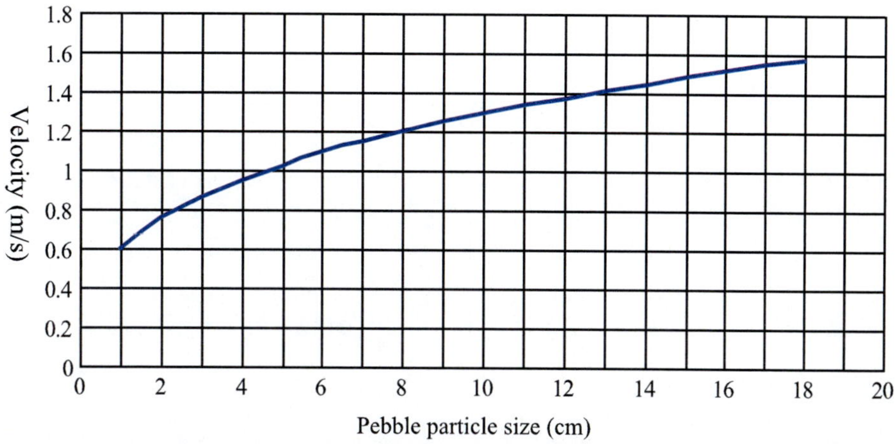

Fig. 5.36 Pebble particle size versus starting speed

Yangtze River Tunnel will not have the accidents of pebble precipitation, discharge and pipe blockage.

Mud circulation system is a key factor in construction, which affects the density index of mud and has an important influence on mud film formation and excavation surface stability. The circulation system screening slag removal effect should be controlled properly. If the fine particles less than 75 μm in the formation are removed by the screener more than 30% (especially the particles less than 50 μm), the density of the mud will be low and the properties of the mud will be affected. If the screening machine cannot screen most particles with particle size greater than 50 μm, large particles will circulate to the slurry returning tank, resulting in a large amount of precipitation in the slurry returning tank, affecting the normal production of slurry. Through calculation and field investigation, the slag carrying capacity and circulation system of sludge discharge pipe can meet the construction requirements when slurry shield tunneling in sandy gravel composite stratum. It is suggested that the setting parameters of the system should be properly adjusted according to the actual working conditions of the site when the shield passes through the gravel composite stratum to improve the efficiency.

Chapter 6
Long-Distance Tunneling of Shields and Replacement of Cutting Tools in the River

During the long-distance shield tunneling, it is very important to effectively solve the tunneling problems such as instability of excavation face, excessive surface subsidence, and dislocation of segment assembly. Among them, the problem of frequent shutdown of shield caused by obstacles or excessive wear of cutters has become an important problem in shield construction in China. Opening maintenance is the main means to eliminate shield faults and remove obstacles. Due to the position of forced shutdown of shield is often in the deep buried depth, complex hydrogeological conditions or the bottom of the river and other adverse environment, under the action of soil pressure and water pressure of the stratum, there is a great safety risk in opening the cabin. Therefore, how to carry out safe cabin opening has become a key technical problem of shield tunnel industry in China. This chapter introduces the long-distance tunneling technology through the adaptability of shield cutter, and summarizes the common faults and corresponding detection technology of shield. At the same time, combined with the mud film air-closure method, the shield cutter replacement technology with pressure is introduced.

6.1 Shield Gravel Formation Long Distance Boring Tool Replacement Technology

In the construction of urban rail transit projects, shield method is a widely used construction technology. Sandy soil has the characteristics of small cohesion, large internal friction angle, poor flow plasticity and large permeability coefficient. When shield tunneling in sandy stratum, it is prone to wear of cutter head, cutter and screw conveyor, increase of torque and thrust of cutter head, abnormal tunneling parameters, difficult excavation and difficult control of land subsidence. Serious accidents will occur. In recent years, more and more attention has been paid to the problems caused by shield tunneling in sandy strata, especially long distance tunneling.

J. Chen et al., *Construction Technology of Large Diameter Underwater Shield Tunnel*, Key Technologies for Tunnel Construction under Complex Geological and Environmental Conditions, https://doi.org/10.1007/978-981-16-5896-9_6

6.1.1 Shield Tool Type and Cutting Principle

6.1.1.1 Shield Tool Type

The performance of shield excavation is mainly guaranteed by the selection and arrangement of cutters. Different types of cutters are arranged on the cutter head of the shield, so that the shield can adapt to the excavation of various strata from soft soil to hard rock. Mastering the correct selection and use method as well as reasonable arrangement on the cutterhead is crucial to improve the tunneling efficiency of shield and create good economic benefits.

In terms of tool shape, Fig. 6.1 shows the front view and side view of the four commonly used digging tools (tooth cutter, roof cutter, mosaic cutter and disc cutter). Among them, tooth cutters and roof cutters are mainly used for excavating weak strata such as sand, silt and clay. The mosaic cutter and disc cutter are mainly used for excavation of hard strata such as gravel layer and rock stratum.

The tool can also be classified according to the purpose of the cutting and setting position, as seen in Table 6.1.

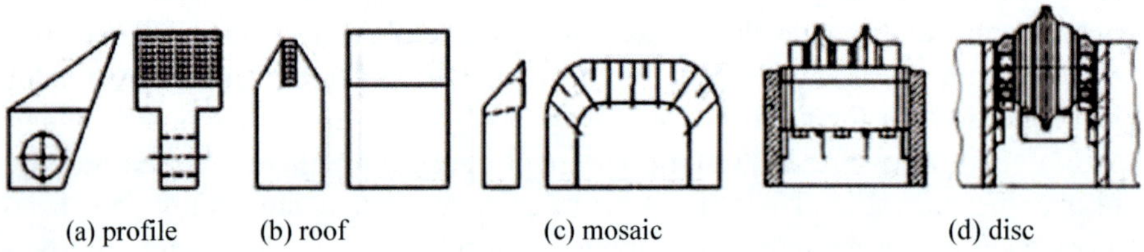

Fig. 6.1 Excavation cutting tool

Table 6.1 Tool type, set position and purpose

Type name	Purpose	Set position
Fixed cutter	Excavation and cutting face	Front panel
Rotational cutter	Excavation and cutting face	Front panel
Advanced cutter	Advanced excavation	Front panel
Guide bit	Break formation	Front panel
Outer edge protection cutter	Protect the outside edge of the knife disc	Front panel extension
Knife tool	Reduced propulsion resistance	Panel extension
Back protection tool	Protect back panel	Back panel
Mud nozzles protection tool	Protect the mud mouth	The mud mouth
Cutting and digging obstacles tool	Cut and dig obstacle	Panel front periphery

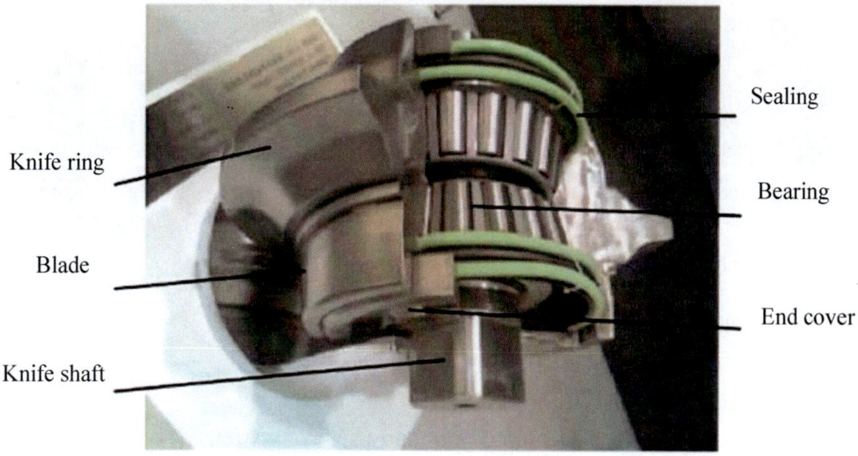

Fig. 6.2 Composition of hob

Shield cutting tools can be divided into two categories: rolling tools and cutting tools. Rolling tools include single edge hob, double edge hob, three edge hob, etc. Cutting tools include cutters, tooth cutters, scrapers, advance cutters, profiling cutters, central fishtail cutters, shellfish cutters, etc. For the excavation of different strata, the cutters of shield usually adopt different forms: when the excavation stratum is hard rock, the disc cutter is used. When the stratum is soft rock, the tooth knife is used. When the stratum is soft or broken soft rock, the cutter or scraper can be used.

1) Hob

In hard rock excavation, the use of rock cutter. The feature of rock breaking by hob is to achieve the purpose of rock breaking by the impact crushing and shear crushing generated by the rolling of the cutter. The hob is mainly composed of cutter ring, cutter body (sleeve), cutter shaft, bearing, seal, end cover and other components (Fig. 6.2).

According to different shapes, hobs can be divided into disc hobs, ball-tooth hobs and wedge-tooth hobs. According to different installation positions, it can be divided into central hob, positive hob and profiling hob. According to the material of blade, it can be divided into all-steel hobs and cemented carbide inserts. According to different structure forms, it can be divided into single-edged hob, double-edged hob, three-edged hob, four-edged profiling hob, two-joint hob, three-joint hob and eccentric hob. Some hobs for shield are shown in Fig. 6.3.

Three-edge and four-edge hobs are only used in the center of the cutterhead to adapt to the small radius rotation of the cutterhead center. The double-edged hob is usually used in mudstone, shale and other soft rock, which requires small blade pitch. The double-edged hob can meet the geological requirements. When the strength of rock mass is high, too small blade distance will lead to low tunneling efficiency on the one hand, and on the other hand, the double-edged forced synchronous rotation will aggravate wear under the condition of non-ideal leveling of the excavation face,

(a) ball gear hob (b) single-edge hob (c) double-edge hob (d) four-edge copying knife

Fig. 6.3 Shield hob

and one of the blade wears and the whole blade fails. Therefore, the double-edged hob is not recommended in the medium and hard rock strata.

2) Cutter

The cutter is a soft soil cutter, which is installed on both sides of the cutter head opening groove to cut the soil layer on the palm surface. When the cutter rotates with the cutterhead, the axial shear force and radial cutting force are generated on the soil of the excavation face. The cutter head and blade are inserted into the stratum, and the soil of the excavation face is continuously cut down. The cutter is generally applicable to loose strata such as sand, pebble and clay with particle size less than 40 mm. The cutter arrangement on the cutterhead generally has concentric circles and Archimedes spiral arrangement. At present, Archimedes spiral arrangement method is widely used.

3) Tooth cutter

When the shield tunneling in the case of soft rock, due to the lack of effective adhesion between the disc cutter and the rock face, the disc cutter cannot roll, resulting in string grinding. If the disc cutter cannot roll, it will lose the effective rock breaking function. Therefore, some disc cutters should be replaced by replaceable tooth cutters in soft rock strata, and the rock breaking trajectory is the same as that of the hob. The rock breaking principle of shield cutter is similar to that of cutter, and there are two cutting edges on the cutter, and the cutter head can break rock when the cutter head is rotating or reversing. Some shield cutters are shown in Fig. 6.4.

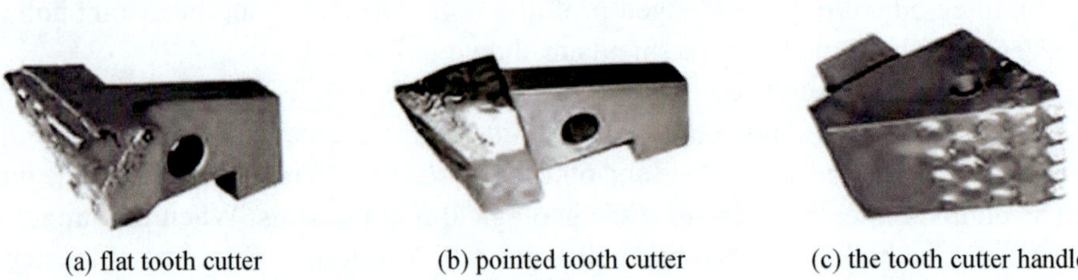

(a) flat tooth cutter (b) pointed tooth cutter (c) the tooth cutter handle

Fig. 6.4 Tooth cutter

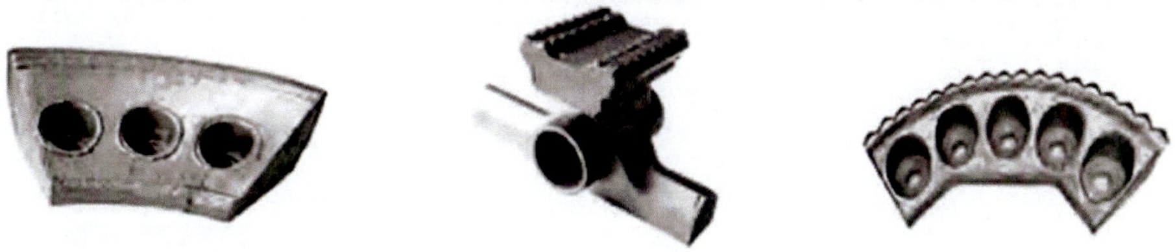

(a) Three-hole scraper (b) Claw knife(double-ended scraper) (c) Edge scraper

Fig. 6.5 Scraper

4) Scraper

The scraper is a tool for cutting soft rock and soil layer or scraping slag soil, and its main role is to clean up rock debris and edge slag soil. The cutting mechanism of the scraper is similar to that of the cutter, and the soil under cutting is scraped into the soil cabin. According to the different installation position, the scraper can be divided into positive scraper and edge scraper (peripheral scraper, diameter cutter). The disadvantage is easy to wear failure. Some scrapers for shield are shown in Fig. 6.5.

5) Fishtail knife

The fishtail cutter is installed at the center of the shield cutterhead, which is usually used in sandy cobble stratum or clay stratum with high strength. It is used to first scrape the soil at the center of the cutterhead, improve the fluidity of the soil at the center, prevent it from consolidating into mud cake, and reduce the cutting resistance of other scrapers and reduce wear. The central knife can be divided into fishtail knife and horn knife, and the fishtail knife can also be divided into integral and combined types. The fishtail knife is shown in Fig. 6.6.

The size of the fishtail knife is large, generally 1200–1500 mm long, 400–500 mm high, 200–300 mm higher than the ordinary scraper. The design and arrangement principle of the fishtail cutter is that the fishtail cutter is not on a plane with other cutters, so the fishtail cutter first cuts the soil and loosens the original stratum. The root of the fishtail cutter is designed as a cone, which adds a turning motion to the

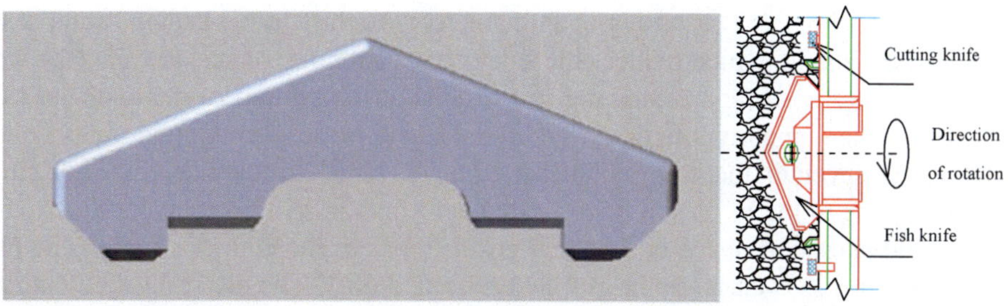

Fig. 6.6 Fishtail knife

Fig. 6.7 First cutter

soil cut by the fishtail cutter when the cutter rotates. This not only solves the cutting problem of the central part of the soil, but also improves the fluidity of the cutting soil.

6) First knife (leading knife)

It is arranged between the panel and the spoke, and the cutting trajectory of the scraper. The first knife is usually 40–50 mm higher than the scraper or the cutter, which does not play a direct cutting role. It contacts the stratum before the cutter, ploughing soft rock and soil layer. The cutting tool first cuts the soil before cutting the soil, and cuts the soil into blocks to create good cutting conditions for the cutting tool. The use of advance cutter can generally significantly increase the flow of cutting soil, greatly reduce the torque of the cutter, reduce the cutting resistance of the scraper, improve the cutting efficiency of the cutter, and reduce the wear of the cutter. First knife is divided into replaceable and non-replaceable two types. The leading knife is shown in Fig. 6.7.

7) The rim bell knife

It can be regarded as the peripheral advance cutter, which is used for the right angle cutter head around. The cutting outer diameter is slightly larger than the outer diameter of the shield, which reduces the shield propulsion resistance, prevents the wear of the cutter head, and cleans the soil residue at the bottom of the cutter head to improve the tunneling efficiency. When the shield passes through the sandy cobble stratum, especially the sandy cobble stratum with large particle size, if the hob cutter is used, due to the loose soil debris, large deformation will occur under the excavation and extrusion of the hob, which greatly reduces the cutting effect of the hob, and sometimes even loses the cutting crushing ability. The disc ring shell cutter is arranged on the front face of the disc ring of the cutter, which is dedicated to cutting sandy pebbles. The problem of cutting soil (sandy cobble) by shield machine can

Fig. 6.8 Shell knife with circle

be well solved by using circle scallop cutter. The circle scallop knife is shown in Fig. 6.8.

8) Copying knife (super-cutting knife)

Shield generally designed two profiling cutters (one spare), arranged at both ends of the spoke. During construction, according to the requirements of over-excavation amount and over-excavation range, the profiling cutter can be extended and retracted from both ends of the spoke radially to achieve the purpose of over-excavation cutting. The maximum extension of profiling knife is generally 80–130 mm. The profiling cutter is designed for curve propulsion, turning and rectification. It is installed on the edge of the cutter head, and the extension of the cutter is controlled by a hydraulic cylinder. The ordinary section does not work, and the cutter head extends during curve propulsion, turning and rectification. Some profiling cutters for shields are shown in Fig. 6.9.

6.1.1.2 Shield Cutter Cutting Principle

1) Cutting principle of the hob

The research on the cutting mechanism of the hob was first started from the mining and coal breaking. It was considered that the broken line of the coal seam after the hob cutting was circular, and the cutting force of the hob was calculated based on this assumption. The main ways of rock breaking by the hob were extrusion failure, shear failure and tension failure. The extrusion rock breaking of the hob mainly occurs on the path of the hob. The rock mass is directly subjected to compressive stress through the blade surface of the hob. The rock mass surface is subjected to excessive compressive stress, which leads to damage and produces fine broken rock particles. The shear rock breaking mainly occurs outside the trajectory line of the hob blade, and the rock mass is damaged due to the shear stress exceeds its shear strength. The main rock breaking method of hob is tension failure, and the failure of rock mass conforms to the tension failure theory. The roller exerts compressive stress on the rock mass, which makes the rock mass broken and cracks appear inside the rock

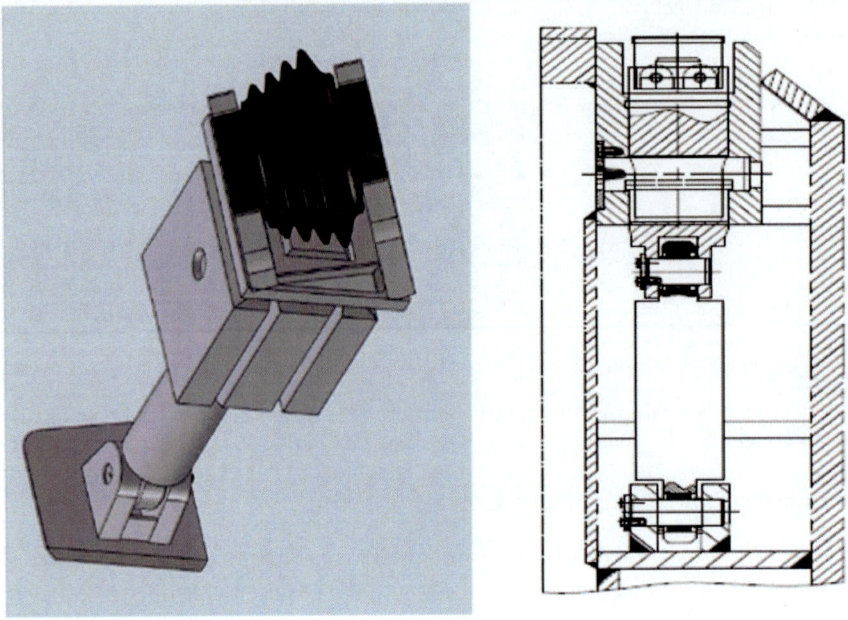

Fig. 6.9 Copying knife

mass. Then, due to the elongation and coalescence of cracks, the rock mass produces excessive plastic deformation and finally tensile failure, as shown in Fig. 6.10.

By comprehensively considering the three rock breaking methods of tension, extrusion and shear of the hob, and combining with the tool selection, cutting parameters, rock type and linear cutting machine experimental data, Ozdemir et al. of Colorado University of Mining and Technology proposed the CSM model of semi-theoretical and semi-experimental, which was used to calculate the force of the hob (Fig. 6.11), providing a basis for the selection of the hob. However, the model only considers the normal force and rolling force of the cutter, and does not consider the

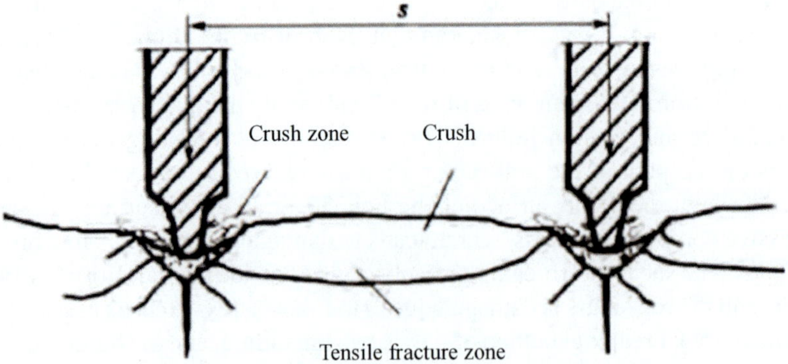

Fig. 6.10 Schematic diagram of rock breaking by hob

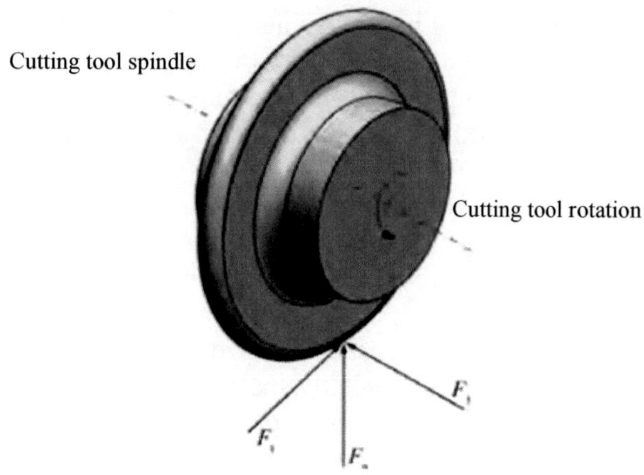

Fig. 6.11 Hob force diagram

lateral force of the cutter. Therefore, the force of the cutter calculated based on the model is smaller than the experimental test value:

$$F_t = \int_0^{\phi} TRPd\theta = \int_0^{\varphi} TRP^0 + d\frac{\theta}{\phi}n^{\psi}d\theta = \frac{TRP^0\phi}{1+\psi} \tag{6.1}$$

$$\phi = \mathrm{arccosd}\frac{\mathrm{R}-\mathrm{p}}{\mathrm{R}}n,\ P = P^0\mathrm{d}\frac{\theta}{\phi}n^{\psi}$$

In the formula

F_t—the resultant force of the hob.

φ—the contact angle between the hob and the rock.

T—the width of the hob tip.

R—hob radius.

P—the pressure in the crushing zone.

P_0—the basic pressure of the crushing zone directly under the hob, and the strength of the rock (rock compressive strength σ_c and rock shear strength σ_t), hob size (hob radius R, tool tip spacing S and one excavation amount p), the blade shape (the hob tip width T, etc.) is related, that is, $P_0 = f\ (\sigma_c, \sigma_t, S, T, R, p)$.

ψ—the pressure distribution coefficient of the tool tip, $\psi = -0.2$ to 0.2 (if the hob is V-shaped and sharp, $\psi = 0.2$; if the width of the hob tip is large, $\psi = -0.2$; in general, $\psi = 0.1$).

θ—blade angle.

However, the single-blade hob and the double-edged hob are not exactly the same in the same kind of rock. The single-blade hob is destroyed by multiple crushing of the rock until the tensile crack occurs in the rock mass. The more regular cracks, the larger the lump block is; for the double-edged hob, the rock is broken by the synergy between the hobs. The hob has a greater pressing force on the rock mass, and the rock is easily broken and produces fine gravel and rock shovel. The energy consumption of rupture is high, and only complete lenticular rock ridges appear in extremely hard rock. Therefore, in the general rock mass, the single-edged hob should be preferentially arranged to reduce the energy consumption of rock breaking; and the double-edged hob should be arranged in the extremely hard rock to strengthen the pressing force of the hob and improve the rock breaking efficiency.

Different rock properties also affect the cutting mechanism of hob. The discrete element method is used to simulate the cutter cutting process. It is found that the rock joint characteristics have a significant impact on the cutter rock breaking. The smaller the joint spacing is, the greater the impact on the penetration of the cutter is. The cracks produced by the rock are easy to develop along the structural plane of the rock mass. Similarly, the rock confining pressure also has a significant impact on the cutting performance of the hob. The increase of confining pressure hinders the development of cracks to a certain extent. In the process of rock intrusion by the hob under different confining pressures, it is found that the broken form of the rock gradually changes from a single lenticular rock mass to three irregular rock fragments.

2) Cutting principle of the scraper

The early research on scraper cutting mechanical model mainly referred to the experience of soil cutting processes such as bucket teeth of excavators in earthwork engineering and plough teeth in farming and forestry. Among them, the most influential one was the flow model of Japanese scholars Yano Shintarou and Tomiaki Jirou. The soil excavated by cutting tools has different flow patterns, which are mainly related to cutting angle, speed, thickness and soil properties. There are four common flow patterns: flow pattern, fracture pattern, shear pattern and exfoliation pattern, as shown in Fig. 6.12. Flow cutting mainly occurs in low-intensity strata such as silt stratum, water-rich clay stratum and silty soil stratum. With the movement of scraper, the soil generates continuous shear deformation from the blade. Shear cutting mainly occurs in the clay and silty soil layers with relatively high strength. When cutting, the soil first generates compression deformation, and then the cutting soil moves along the blade surface of the tool until shear failure occurs. Fractured cutting occurs in sand stratum. The soil generates compression deformation at the front of the blade and maintains stability for a certain time. With the cutting progress, the soil at the blade generates cracks and destroys, and the formed slag soil is small block. The stripping cutting occurs in the ordinary gravel and sandy cobble stratum, and the scraper plays the role of stripping sandy cobble.

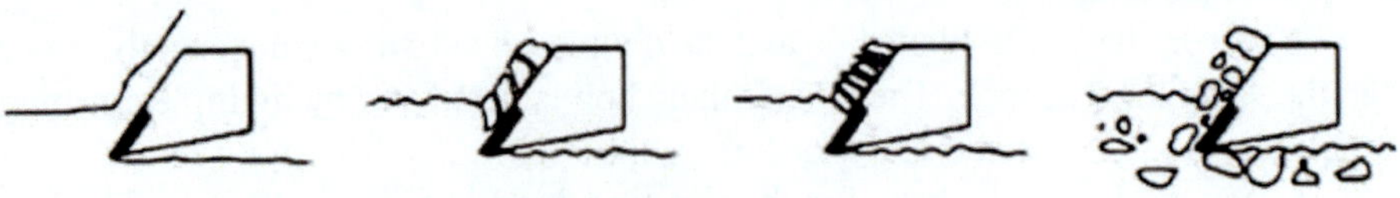

(a) Water type (b) Shear (c) Fracture (d) Flake type

Fig. 6.12 Flow model of scraper cutting soil

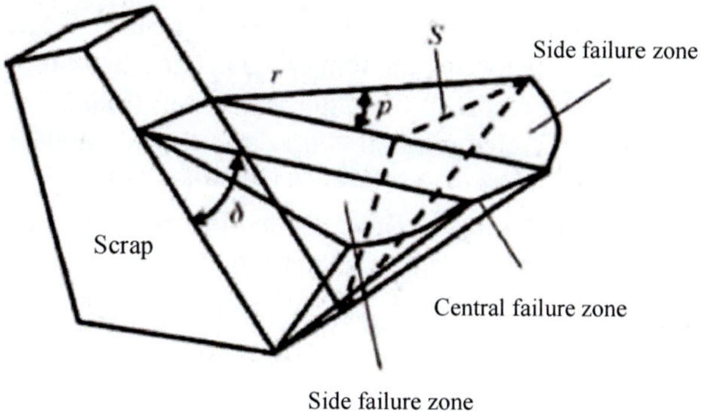

Fig. 6.13 Mckyes-Ail theoretical soil failure

As the most commonly used force model in scraper cutting, Mckyes-Ali model divides soil failure into central failure zone and two half-moon failure zones, as shown in Fig. 6.13. The logarithmic spiral failure line is simplified to a straight line, and the cutting force of the tool is derived. Domestic research on scraper cutting model is mostly based on Mckyes-Ali model, by assuming that the scraper vertical cutting soil or ignoring the weight of soil, so as to improve the calculation model.

6.1.2 Analysis of Geological Adaptability Design of Cutterhead and Cutter

The cutterhead is one of the core components of shield tunneling. In general, the cutterhead has the following three functions: cutting rock and soil mass under the combined action of cutterhead torque and propulsion. Support the excavation face during excavation and shutdown. Improvement of rock and soil during excavation.

In many projects, due to the inappropriate design of head is not suitable for the location, there are serious wear of cutter head and surrounding cutters, and large-scale loss and failure of cutter alloy. In this section, combined with the design of existing cutter head of Nanjing Yangtze River Tunnel and the analysis of the advantages and disadvantages of the adopted process, the technical points and parameter selection of geological adaptability design of large-diameter slurry shield cutter head in gravel or sandy cobble stratum are proposed.

6.1.2.1 Analysis of Geological Adaptability Design of Cutterhead

There are usually three types of cutterhead structure, namely spoke type, spoke panel type and panel type, and the corresponding opening rates are usually about 65, 35 and 20%. According to the analysis, the shield cutter head in sandy cobble stratum should adopt the spoke cutter head with large opening rate, which is beneficial to the flow of slag and soil and the rapid excavation, reduce the tool wear and improve the secondary excavation distance of the tool. However, for the large-diameter slurry shield, considering the tunnel scale, cutterhead function, slag and soil properties and other factors, the spoke cutterhead is not applicable. The large-diameter slurry shields in China and abroad all choose the spoke panel type or pure panel type cutterhead structure without exception. The cutterhead opening rate is 15–40%. The cutterhead of Nanjing Yangtze River Tunnel Project adopts the spoke panel type structure, and the cutterhead opening rate is about 35%. The analysis is as follows:

(1) The large-diameter cutter head adopts a spoke type structure, and the structural rigidity is difficult to be ensured, and the cutter disc steel ring is easily deformed or even unstable.
(2) Large-diameter slurry shield is generally used in highway tunnels or railway tunnels. The tunnel is large in scale and long in mileage, and most projects are more than 2 km long. When the excavation section contains sand layer or sandy cobble stratum, it is necessary to develop a medium-distance tool change plan. The relatively moderate opening rate of the cutterhead is conducive to the stability of the auxiliary excavation face when the pressure tool change is applied.
(3) When the project is a large diameter slurry shield at the bottom of the water, if the support pressure is improperly controlled in the tunneling process, the slurry is likely to occur over-pressure splitting roof fall failure or pressure collapse failure. The panel can play the role of temporary support excavation face, and win time for taking corrective measures to avoid accidents.
(4) The residual soil of slurry balance shield is in the form of dilute slurry with a specific gravity of about 1.35, and there is no poor flow plasticity of residual soil, which does not affect the tunneling speed.

6.1.2.2 Tool Geology Adaptability Design Analysis

The leading cutter and scraper on the cutterhead are the main cutters, and the center cutter, overbreaker and peripheral scraper have fixed positions on the cutterhead. Therefore, the layout design of the cutter mainly refers to the layout of the leading cutter and scraper.

1) First knife design

The advance cutter of shield tunneling in sandy cobble stratum is the primary position, which is widely used in earth pressure balance shield with mature technology. According to the working principle of the leading knife and scraper, the content of the tool layout design includes the design of the leading amount of the tool and the spacing of the leading knife. The cutting tool advance is the height difference between the cutting tool and the scraper, which depends on the tool wear length and the rock type, and should not be less than the tool wear length. Whether the cutting tool design is reasonable or not directly affects the cutting effect of the cutter head, the stirring effect of the cutter on the slag soil and the life of the scraper. For sand stratum, the fine particle content is high, and the slag soil is easy to be improved and

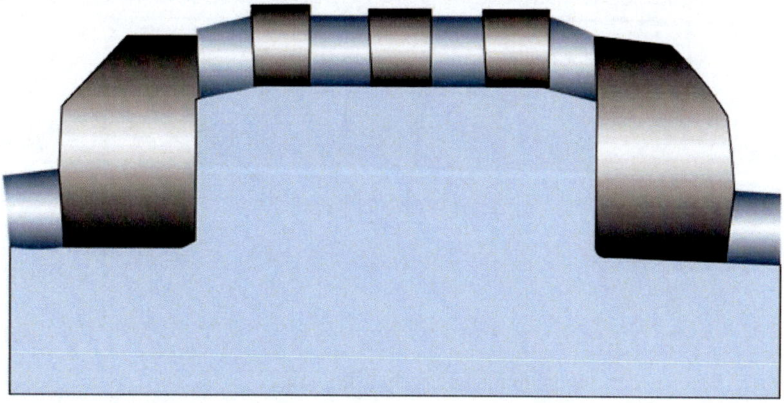

Fig. 6.14 Best advance cutting for sand and pebble formation

cut. Therefore, the length of the leading knife is shorter than that of the scraper, and the leading amount is generally 10–15 mm. In Nanjing Yangtze River Tunnel Project, the cutter head is equipped with three rows of alloy columns only in the middle of the cutter body. After the wear of the cutter body on both sides, the alloy columns fall off, which leads to further wear of the cutter body as a whole. It is completely unsuitable for the serious wear characteristics of gravel sand stratum. In practical engineering, the upper part of the cutter head is completely worn. According to the surfacing traces, the wear is estimated to be about 90 mm, and the remaining height is about 165 mm, which is lower than the height of the scraper and completely invalid (less than 300 m in the gravel sand stratum of some sections). The leading knife with the best application effect in sandy cobble stratum is shown in Fig. 6.14.

2) Scraper design

The scrapers (fixed scrapers and replaceable scrapers) are installed in the left and right shields of Nanjing Yangtze River Tunnel. When tunneling in the middle and short distance of gravel stratum, the phenomena such as alloy cracking, overall shedding and groove turnover are common. The scrapers used in this project are comprehensively analyzed. Compared with the design of shield scrapers in sandy cobble stratum in China, the design parameters and principles of scrapers suitable for sandy cobble stratum are proposed.

(1) Alloy materials. A large area of tool change was carried out on the left line after heading to the 578 ring. Some tool alloys showed normal wear. The wear amount was taken for the inverse calculation of the wear coefficient, and the alloy performance was compared with the existing empirical wear coefficient. Through comparison, it is found that the reverse wear coefficient of the normal wear of the surrounding tool is larger than that of the empirical wear coefficient, which may be due to the failure of the scraper and the fixed knife around the same trajectory in advance, and the actual number of effective tools is less than the number of calculated tools. The discreteness of the reverse wear coefficient

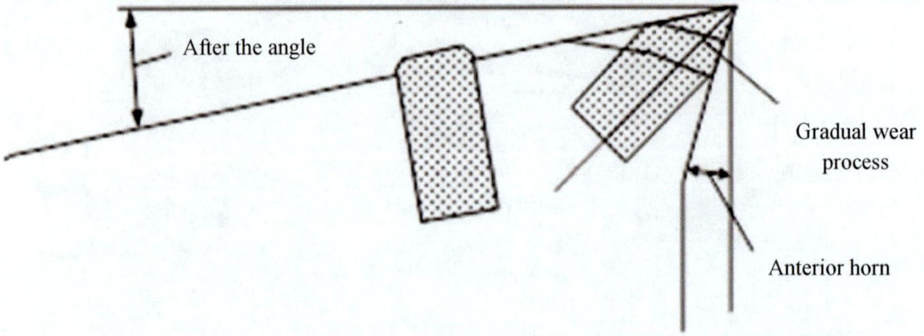

Fig. 6.15 Blade parameter definition and secondary wear

of most tools is small, indicating that the hardness of the alloy is normal. Tool cemented carbide (blade) has two performance evaluation parameters, namely hardness and bending strength. These two parameters restrict each other. Increasing hardness is beneficial to wear resistance, but bending strength will decrease. Shield cutter in sandy cobble stratum is prone to alloy cracking. Therefore, appropriate sacrifice of hardness, improving bending strength and reducing the probability of alloy cracking are the correct ideas to select alloy performance parameters. It is suggested that the hardness of cemented carbide for cutting tools in sandy cobble stratum should be HRA bigger than 86, and the bending strength should be greater than 2000 N/mm^2.

(2) The front and rear angle parameters of scraper are taken. The front and rear angles of shield cutters are usually 5°–15°, and the specific values are determined according to different geological conditions. The secondary wear on the back of the cutter in sandy cobble stratum is serious as shown in Fig. 6.15, so as to reduce the embedded depth of the alloy. In order to delay this trend, the back angle should not be too small, and it is recommended to take 15°. The scraper in gravel or sandy cobble stratum is mainly peeled and stirred. In order to reduce the alloy cracking, the rake angle should not be too large, and it is recommended to take 5°. The rake and rake angles of the scraper in this project are 10°, which can be further optimized.

(3) Selection of bonding mode, section size and chamfer radius of blade. Scraper cemented carbide and cutter body usually have two kinds of combination, surface-mounted and embedded, both have their advantages and disadvantages. The main problem of scraper in sandy cobble stratum is to prevent alloy cracking, and the direct cutting function requirements are relatively low, so the embedded application is more, but the engineering effect is poor in harsh sandy cobble stratum. Therefore, the application of surface-mounted combination is gradually increasing.

The definition of cemented carbide section size (width × height) is shown in Fig. 6.16. When the surface bonding method is used, too small width will lead to weak impact resistance, and too small height will lead to all cemented carbides

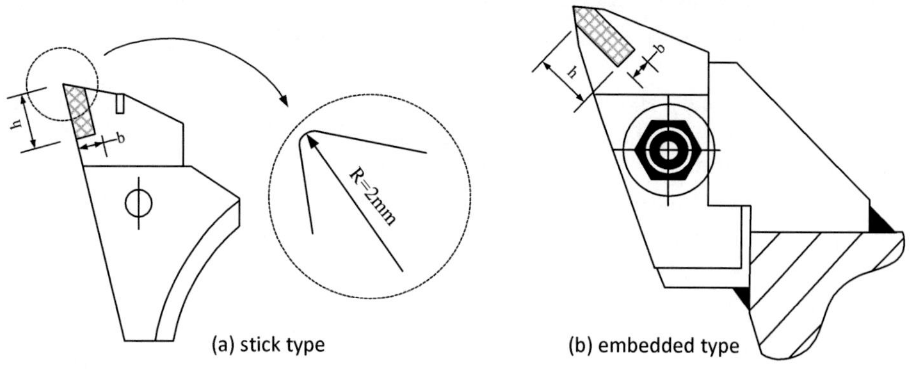

Fig. 6.16 Schematic diagram of two combination modes of alloy and knife body and blade radius

entering the stratum during tunneling, which aggravates the wear of the tool body at the bottom of the alloy and causes the alloy to fall off. When the embedded bonding method is adopted, the alloy width can be taken as the conventional value (≥15 mm), but the height cannot be too small. Otherwise, the slight wear of the cutters on both sides of the alloy leads to insufficient embedded depth and alloy shedding.

The chamfer radius of the blade reflects the sharpness of the tool, usually $R =$ 2 mm. In general, increasing the blade radius will cause the tool to become blunt and reduce the cutting ability. However, in practical engineering, the blade in sandy cobble stratum is too sharp, which often occurs cracking in the initial tunneling. As mentioned above, the sandy cobble stratum is spalling failure, and the requirement for cutting function is not high. In practical engineering, the blade angle radius of the newly improved scraper in Beijing diameter line reaches 10 mm, and the engineering application effect is good. Therefore, it is recommended that the blade chamfer radius of the cutting knife in sandy cobble stratum should be 5–10 mm, and the low value should be taken when the interval contains clay stratum.

Through the above analysis, the design principle of the geological adaptability of the scraper in the gravel sand and gravel stratum is proposed: reduce the rake angle, increase the back angle, increase the alloy, and passivate the blade.

6.1.3 Field Excavation Test Study

In the process of shield construction, different cutters should be scientifically and reasonably selected according to different geological conditions to prolong the service life of cutters, which will have an important impact on the progress and benefit of the project. Combined with Nanjing Yangtze River Tunnel Project, this section mainly optimizes the design of scraper and summarizes the long-distance construction experience of gravel stratum.

6.1.3.1 Replaceable Scraper Improvement Program

Based on the analysis of the insufficient design of the initial scraper, the engineering performance of the scraper corresponding to different improvement schemes is studied, and the geological adaptability design points of the refined scraper in the gravel sand stratum are further verified. The influence of tool parameters and tunneling parameters on the torque of the cutterhead is analyzed. Three sets of scraper improvements are shown in Figs. 6.17, 6.18 and 6.19 and Table 6.2.

Compared with the original scheme, the three schemes are basically in line with the principle of "reducing the front angle, increasing the rear angle, increasing the alloy section, and passivating the alloy blade". At the same time, the width of the alloy block is increased, and the impact resistance is enhanced from the size and specification. In order to be suitable for the alloy section form and increase the area of the alloy and the blade body, the combination mode of the alloy and the blade body is changed to the surface type. Scheme I embeds a small alloy behind the main alloy, which can effectively delay the secondary wear of the rear corner. Scheme II peripheral cutter head is designed as parallel double edge, so that it has both forward and reverse cutting ability. Scheme III blade radius is directly passivated to $R = 10$ mm, which is the first application in China (Beijing railway diameter line scraper is improved to the second application). The above is the design innovation of the

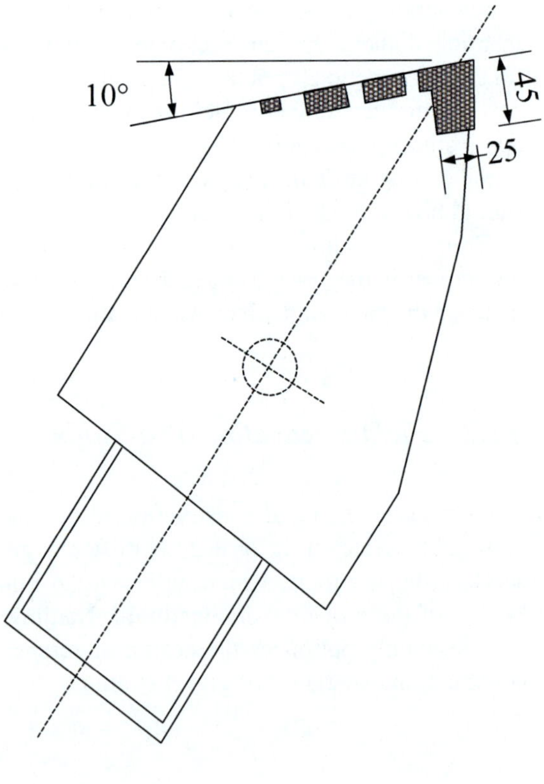

Fig. 6.17 Changing the squeege improvement scheme

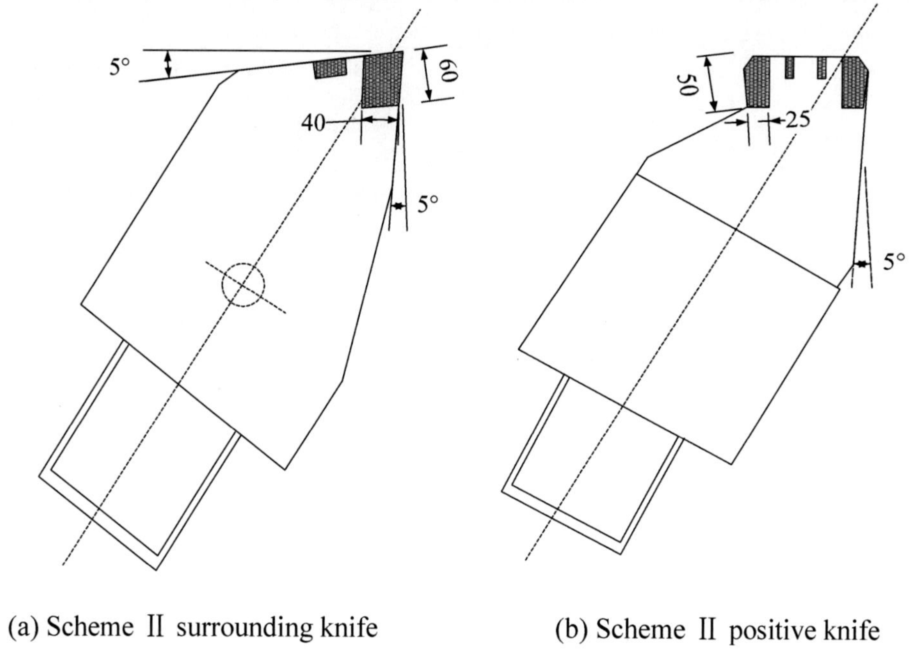

Fig. 6.18 Changing the squeegee II improvement scheme

Fig. 6.19 Changing the squeegee III improvement

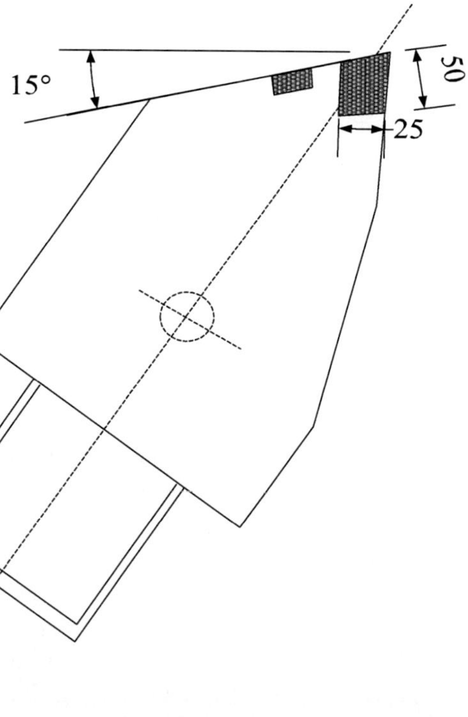

Table 6.2 Scraper parameters and design description of different schemes

Scheme	Anterior angle (°)	Relief angle (°)	Alloy section size (mm × mm)	Alloy width (mm)	Blade radius (mm)	Combination mode of alloy and knife body
Original scheme	10	10	15 × 35	20	2	Flush bonding
Improved scheme I	0	10	25 × 45	48	Quadrilateral processing	Face-mounted
Improved scheme II around knife	5	0	25 × 50	40	Chamfer angle processing	Face-mounted
Improved scheme II positive knife	0	15	25 × 50	40	Chamfer angle processing	Face-mounted
Improved scheme III	5	5	40 × 60	55	10	Face-mounted

three schemes, and the specific engineering performance must be tested in the actual excavation.

6.1.3.2 Scraper Replacement Experiment

Three sets of scraper improvement schemes have distinct characteristics in the design, but their applicability and process reliability still need to be tested by practical engineering. The improved scraper is first applied to the left line tunnel. The improved scraper and the original scraper are installed at the same time at different cutter positions of the cutter head. In the tunneling process, the improved scraper is gradually adjusted according to the tunneling effect and the performance of the cutter, and the left line shield is successfully penetrated.

6.1.3.3 Statistical Comparison of Normal Wear Coefficients

The normal wear monitoring value (the wear value of uncracked alloy tooth) obtained during the whole tool changing process before and after the improvement of the scraper was used to deduce the wear coefficient. According to the proportion of various strata in the tool changing interval, the wear coefficient of the cutter in different strata was calculated. The wear resistance of different types of scrapers was tested, and the calculation results were used for reference in future projects.

The regression of wear coefficient in different intervals is shown in Figs. 6.20, 6.21, 6.22, 6.23, 6.24, 6.25, 6.26, 6.27 and 6.28. The starting mileage in Figs. 6.21, 6.22 and 6.23 is ring 647 due to the alternate tool change of different tool numbers. The proportion of various strata in each tool change interval is shown in Table 6.3.

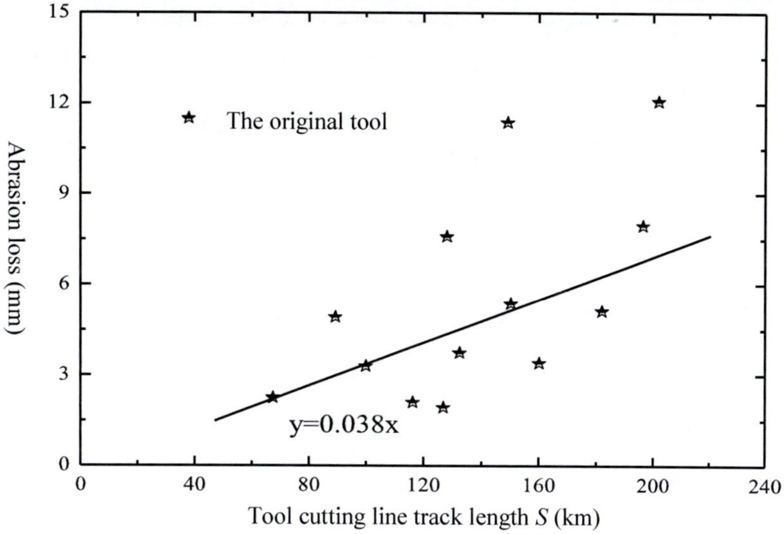

Fig. 6.20 578–647 ring original cutting tool

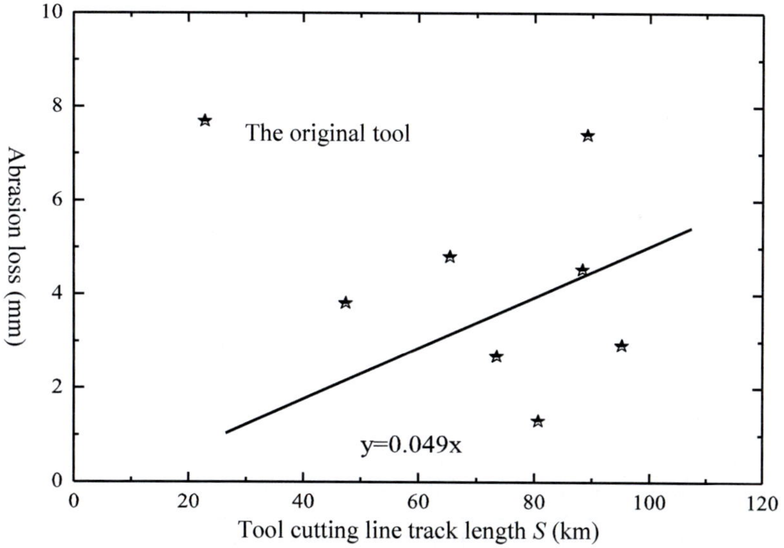

Fig. 6.21 647–680 ring original cutting tool

According to the proportion of different types of strata in different cutter replacement intervals, combined with the statistical results of 0–578 rings in Table 6.3 and Figs. 6.20, 6.21, 6.22, 6.23, 6.24, 6.25, 6.26 and 6.27, the normal wear coefficient of each type of scraper in different strata is calculated. The effective statistical intervals of improved No. I and improved No. II are only one and two, respectively, and the

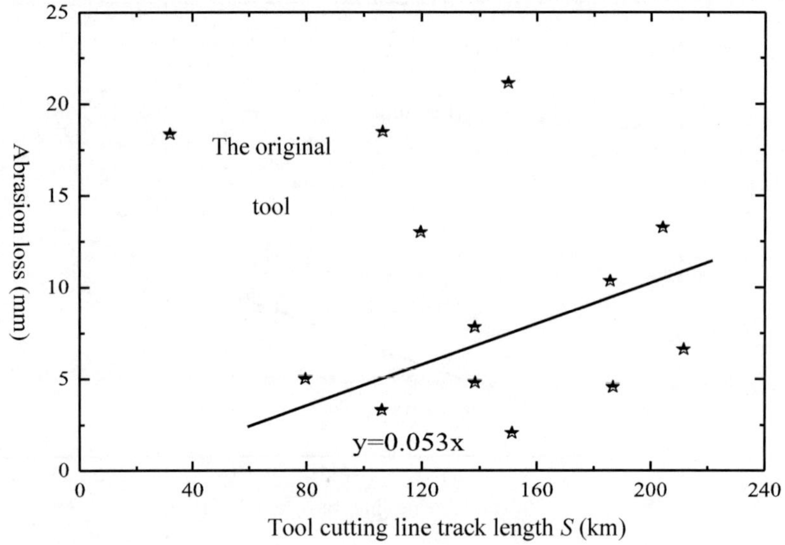

Fig. 6.22 647–719 ring original cutting tool

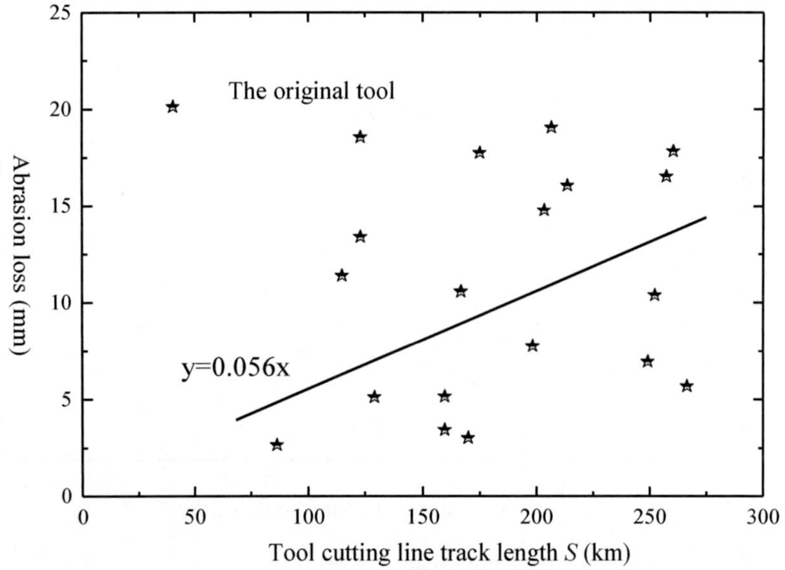

Fig. 6.23 647–742 ring original cutting tool

wear coefficient of this type of scraper in the corresponding stratum cannot be calculated. Since the alloy blades of the improved No. I and the original cutter are both sharp designs, the improved No. II and the improved No. II are both sharp designs. No. III is designed for passivation treatment, so it is assumed that the wear coefficient of the improved No. I and the original tool is equal in the fine sand stratum, and the

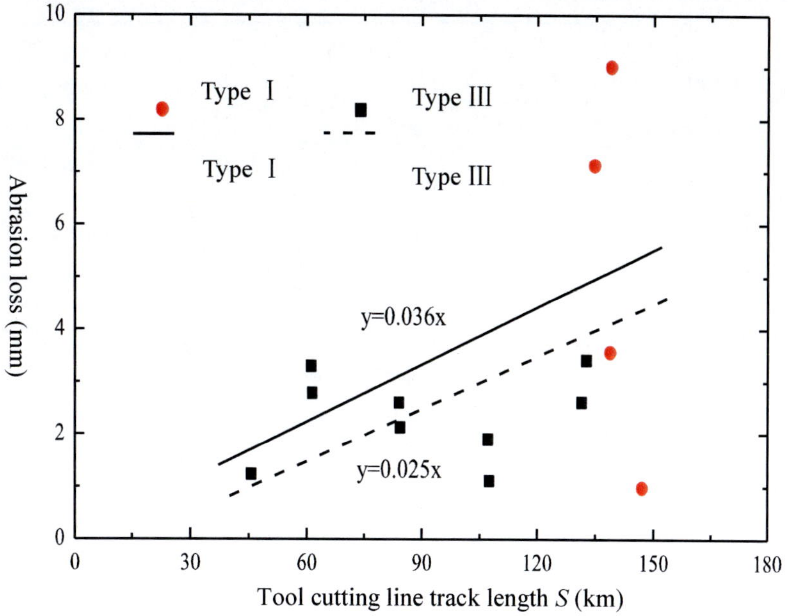

Fig. 6.24 742–792 ring coefficient of wear

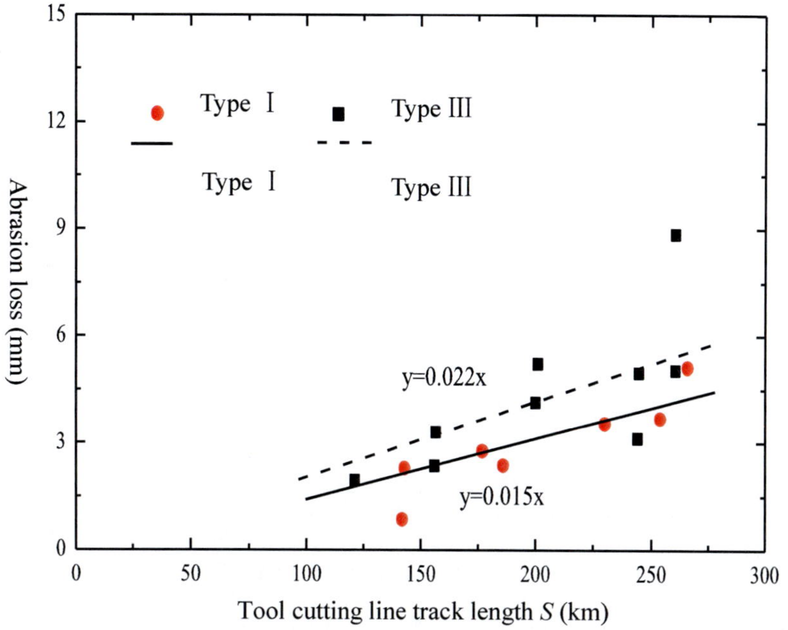

Fig. 6.25 792–885 ring coefficient of wear

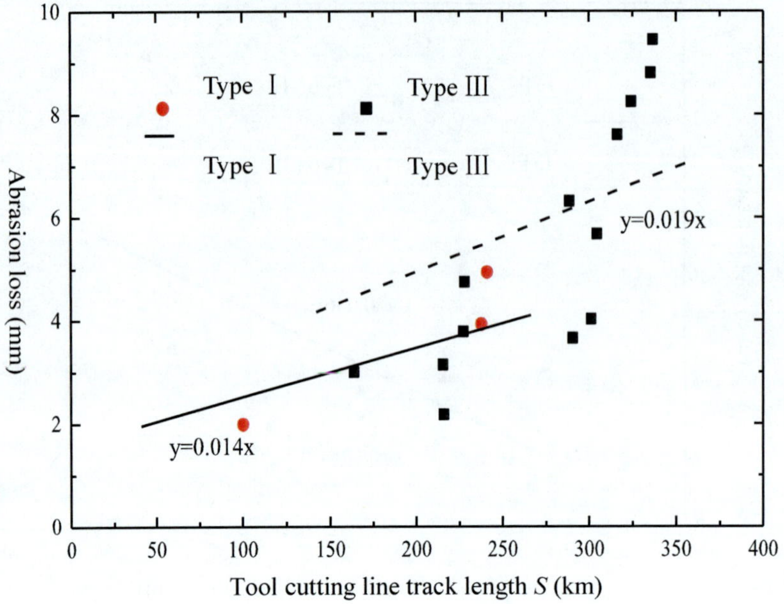

Fig. 6.26 792–885 ring coefficient of wear

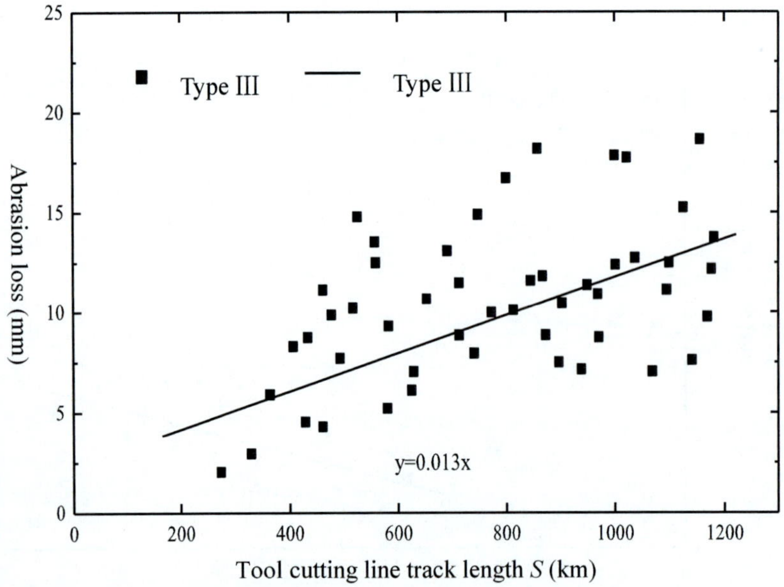

Fig. 6.27 1006–1513 ring coefficient of wear

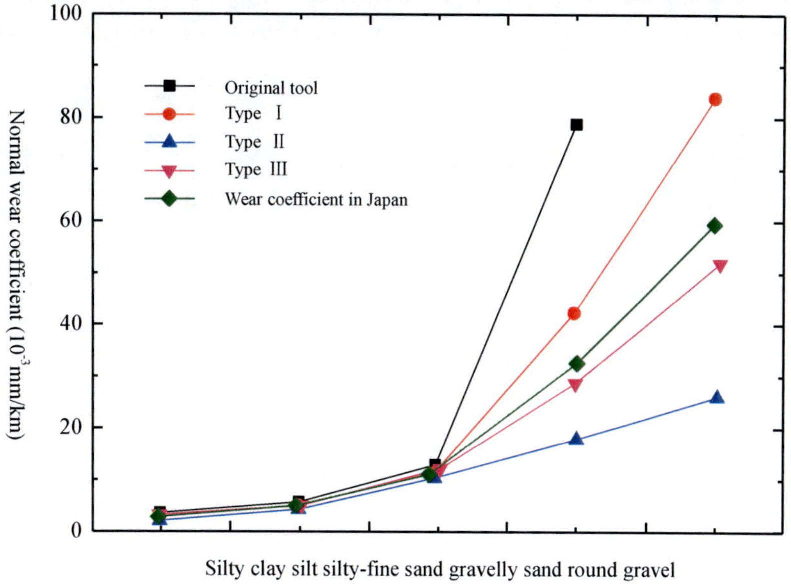

Fig. 6.28 Comparison of wear coefficient of each type of scraper in different stratum

Table 6.3 The proportion of all types of strata in the tool changing interval

Tool change interval	Silt-fine sand (%)	Gravelly sand (%)	Round gravel
578–647 rings	61.5	38.5	–
647–680 rings	45.6	54.4	–
647–719 rings	38.9	61.1	–
719–942 rings	34.9	65.1	–
742–792 rings	35.1	54.1	10.8%
792–885 rings	41.7	58.3	–
885–1006 rings	61.5	30.2	8.3%
1006–1513 rings	91.9	6.0	2.1%

wear coefficient of the improved No. II and the improved No. III is equal. According to experience, it is assumed that the wear coefficient of the improved No. I in the round gravel layer is twice its wear coefficient in the gravel layer. On the basis of the above two assumptions, the classification wear coefficient of each type of scraper in

Table 6.4 Statistical results of wear coefficient of each type of scraper in different stratum ($\times 10^{-3}$ mm/km)

Type	Silty clay	Slit	Silty-fine sand	Gravelly sand	Round gravel
Original scraper	2.5	4.5	12	79	–
Improved type I	–	–	12	42	84
Improved type II	–	–	11	18	27
Improved type III	–	–	11	29	52
Japanese statistics experience value	4	4	14	32	60

different strata is calculated. According to the current situation of the advance cutter in the right-line cabin opening, it can be concluded that the left-line advance cutter is also completely invalid. Therefore, all statistical calculation results are based on the failure of the advance cutter, which can be used as a reference for the prediction of tunneling wear of similar projects without the advance cutter (Table 6.4).

It can be seen from the statistical data that the wear coefficient of the original scraper in silty clay and fine sand stratum is close to the corresponding empirical statistical data, but the wear coefficient increases rapidly to 0.079 mm/km in gravel sand stratum, which is about 2.47 times of the corresponding empirical wear coefficient. The wear coefficient of the original scraper in the fine sand stratum is normal, indicating that the hardness parameters are normal. In the gravel sand stratum, not only the cracking and overall shedding are serious, but also the normal wear coefficient is too large. The fundamental reason is that the alloy blade is too sharp, and the normal wear process has been accompanied by fine cracking, which also proves that the alloy is too sharp in the gravel or pebble stratum.

The improved No. I has improved the design of the alloy blade. The normal wear performance of the alloy blade in the gravel layer is better than that of the original tool, but the wear coefficient in the round gravel layer is as high as 84×10^{-3} mm/km, which is about 1.4 times of the corresponding empirical wear coefficient. The fundamental reason is that the design of the alloy blade takes too much consideration of its cutting performance and the passivation degree is not enough, which is not conducive to crack prevention and normal wear, and the blade tip is quickly flattened after short-distance tunneling.

The normal wear coefficient of the improved No. II is the smallest in all types of cutting tools, and is far less than the empirical wear value, which shows the rationality of the alloy edge design and the high hardness of the alloy. At the same time, the cutting tool also shows the characteristics of easy cracking, indicating that the high hardness of the alloy leads to its brittleness and insufficient bending strength. It does not conform to the principle of selecting the alloy performance parameters in the gravel stratum. It is suggested to reduce the hardness of the alloy and improve the bending strength.

The wear coefficient of the improved No. III is moderate, which is slightly less than the empirical wear coefficient, and about 70 cutters are used in the whole tool

Table 6.5 The suggested wear coefficient of sludge-shield scraper under the condition of no advance cutter k ($\times 10^{-3}$ mm/km)

Clay	Silty clay	Silt	Silty-fine sand	Gravelly sand	Round gravel
2	3	5	10	20	50

change process. Only 4 cutters appear alloy cracking, and the rest are normal wear. In the gravel layer, the general tool change distance reaches 200 ring, and the longest reaches 263 ring, which is much higher than the average 50 ring of other types of scrapers. The replaced cutters of the improved number III maintain good shape, and can still be used in practice. The improved No. III tunneling process has the most application and stable performance, which not only directly verifies the good geological adaptability of "blunt knife" in gravel and round gravel strata, but also shows that the material selection of this type of tool alloy is reasonable in taking into account both hardness and bending strength.

According to the engineering performance of different types of improved cutters, it is suggested that the design and material selection of similar engineering scrapers can refer to the improved NO. III, and based on the statistical value of wear coefficient in sandy soil, and referring to the statistical value of wear coefficient of original cutters in clay and silt, the recommended value of wear coefficient under similar engineering conditions is proposed, as shown in Table 6.5.

6.1.3.4 Evaluation of Each Improved Blade Design

Although the engineering performance of the improved scraper I and II is better than that of the original tool, there are still serious alloy cracking and overall shedding in practical engineering applications. The overall performance of the improved scraper III is excellent, but there are also improvements. The reasons are analyzed, and the shortcomings of the design and process are found. It has a direct reference for the design of the scraper in gravel-sand cobble formation. Break probability statistics of three types of scrapers in each tool change interval are shown in Table 6.6.

It can be seen from Table 6.6 that the fracture of improved scraper I alloy is common, but there is no overall shedding, indicating that the brazing quality control is greatly improved compared with the original tool. The fracture of the improved scraper II alloy is better than that of the improved scraper I, but some of the tools have the whole alloy falling off, and the brazing process and structural design have defects. In previous inspections, only four blades of the improved scraper III appeared alloy cracking, and the good rate was above 90%. The gravel sand gravel stratum had strong geological adaptability, but the wear condition showed that the structural design could be further optimized. According to the typical shapes of three kinds of tools after wear or fracture, the shortcomings of tool design or process are analyzed. On this basis, the optimization design is proposed.

Table 6.6 Improved blade collapse design

Tool-changing ring number	Improved type I			Improved type II			Improved type III		
	Split number	Check total	Split probability	Split number	Check total	Split probability	Split number	Check total	Split probability
742 ring	–	–	–	0	1	0	–	–	–
792 ring	3	4	75.0%	1	3	33.3%	0	9	0.0%
850 ring	2	3	66.7%	4	9	40.0%	0	2	0.0%
885 ring	1	2	50.0%	2	4	50.0%	0	15	0.0%
942 ring	0	2	0.0%	0	1	0.0%	0	2	0.0%
1006 ring	1	2	50.0%	4	6	66.7%	1	13	7.7%
1133 ring	–	–	–	–	–	–	2	14	14.3%
1513 ring	–	–	–	0	1	100.0%	1	54	1.9%

6.1.3.5 Scraper Optimization Design

According to the practical engineering application performance of scraper, the scraper is further optimized based on the improved No. III. The front angle is changed from 5° to 0°, and the rear angle is changed from 5° to 15°, so as to reduce the tool wear. The main alloy tooth is changed from rectangle to L shape, and the amount of residual alloy is reduced, so that it not only has excellent engineering performance, but also meets the principle of saving and application. The proposed improved tool has not been used in engineering practice, and the actual engineering effect needs to be tested.

6.1.4 Tool Cutter Wear and Its Influencing Factors

Tool wear types are mainly divided into normal wear and abnormal wear. The normal wear is due to the friction between the tool and the rock and soil layer during the shield tunneling process, and the alloy and surfacing area on the tool surface are continuously subjected to friction loss. The abnormal wear is mainly due to the fracture of the alloy tooth or the partial wear of the hob during the cutting process, and the tool cannot be used again.

6.1.4.1 Tool Wear Analysis

Tool normal wear calculation formula:

$$\delta = \frac{L \cdot K_n \cdot N \cdot \pi \cdot D}{V} \tag{6.2}$$

δ—the amount of tool wear (mm);

L—drilling distance (km);

K_n—the comprehensive wear coefficient of the tool ($\times 10^{-3}$ mm/km) when the tool is placed in the same way as the trajectory;

D—tool outer diameter (m);

N—cutter speed (r/h);

V—advance speed (mm/min).

The relationship between the wear coefficient of the same cutting path layout knife and only one knife is:

$$\mathrm{K_n = K \cdot n^{-0.333}} \tag{6.3}$$

The formula is used to calculate the tool wear in Nanjing Yangtze River Tunnel (Wei Qi Road) project. In the formula, the tool alloy value is Japanese standard E3 cemented carbide (both toughness and hardness are moderate, which is suitable for

Table 6.7 Tool wear experience calculated value

Tool number	Tunneling distance L (km)	Cutterhead speed N (r/min)	Tunneling speed V (mm/min)	Calculated tool wear experience value K_n ($\times 10^{-3}$ mm/km)	Tool wear factor	Cutting tool outside diameter D (m)	Tool wear calculated value (mm)
28	1.3	0.75	25	12	1.39	14.96	3.43
32				9	1.53	12.99	3.23
26				3	2.21	10.59	3.86
30				3	2.21	8.19	2.97 m

sandy gravel stratum). The wear coefficient of scraper in clay silty fine sand stratum is 2.5×10^{-3} mm/km in slurry shield with the protection of the leading cutter, and 14×10^{-3} mm/km in sandy gravel stratum (the wear coefficient is the empirical data corresponding to the single knife with the same trajectory). At present, the project is in muddy powder. The excavation depth in clay is about 1100 m, and that in gravel is about 200 m. The weighted wear coefficient is calculated as follows:

$$\mathrm{K} = \left(2.5 \cdot 10^{-3} \cdot 11 + 14 \cdot 10^{-3} \cdot 2\right)/13 = 4.27 \cdot 10^{-3}\,\mathrm{mm/km} \tag{6.4}$$

Calculate the tool wear at different cutting radii as shown in Table 6.7.

The abnormal wear of the tool is due to the collision and abnormal cutting between the tool and the complex mixed strata such as hard rock or gravel layer with high abrasiveness in the cutting process, which leads to the failure of the alloy block of the tool before its wear resistance is fully played, so that the tool loses its cutting function. The abnormal wear of general tools includes the eccentric wear of hobs and scrapers, the fracture of alloy teeth, and the bearing fracture of hobs. Compared with normal wear, abnormal wear has obvious suddenness and irregularity.

6.1.4.2 Factors Affecting Tool Wear

In most shield projects with complex geological conditions, there are varying degrees of tool wear. In view of the various wear phenomena of the tool, combined with the strata where the tool wear occurs, the reasons for the occurrence are summarized and analyzed, and reasonable tool configuration and selection measures are carried out, so as to reduce the tool wear phenomenon and increase the effective cutting distance of the tool.

The normal wear of the tool is the result of the normal use of the tool. The wear is mainly due to the contact and sliding friction between the blade surface of the tool and the rock and soil body, which makes the blade surface of the tool lose continuously. The main phenomenon is the uniform wear of the alloy block.

There are many kinds of abnormal wear of cutting tools. The main reasons are improper tool configuration, improper tool selection and low quality of financial embedding. The improper tool configuration means that the selection of tool type is not suitable for the current formation. Such as Guangfo subway, hob cutting soft soil layer and produce eccentric wear. The improper selection of cutting tools is due to the section shape of the cutting tool is not suitable for cutting strata, such as Nanjing Wei Qi Road Yangtze River Tunnel crossing sandy cobble strata, due to the sharp blade of the scraper, resulting in a large number of fracture of the alloy block of the cutting tool. In the following, the causes are analyzed through the abnormal wear form of the tool. The eccentric wear of the hob is due to the insufficient friction force given to the hob in the soft soil layer, so that the hob does not rotate, so only one side of the hob cuts the stratum, resulting in the arc on the cutter ring being flattened and the eccentric wear occurs, as shown in Fig. 6.29. After the hob is eccentrically worn, the cutter ring is ground into a plane, resulting in smaller contact area between the palm surface and the hob, thus giving the hob less friction, and the plane is more difficult to rotate than the arc surface, resulting in more serious abnormal wear. In addition, when the shield tunneling in the soft and hard uneven strata, the cutter ring of the hob collides in the rock layer of the soft and hard interface. Because the cutter ring is brittle and hard alloy material, it is prone to alloy cracking, which will seriously lead to the fracture of the bearing due to excessive bearing torque. If the damaged hob is not replaced in time, the adjacent hob will bear more load, resulting in more serious damage.

The abnormal wear of scraper mainly occurs in hard rock and uneven soft and hard strata. The scraper collides with the rock in the palm surface, which will lead to tool damage such as the fracture and fracture of alloy teeth, as shown in Fig. 6.30. In

Fig. 6.29 Partial grinding of hob and fracture of knife ring

(a) The alloy teeth of the scraper (b) Excess wear of blade alloy teeth

Fig. 6.30 Blade's abnormal wear

the tunneling process, alloy teeth also fall off due to excessive wear in the surfacing area. Once the alloy tooth is damaged, the wear resistance of the scraper will be greatly reduced, and the subsequent tunneling will lead to excessive wear or partial wear of the cutter. In extreme cases, due to the failure of large rocks to enter the pressure chamber smoothly and to stop at the scraper, the scraper repeatedly cuts rocks, resulting in excessive wear or partial wear of the blade body.

6.2 Shield Fault and Detection Technology

Shield is the most advanced super-large special equipment for tunneling in the world. Today, with increasing emphasis on the development and utilization of underground space, shield construction has a very broad market prospect. In order to ensure the smooth progress of shield construction, it is necessary to continuously inspect and monitor the equipment status of the shield during the shield construction, master the status of the equipment, and provide guidance or prediction for the maintenance of the shield. It will avoid the sudden damage of the shield due to the lack of prior preparation, which will lead to the long-term shutdown of the shield to be repaired, affect the smooth progress of the construction and bring safety hazards. At the same time, due to the complex structure of the shield, it involves many specialties such as machinery, electricity and fluid. Once a fault occurs, how to analyze and judge the cause of the fault as soon as possible and eliminate the equipment fault as soon as possible is also an urgent problem to be solved in shield construction.

Due to the large number of shield equipment and strong correlation, the failure of any single set of equipment will affect the tunneling progress of the shield to varying degrees. The damage of key parts will lead to the stagnation or even paralysis of the whole shield, which will affect the progress of the whole project and greatly increase the engineering cost. Therefore, it is necessary to develop a set of shield maintenance procedures, carry out the research on the technology and method of condition monitoring and fault diagnosis, and quickly determine the fault location, so as to effectively shorten the maintenance time and improve the utilization rate of equipment. This section summarizes the common faults and maintenance methods of shield, and introduces the shield fault detection technology based on virtual instrument technology and oil analysis technology combined with Nanjing Yangtze River tunnel project.

6.2.1 Shield Tunnel Common Faults and Maintenance

6.2.1.1 Common Faults of Shield Tunneling and Treatment Methods

1) Shield manufacturing is unified, but in the face of various environmental challenges, unified configuration is always flawed and cannot cope with all environments. So the shield cannot move forward is generally caused by environmental conditions.

If the water content in the construction site is large, it will lead to the increase of soil viscosity. There are several serious problems when the shield is moving forward: the shield may squeeze the cohesive soil into a wall that is difficult to excavate and has a large density, which hinders the shield from moving forward. The soil surface with high moisture content will also be relatively smooth, resulting in the failure of the shield cutterhead to find the digging point, which makes the cutterhead empty and white loss of energy power, and even causes the cutterhead to fall off due to empty rotation. When the moist soil enters the soil cabin, due to the large viscosity, it may even cause the soil cabin to be unable to discharge and clean up the internal soil in time, making the soil cabin blocked, and the shield cannot carry out the construction operation normally, so it is forced to stop moving forward. The earth pressure is too large, which makes the shield load exceed the standard, cannot be effectively recycled, and also makes the shield unable to move forward.

If the soil is hard and contains a large number of rock fragments, resulting in shield cutterhead working pressure, and may even damage the cutterhead, which is one of the important reasons for shield cannot continue to move forward. In dealing with such problems, it is generally considered that the characteristics of different geographical environments are taken into account in the construction of shields, and the corresponding maintenance and use equipment will be served for shield support to help shields face and deal with tunnel construction problems under various geographical conditions. For example, when wet soil is encountered, the artificial construction team strength can be added to help quickly clean up the wet soil with large viscosity, and help shield obstacles, so that normal work can be carried out. Some chemical agents can also be added to make the moist soil solidified, so that the shield can quickly and easily solve the problem of excavation obstacles. When facing the situation of hard soil, aiming at the improvement and innovation of cutterhead, the cutterhead is more solid, which can resist the problem of hard rock that is difficult to be broken. Stone crusher can also be used to assist the normal operation of shield.

2) Hydraulic system

Hydraulic technology is one of the most important operation links in shield. However, it is also the most prone to problems and faults. The most fatal problem is the contamination of oil in hydraulic components, which can lead to a series of component failures. Oil is tainted, and the temperature of the first tank filled with oil will rise sharply during operation, reaching the limit of overload, the tank may be scrapped.

Oil pollution may contain some particles, which will increase the friction between the components. When the friction is too fast, it may lead to burning and failure of the components. Oil pollution may lead to oil agglomeration, which will block the channel and filter of shield hydraulic system, so that the oil cannot flow smoothly, resulting in the hydraulic system can not work normally, and even cause damage and operation obstacles of other systems. In the face of this problem, the most fundamental operation is to timely clean up the oil, clean up the components, and replace the oil. Avoid oil pollution, from the most basic, the most fundamental place to solve the most serious problems. When cleaning components, the same oil is used to rinse them to ensure that lubrication between components prevent excessive friction damage. When replacing oil, you must use the same kind of oil to fill and supplement to prevent the possibility of chemical reactions between different oils. In the hydraulic system, if some parts are damaged, they must be repaired or replaced in time to prevent secondary pollution after oil supplement, resulting in paralysis of the hydraulic system and unable to be used again.

6.2.1.2 Maintenance of Shield Machine

The maintenance of shield machine follows the principle of paying equal attention to maintenance and putting prevention first, and adopts the combination of daily maintenance, regular maintenance and forced maintenance. Main contents of maintenance are: cleaning, fastening, adjusting, lubrication, anti-corrosion, etc., also known as "10 word practice".

(1) For the shield used for the first time, a comprehensive inspection and maintenance should be carried out after the completion of 100 m test driving. The hydraulic system filter should be replaced, and the gear oil driven by cutter head should be replaced. During the test period, the inspection of equipment

should be strengthened, the frequency of inspection should be increased, and the precursor of possible failure in the first load test should be found in time to avoid unnecessary losses.

(2) The mechanical and electrical engineers of the project department should inspect the operation status of the shield and the supporting equipment every day, check and confirm the "Operation records of shield machine" filled out by the shield operators and the "Daily maintenance records of shield machine" filled out by the maintenance personnel item by item, find out the hidden trouble in time, and do a good job of preventive maintenance to keep the shield in good condition.

(3) It is forbidden to dismantle and change the components on the shield without authorization. It is forbidden to change the components for different purposes at will. It is forbidden to dismantle sleeves for hydraulic and electrical control components. It is forbidden to dismantle safety and indicative sensors, safety valves, instruments, relief valves, etc. If damaged or inaccurate, it should be found and replaced in time.

(4) The maintenance of shield must be carried out under the premise of ensuring safety. Only when the shield stops running, can the maintenance of shield be carried out. In electrical maintenance, the electrical switch should be disconnected, and the warning sign of "prohibiting closing" should be suspended. After the end of maintenance, the warning sign is confirmed by the electrician who suspends the warning sign and removes the warning sign to close the switch. High voltage electrical equipment maintenance should discharge first. Before the maintenance of the hydraulic system, the relevant valves must be closed and the pressure must be reduced to prevent the hydraulic cylinder from retracting and the accidental operation of the hydraulic motor. Hydraulic system maintenance must pay attention to cleaning, strictly prohibit the use of cotton yarn and other easy to wool articles cleaning pipe joint inner wall, grease bucket, pipe, etc.

(5) The maintenance of shield must first be familiar with its working principle and the relationship between electrical, hydraulic and mechanical, as well as the role of various parts in the equipment system, so as to solve the problem quickly and accurately. Equipment maintenance should follow the principle of "looking" first, "thinking" later, and "doing" again. When a machine fails, it is necessary to first see what fault is, and do not blindly do so, so as not to affect the maintenance speed or even expand the problem, and find the root of the problem before doing it. Equipment maintenance should be external and internal, such as in electrical maintenance, first check whether the wire is broken, broken, shedding and other obvious problems, often simple and obvious problems are most likely to be ignored. Do a good job of equipment fault handling records and summarize to improve the speed of similar fault handling.

(6) The daily maintenance and maintenance of equipment are also very important. Proper maintenance can reduce the failure rate of equipment, improve the service life of equipment, and increase production and efficiency.

(7) When the shield machine is suspended for a long time, necessary inspection and maintenance should be carried out, and the following work should be done: the storage site should be hardened, and there should be a roof for rain and sun protection. Apply grease on the joint surface exposed to air, attach it with felt and cover it with rainproof cloth. Regular lubrication maintenance. Winter to do a good job of cold and antifreeze.

6.2.2 Shield State Monitoring and Fault Diagnosis Technology Based on Virtual Instrument Technology

In shield construction, in order to facilitate operators to grasp the real-time status of the whole machine, the shield generally integrates a set of complex monitoring systems, and sometimes thousands of monitoring points, including propulsion system, assembly system, hydraulic system, grouting system, slurry system and lubrication system. These monitoring parameters are generally set the alarm threshold, such as the filter pressure differential alarm, sewage tank level alarm, and most of the parameters are simply feedback, display, by the operator with experience to determine its state. It is difficult for some key parts to judge their health status and make early prediction only by these feedback parameters. Such as the detection of hydraulic pump, shield only feedback its output pressure, only from the pressure observation, can not accurately determine its state, especially in the early stage of the fault. Therefore, it is necessary to develop a set of precision detection instruments to make up for the deficiency of the shield self-monitoring system. This section introduces a shield state detection and fault diagnosis technology based on virtual instrument technology combined with Nanjing Yangtze River Tunnel project.

6.2.2.1 Virtual Instrument Technology

The concept of virtual instrument was first proposed by National Instruments Corporation in 1986. The so-called virtual instrument (VI) combines the existing mainstream computer technology with flexible and easy-to-use software hardware to establish a powerful, flexible and changeable computer-based test measurement and control system, that is, to realize the function of traditional hardware with software.

The introduction of virtual instrument technology into the development of the system, not only saves the cost, but also greatly reduces the system development cycle, hardware and software system development time is less than two months. Through the modular management mode, the upgrade of the detection system is more flexible, which is convenient for the integration of new functions in the future.

6.2.2.2 Overall System Design

1) Test object

Carry out state detection, first determine the detection object. The structure of shield is complex, the probability of failure in each part is different, and the influence on the construction is different. Therefore, the system analysis of each part is carried out before determining the detection object, and the components with high risk, large fault loss or difficult repair after damage are selected as the key detection objects of this system. Vibration detection technology is most suitable for rotating equipment detection, and the key parts affecting shield tunneling are rotating parts, such as cutterhead drive system, propulsion system, slurry circulation system, etc. These components can be classified into four categories, namely motor, gearbox, hydraulic pump and bearing. The statistics of monitoring objects are shown in Table 6.8.

2) Measuring point selection

The quality of the measurement point determines the quality of the acquisition signal, so you need to do this step carefully. The selection of screening and measuring points follows the following five principles:

Table 6.8 Vibration detection object of shield

Type	Device	Quantity (number)
Electrical machine	Main drive motor	15
	Hydraulic pump motor	8
	Mud pump motor	7
	Mortar filling machine	4
	Blowdown pump motor	1
	Fan motor	1
	Gear oil cooling motor	2
Gearbox	Main drive gearbox	15
	Mortar filling gearbox	4
Oil hydraulic pump	Pushing hydraulic pump	2
	Assembly hydraulic pump	2
	Auxiliary hydraulic pump	2
	Grouting pump hydraulic pump	2
	Servo valve hydraulic pump	1
	Copying knife hydraulic pump	1
	Hydraulic crusher pump	1
	Bentonite nozzle hydraulic pump	2
Bearing	Mud pump bearing	2
	Motor bearing	10
	Mud pump bearing	78

Table 6.9 Vibration monitoring scheme

Monitoring method	Period	Remark
Simple measurement vibration	50d or 50 ring	Find abnormal, shorten the detection period, and focus on precision detection
Precision measurement of vibration	15d or 150 ring	

(1) The surface of the measuring point should be as smooth as possible and the sensor should be installed securely.
(2) The measuring point should be as close as possible to the measured vibration source, so that the vibration signal has the least attenuation during the propagation process.
(3) Prefer the position near the bearing housing.
(4) More points should be arranged near the faulty multiple components.
(5) Low frequency (below 1000 Hz) takes into account three directions of horizontal, vertical and axial directions; high frequency (1000 Hz or higher) can be used in one direction because the direction sensitivity of high frequency vibration signals is poor.

The measuring points of each bearing are the horizontal radial, vertical radial and axial positions of the bearing. The bearings inside the equipment, such as motor bearings and gearbox bearings, are not monitored separately, and the inspection of the corresponding equipment is the main.

3) Testing plan

It can be seen from Table 6.8 that the number of detected objects of shield is large and the position is dispersed. In order to reduce the cost of detection system and the workload of detection, the system adopts portable mobile measurement and cooperates with a simple vibration measuring instrument for inspection. The specific detection scheme is shown in Table 6.9. The data of vibration detection need to be systematically managed, especially the simple vibration measuring instrument, which needs to record the data manually and enter into the computer for preservation. In practical application, it is necessary to use infrared thermometer, oil analyzer, mechanical fault hearing device and other simple equipment to grasp the health status of the equipment through various means.

6.2.2.3 System Hardware Configuration

The hardware part of the shield vibration detection system includes vibration acceleration sensor, signal conditioning module, signal acquisition card and notebook computer.

The sensor is used to convert real physical signals into electrical signals. Because the electrical signals output by the sensor are usually weak and accompanied by a large number of noise signals, it is necessary to use signal conditioning devices to

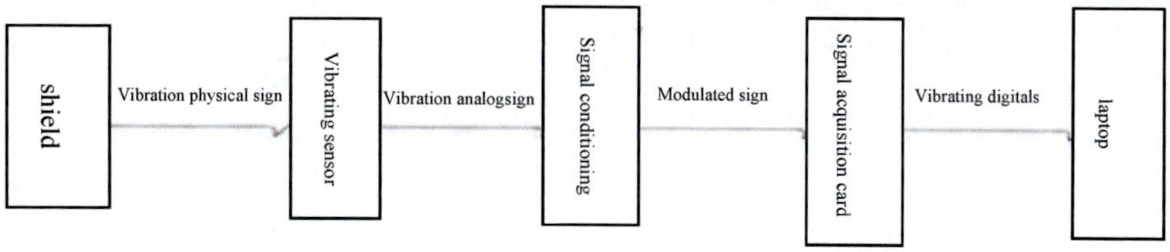

Fig. 6.31 The hardware structure of shield vibration detection system composition

Table 6.10 Vibration frequency and amplitude range of each equipment of shield

Device type	Normal		Abnormal (five times normal level)	
	Frequency (Hz)	Amplitude (g)	Frequency (Hz)	Amplitude (g)
Gearbox	20–600	0.15–1.5	100–3000	0.75–7.5
Electrical machinery	10–500	0.10–2.0	50–2500	0.50–10.0
oil hydraulic pump	25–1500	0.50–3.5	125–7500	2.5–7.5
Bearing	1–500	0.10–2.0	5–2500	0.50–10.0

regulate them. The functions of signal conditioning devices generally include amplification, isolation, multiplexing, filtering, excitation and linearization. The acquisition card sends the modulated signal to the computer, and then the computer processes, analyzes and displays it.

Figure 6.31 shows the hardware structure of the shield vibration detection system. The frequency of each equipment is calculated according to the data detected by the simple vibration tester. The statistics of the vibration frequency and amplitude range of the key equipment of the shield are shown in Table 6.10. Considering the surrounding environment and system cost of the detection object, the magnetic seat installation method is adopted.

The hardware of the shield vibration detection system is shown in Table 6.11. The overall hardware of the system is compact and portable, and it can be operated by one person without external power supply. The battery runs for more than 4 h, which is sufficient to complete a complete detection. At the same time, the hardware has shielding function, and has certain waterproof, dustproof and seismic ability, which is suitable for complex environment.

6.2.2.4 Software System Development

LabVIEW is a powerful and flexible instrument and analysis software application excavation tool, which is an important part of virtual instrument technology. In the software part of the system, LabVIEW is used as the software platform and combined with the design and development of Access database. The functions of data acquisition, waveform storage, signal processing, data analysis and report generation are completed. Using the characteristics of LabVIEW itself, the modular design and

Table 6.11 Shield vibration detection hardware system

Name	Quantity	Type	Major parameter
Vibration acceleration sensor	3	LC0104T	Measuring range: 50 g; frequency: 0.5–9000 Hz; accuracy: 200 mV g^{-1}
Signal conditioning module	3	CM3501	Fourth order anti-mixing filtering; built-in lithium battery
Signal acquisition card	1	DAQCard-6024E	16 single-ended input channels; Maximum sampling rate is 200 kb s^{-1}
Laptop	1	Dell D5400	PCMCIA slot; nine core batteries
Magnet base	3	GZ-4	Suction is no less than 8 kg

development of software are realized, and the efficiency of program development is greatly improved. The functions of the software part of the system are shown in Fig. 6.32.

6.2.2.5 Engineering Application

The system has been applied in Nanjing Yangtze River Tunnel since February 2009 to guide the full-line operation of the two tunnels on August 22, 2009. During the monitoring process, too many misjudgments were eliminated, and the expensive shutdown was avoided. The detected objects were in good condition and no obvious abnormalities were found. For example, on March 17, 2009, a high temperature of P1.1B shaft sleeve bearing occurred, which exceeded the daily temperature by about 20 °C. Foreign experts in field work preliminarily judged that the motor was eccentric or different from the load. The system is used to detect the motor. Table 6.12 is the radial vibration time domain statistics of the bearing position at the output end of the motor. By comparing with the vibration value of the motor under normal state, it is found that the amplitude of the statistics increases little, some also appear negative growth, and there is no abnormal increase in statistics.

The spectrum diagram is further observed, as shown in Fig. 6.33. If the motor has eccentricity or different load centers, the double frequency of the motor will increase with sideband, and the double grid frequency will also increase significantly with sideband. When the speed of the motor is 398 r/min, the rotating frequency is 6.63 Hz, compared with the spectrum of the motor under normal condition, it can be found that the frequency change of twice the grid is not obvious, and the amplitude change of twice the rotating frequency is not large. Later investigation is due to too much lubricating oil injected into the shaft sleeve, resulting in high temperature bearing.

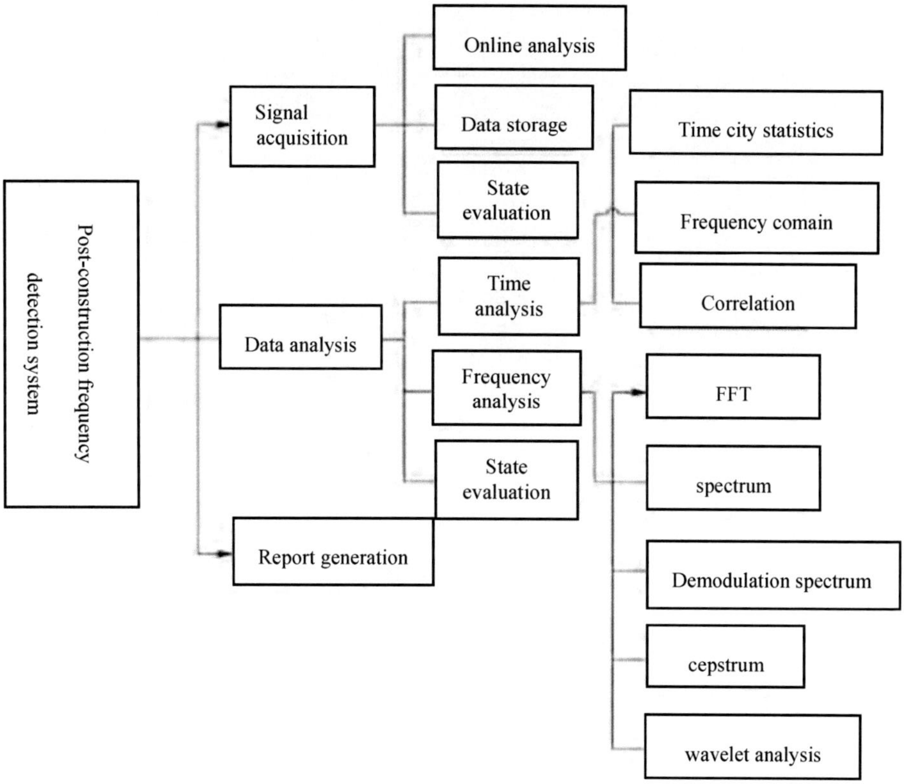

Fig. 6.32 Vibration detection software function of shield

6.2.3 Shield Fault Diagnosis Technology and Application Based on Oil Analysis

In order to ensure the normal construction of shield tunneling, through the analysis of hydraulic oil and gear oil, without disassembly of equipment and without affecting the normal construction, the fault prediction of hydraulic system and main drive gear system is realized, which provides scientific basis for the maintenance of relevant departments.

6.2.3.1 Sampling

The oil sample is the main basis for oil analysis, and all fault information is integrated into it. Therefore, whether the oil extraction specification directly determines the monitoring conclusion of the equipment.

Table 6.12 P1.1B motor vibration time domain statistics

State	Dimensional						Dimensionless		
	Accelerated speed (m^2/s)	Speed (m/s)	Displacement (mm)	Maximum (m/s)	Minimum (m/s)	Mean square root (m^2/s)	Kurtosis	Vibration severity	Peak metric
Normal	3.115	0.181	0.007	13.958	−10.965	3.693	−0.382	0.287	3.777
Abnormal	2.5896	0.423	0.001	13.418	−12.793	3.927	−0.126	0.419	3.417
Amplification	−0.519	0.242	0.003	−0.530	−1.828	0.234	0.556	0.132	−0.360

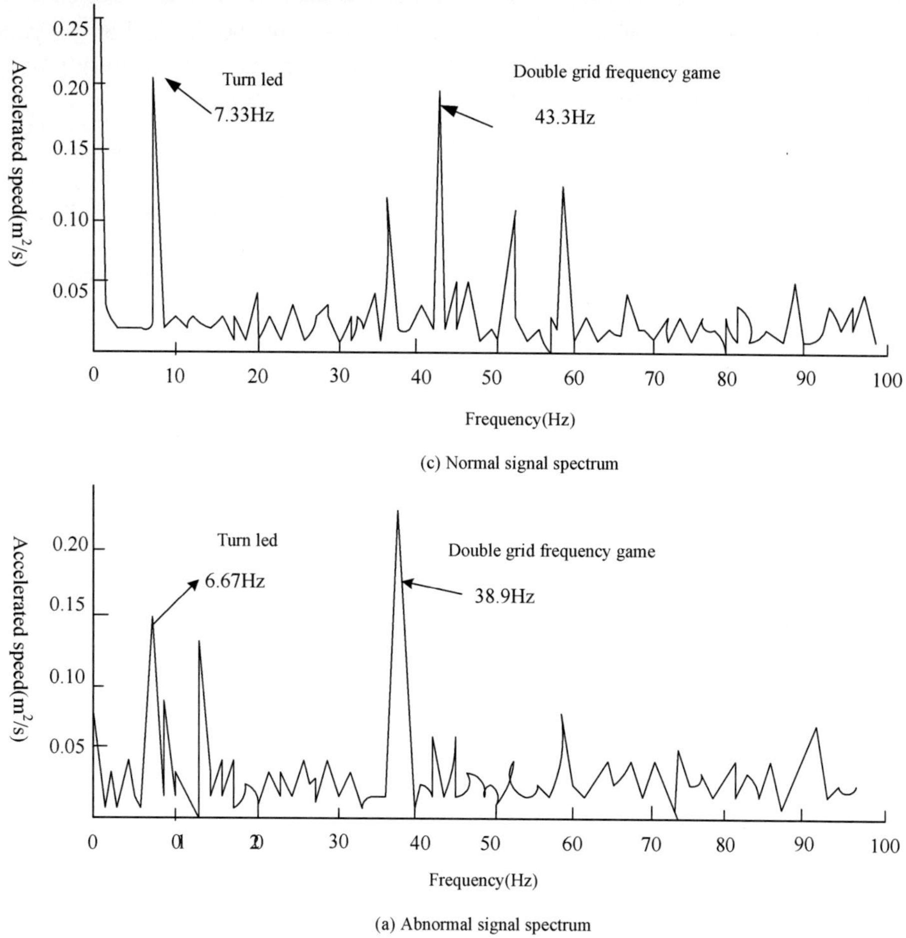

Fig. 6.33 Spectrum of motor vibration

1) Sampling conditions

Abrasives and impurities entering the lubricating oil mostly exist in the form of organic compounds, colloidal suspensions and free particles, some hanging in the inner wall, some in the oil, and some deposited in the oil bottom when the lubricating oil is static. Therefore, the oil extraction must be in the shield tunneling state or just stop tunneling, to ensure that the abrasive particles in a just uniform suspension state.

2) The sampling location

Reasonable location should be selected after the hydraulic system and the friction pair of the main drive gear system and before the filter.

3) Sampling tools and sampling quantities

Use the special tools for sampling to take oil. The sample bottles and oil collection pipes are disposable and discarded after use. There are many items considering oil analysis, and the sampling amount is set to 200 mL each time.

4) Sampling cycle

The oil sampling period is one month. If the equipment has abnormal wear, the sampling period should be shortened as appropriate. The main drive gear ring and hydraulic system wear rate is relatively low, can be appropriately extended.

6.2.3.2 Oil Analysis

The main items of oil analysis include: kinematic viscosity, iron spectrum, moisture, pollution and spectral analysis, where spectral analysis requires outbound inspection.

1) Kinematic viscosity and moisture analysis

The detection standard of oil kinematic viscosity: the change rate at 40 °C should not exceed 10% of the normal kinematic viscosity. If the viscosity is too low, it will cause lubrication failure and serious wear. Too high viscosity will make the oil temperature rise, resulting in oxidation of the lubricating surface, but also increase power consumption, resulting in energy waste.

Moisture detection standard: moisture content should be no more than 0.1%. Excessive moisture content can cause corrosion of sensitive parts and cause malfunctions.

2) Oil sample iron spectrum analysis

Ferrography is used to separate and detect abrasive particles and debris from lubricating oil samples (greases) by ferrography, so as to analyze and judge the wear type, wear degree and wear position of the surface of the machine moving pair. The direct reading ferrography technology is used to analyze the ferrography of oil samples.

(1) Working principle of direct reading type ferrography.

The wear debris was separated from the oil by a direct-reading ferrograph and deposited on the deposition tube according to the size for analysis by the tester. Magnetic field device is a high strength and high gradient magnetic field. Under the action of magnetic field, large particles first deposit, and small particles deposit with the oil flow to a distance. Photoelectric sensors installed in different locations read out the number of different sizes of abrasive particles.

(2) Direct reading type of ferromagnetic characteristic parameters.

① Large abrasive reading D_l. D_l is the first reading of the photoelectric sensor read by the direct reading spectrometer. It represents the relative concentration of large abrasive particles with a diameter greater than 5 μm and is one of the main parameters for rating mechanical wear.

② Small abrasive readings D_S. D_s is also the original reading directly displayed by the direct reading spectrometer, representing the relative concentration of small abrasive particles (1–2 μm), can also reflect the iron content of the oil sample.

③ Total wear amount Q. Q is the sum of the large abrasive particles and the small abrasive grain readings, $Q = D_t + D_S$. It reflects the relative amount of wear, and its rate of change reflects the rate of development of wear.

④ Wear intensity ls. Ls is the product of total wear and wear severity, i.e. $ls = (D_t + D_{s)}\ (D_t - D_{s)}$, which is related to the severity of wear. Under normal friction and wear conditions, the values of D_t and D_s are close. Only under abnormal wear conditions will the D_t value greatly exceed the D_s value. Therefore, the appearance of the ls peak is a sign of an accident.

⑤ Major abrasive percentages *PLP*. The *PLP* reaction large abrasive grains accounted for the specific gravity of the total amount of abrasive grains, $PLP = [D_t/(D_t + D_{s)}] \times 100\%$. When the equipment is abnormally worn, the number of large abrasive grains will increase sharply, and the *PLP* value will increase significantly. Therefore, *PLP* is very sensitive to the monitoring of large abrasive grains.

⑥ Abrasive concentration *PLP*. $PLP = (D_t + D_{s)}$/sample size, which represents the total amount of wear per ml of undiluted sample. The sharp increase in *PLP* marks the appearance of abnormal wear.

3) Iron spectrum abrasive grain characteristics

The wear process typically produces six basic types of abrasive particles, including combinations of ferromagnetic and non-ferromagnetic particles.

(1) Normal sliding particles. The normal sliding particles are the result of normal wear, and the abrasive particles from the shear mixing layer are thin flakes with smooth surface, the long axis size is 0.5–15 μm, and the thickness is 0.15–1 μm.

(2) Cutting abrasive grains. Cutting abrasive is the result of one surface penetrating another surface, there are two ways to produce this result. First, a relatively hard component may be misalignment or fracture, resulting in sharp edges penetrating into a soft surface. Secondly, the hard abrasive particles in the lubrication system stand out from the soft surface and pierce into another worn surface. If a system is growing large (50 μm long) abrasive particles, it indicates a component failure.

(3) Spherical abrasive grains.

Spherical abrasive particles. The diameter of spherical abrasive particles is about 3 pm, which has bright center and black band imaging characteristics under white light irradiation. Relative movement and rubbing are the characteristics of fatigue wear.

(4) Severe sliding. There are parallel stripes on the surface of severely worn particles, and these stripes are parallel to each other and parallel to the long axis of particles, which is the identification feature. They are generally greater than

15 μm in length and 5–30 μm in thickness. Severe sliding particles sometimes show tempered color, which may change the appearance of particles after heat treatment.

(5) Bearing abrasive grains. These unique particle types are related to the fatigue of rolling bearings. The fatigue debris particles originate from the delamination when the metal surface is dented or cracks expand. During the micro-cracking process, these particles reach the maximum size of 100 μm. The fatigue crack is generally flat, and the main length-thickness ratio is 10:1. The surface is flat and the circumference is random irregular.

(6) Gear wear. Gears involve two types of wear: the fatigue particles of pitch circle are derived from the gear pitch circle, which are very similar to the fatigue particles of rolling bearings. They generally have flat surfaces and are usually irregular shapes. Depending on the design of the gear, the main size and thickness ratio of the particles are between 4:1 and 10:1. The short fat particles are produced by the tensile force on the surface of the gear, so that the fatigue fracture extends into the gear teeth before cracking.

4) Spectral analysis characteristic parameters

The characteristic parameter of spectral analysis is the concentration of various elements in oil. Its quantitative unit is mg/L, commonly known as PPM.

After a large number of tests, certain rules were found from the increase and decrease of element concentration. Wear metal components are: iron, lead, copper, tin, aluminum, nickel, silver, molybdenum, magnesium, zinc, titanium, tungsten, antimony, vanadium, etc. The pollutants are mainly composed of silicon, boron, sodium and potassium. Oil additives are mainly composed of copper, magnesium, calcium barium, zinc, molybdenum, antimony, silicon, boron and phosphorus.

6.2.3.3 Engineering Application

From September 2009 to July 2010, the lubrication and wear status of the main bearing of the left shield in a project department was tracked and monitored by the Equipment Condition Monitoring Institute of Guangzhou Institute of Mechanical Research using oil analysis technology. The project department adopted slurry air cushion balance shield with excavation diameter of 11.92 m. The main bearing lubrication system of the shield adopts BP Energol GR-XP 680 gear oil and adopts centralized lubrication mode.

When sampling and analyzing on November 23, 2009, the viscosity of oil sample increased significantly, far beyond the allowable range of viscosity of lubricating oil in use. Wear metal elements Fe, Cu content beyond the normal range, as shown in Fig. 6.34. Elements Mo, Ca, Na content increased significantly, as shown in Fig. 6.35. According to the results of oil monitoring and analysis, it is preliminarily judged that the oil is polluted by external contamination, which causes viscosity, Mo, Ca content increased sharply.

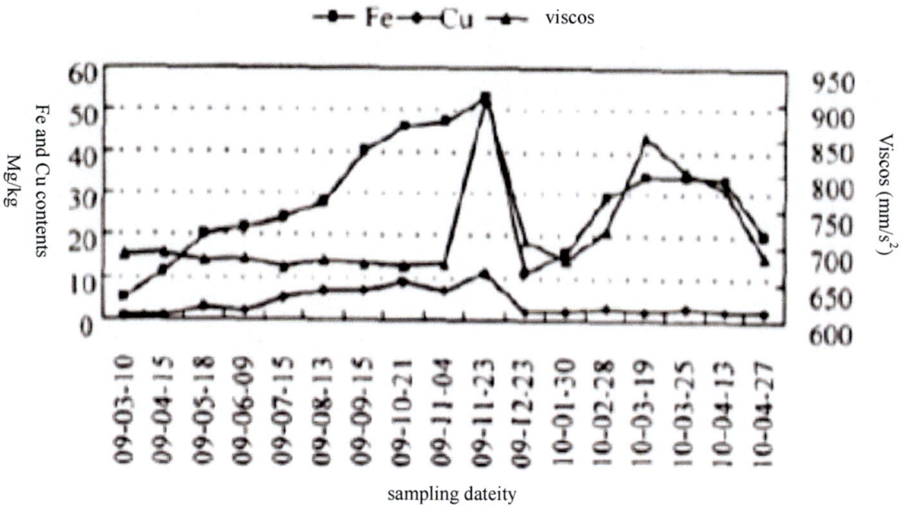

Fig. 6.34 The variation trend of Fe and Cu content and viscosity at 40 °C

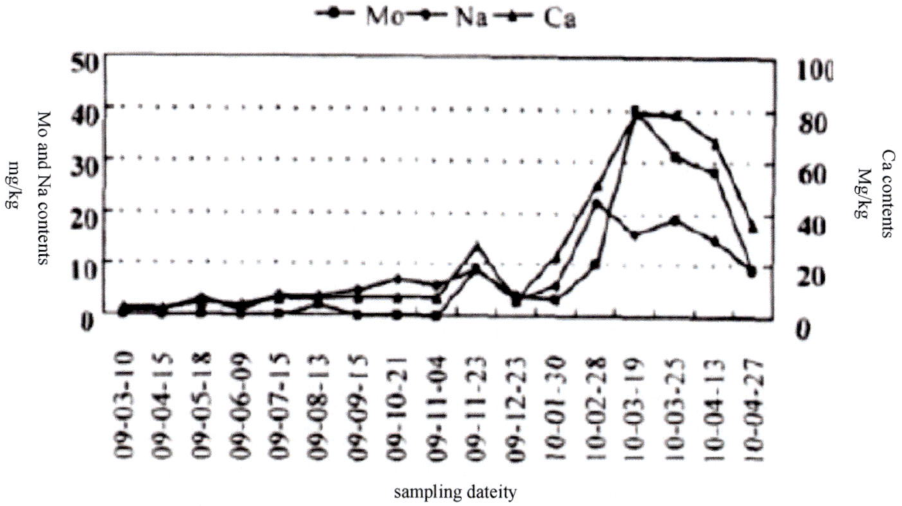

Fig. 6.35 Variation trend of Mo, Ca and Na contents

In order to determine the source of pollution, the possible external pollution sources such as groundwater, slurry water and slag slurry in the construction site were sampled and analyzed, and the BP Energol, GR2XP680 new oil, HBW new and EP2 new grease were detected and analyzed. The results are shown in Tables 6.13 and 6.14.

Table 6.13 Spectral analysis of new oil and pollutants

Element	BP680 new oil	EP2 new fat	HBW new fat	External circulating water	Muddy water	Slag slurry
Fe	0	14	>1200	1	120	>1200
Sn	0	2	5	40	37	40
Si	0	136	>1200	4	321	>1200
Mo	0	232	14	0	0	71
Al	0	26	950	0	67	1126
Mg	0	16	>1200	>1200	>1200	>1200
Na	0	184	38	162	192	205
Ti	0	0	20	0	2	55
Ca	6	33.36	170	433	2479	1954

Table 6.14 Periphery analysis of solid particles of new oil and pollutants

Description of sample	Analysis result
BP680 new oil	NULL
EP2 new fat	A small amount of dust particles
HBW new fat	More uniform, white spherical crystalline particles and flocculent, massively agglomerated fibrous materials
Slag slurry	More dust, different crystal materials and other impurities

Combined with the analysis in Tables 6.13 and 6.14, the main sources of Mo and Ca are EP2 grease, and Na mainly occurs in slag slurry, slurry water and EP2 grease. However, the content of Mg in slag slurry and slurry water is very high, and no increase in Mg content is found in this monitoring. Therefore, it can be inferred that the main reason for the increase in the viscosity of the bearing gear oil and the content of Mo, Na and Ca is the pollution of EP2 grease. After determining the source of pollution, the lubrication system of the main bearing was oil changed, and the oil tracking and monitoring of the unit were strengthened. According to the analysis results, corresponding measures were taken to meet the needs of normal lubrication of the bearing so that the shield can operate safely.

6.3 Replacement Technology of Pressured Cutter in Shield River

With the continuous application and development of shield technology in China, the river-crossing tunnel shows the characteristics of large diameter, long distance, high water pressure and soft and hard excavation surface. Long-distance tunneling in upper soft and lower hard strata puts forward higher requirements for the life of the cutter. It is inevitable to repair or replace the cutter in the tunneling process. However, due to the complexity and variability of the geological conditions and the limitations of the surrounding conditions, it is often difficult to implement atmospheric cabin entry by reinforcing the soil in front of the excavation. With the characteristics of no need for stratum reinforcement, strong geological adaptability, low requirements for the surrounding environment and small influence, the technology of pressurized cabin entry solves the problem of tool change under harsh conditions.

6.3.1 Technical Difficulties in Cutter Replacement of Shield Machine with Pressurized Inlet Cabin

The so-called pressure into the cabin refers to the pressure of compressed gas to replace the mud pressure or soil pressure in the pressure cabin. According to the principle of compressed air construction, the excavation face is supported by air pressure, and the operator enters the pressure cabin under a certain air pressure environment for operation. Although the construction of large-diameter slurry shield in China has been successful, the research on the cutter replacement technology with pressure in ultra-high water pressure (above 0.68 MPa) and strong permeable strata is still in its infancy, and there are still many difficulties in the operation of pressurized cabin entry. In general, there are the following points.

1) Stability of excavation face

The most basic and core issue in shield opening is the stability of the excavation face. In the process of cutter head repair, the high-pressure gas leakage in the working space must be effectively controlled to avoid the damage of environmental pressure fluctuations to the operators and a large number of gas leakages that cause the formation of seepage channels in a certain area of soil layer lead to soil instability, and the gas leakage per unit time exceeds the air supply per unit time of the equipment, resulting in the failure to maintain the spatial pressure. When the high permeability strata such as sand cannot maintain the stability of the air pressure, it is necessary to increase the air closure performance of the strata to complete the auxiliary construction. At present, the most commonly used method to reduce the permeability of soil is to inject mud into the front of the excavation, but when the bottom of the river with high water pressure and high permeability is opened, how to form a mud film with good air tightness is still an important problem in shield opening technology.

2) Personnel safety under pressure

When it is necessary to repair or replace the cutterhead and tools, especially for long-term operations such as welding operation, this kind of pressurized cabin opening generally lasts for several hours, and the excavation surface needs to be supported by compressed gas to form a local relatively stable working space. When foreign countries use compressed gas to maintain the stability of the excavation surface in the cabin opening, due to improper operation, the workers in the tunnel have decompression symptoms, such as ear membrane rupture and hearing impairment. At the same time, when conducting fire operations under compressed air conditions, a large number of harmful gases will be generated, which will threaten the personal safety of staff.

3) Welding repair operation under pressure

In order to ensure the welding quality, it is necessary to protect the molten pool to avoid contact with nitrogen and oxygen in the air, which is difficult to guarantee in high pressure environment. At the same time, compared with the ignition operation under atmospheric pressure, the influence of compressed air on the welding behavior and the possible combustible gas in the cabin should be considered in the welding repair of shield cutterhead in the compressed air environment. In the high pressure environment will cause welding arc column arc pressure increase, arc shrinkage and a series of problems, affect the cutter welding repair operation.

6.3.2 *Pressured Inlet Technology Based on Mud Film Closed Air*

In order to change the cutter for slurry shield with pressure into the chamber, it is usually necessary to reduce the liquid level of the slurry chamber to facilitate the operation of the workers with pressure into the chamber. At this time, the upper and middle soil layers in front of the shield excavation are all supported by gas. For the strong permeable strata, a large number of compressed air will carry fine particles through the pores between the strata to escape upwards. Increasing the pore size of the strata will cause instability of air pressure and cannot balance the water and soil pressure of the excavation face. In order to realize the stability of excavation face in strong permeable stratum, a layer of airtight mud film must be established on the surface of excavation face. This section takes Nanjing Yangtze River Tunnel Project as an example to introduce the pressurized cabin entry technology based on mud film air closure.

6.3.2.1 Air-Closing Technology of Mud Film

In the Nanjing Yangtze River Tunnel Project, after the excavation of the Nanjing Yangtze River Tunnel to the 650th ring, the excavation parameters are significantly increased, in which the torque is increased to 15 kN m, and the excavation speed is reduced to 10 mm/min. Giant granular soil, large blocks of ancient wood and pebbles above 20 cm are found in the discharged slag soil, and the excavation cannot be continued. In order to ensure the safety of equipment and long-distance construction, the shutdown inspection is carried out in the 659 ring. During the tunneling construction before the shield shutdown, the shield has entered the gravel land with strong permeability and high water pressure. In the layer, the polymer mud with high density and viscosity is used, and the mud filtration is less. The mud film is mud-skin mud film. However, the polymer slurry is not stable when it is shut down for a long time, and the slurry is prone to deterioration of quality such as aging, which will further affect the quality of slurry film formation. Therefore, the bentonite slurry with good stability should be used in the opening of the cabin. In other words, before opening the cabin, some adjustments must be made to the formula and basic parameters of the slurry, and the indoor permeability test and air tightness test of the slurry must be carried out to ensure the stability of the excavation face and the successful implementation of the cabin entry operation.

1) Laboratory permeability test of pure bentonite mud

In order to make the mud penetrate a certain distance in the stratum and improve the cohesive force of the soil on the excavation face, the Na grade bentonite was used to swell at the concentration of 3%, 5%, 8% and 12% respectively for 24 h. The specific gravity and viscosity of the mud are shown in Table 6.15.

In this experiment, the gravel sand layer that is the most unfavorable for film formation is selected as the test stratum, that is, the full-face gravel sand stratum is used to simulate the actual stratum. The coarse particles larger than 2 mm account for about 40% of the whole stratum content, and the permeability coefficient of the stratum is about 3.95×10^{-2} cm/s in the test, which is basically consistent with the permeability coefficient of gravel sand stratum in practical engineering.

Test instruments and test methods are shown in Figs. 6.36 and 6.37, and the pressure is 0.3 MPa at one time. The stability of mud infiltration, the amount of filter water and the forming time of mud film are measured as the evaluation indexes of the quality of mud film, and the form of mud film is observed. The smaller the filter

Table 6.15 Summary of basic properties and film forming test results of four groups of bentonite mud

Slurry concentration (%)	Specific gravity	Funnel viscosity (s)	Is mud skin formed	Is there a permeable zone	Formation time of mud film
3	1.03	17.6	No	10 cm permeable zone	15 s
5	1.04	18.9	No	8 cm permeable zone	About 10 s formed mud film
8	1.06	21.3	2 mm mud skin	3 cm permeable zone	About 2 s formed mud film
12	1.08	23.8	1 mm mud skin	2 cm permeable zone	About 2 s formed mud film

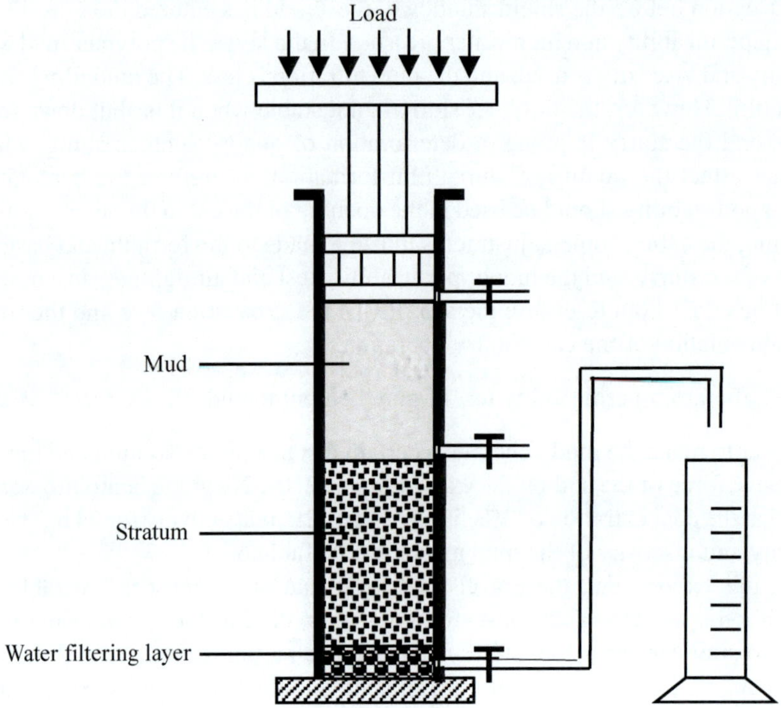

Fig. 6.36 Schematic diagram of mud film forming device

water, the shorter the film-forming time, the denser the mud film formed, and the better the quality of the mud film.

The basic properties of four kinds of mud and the results of film forming test in the formation are shown in Table 6.15.

It can be seen from Table 6.15 that the viscosity of bentonite mud with the concentration of 3%–5% is low and the specific gravity is less than 1.05, which can form a permeable zone mud film in the formation. However, the mud loss is relatively large and the film forming time is long. The viscosity of bentonite mud with 8%–12% concentration is more than 20 s. After the mud penetrates into the stratum for a certain distance, it quickly stabilizes, and the mud loss is less. The mud film of mud skin-permeability zone is formed (Fig. 6.38), and the film forming time is very short.

2) Mud film forming test for increasing specific gravity and viscosity

After forming a mud-skin permeable zone mud film on the excavation surface, the specific gravity of the mud was increased to 1.15–1.20, and the viscosity was more than 25 s. The mud film-forming test was carried out.

Because the proportion of pure bentonite mud is generally small, if the proportion of pure bentonite mud is adjusted to 1.15–1.20, the viscosity will be large and it is difficult to meet the pumping requirements. Therefore, low specific gravity bentonite slurry with high viscosity and natural clay slurry (the excess slurry stored when shield

Fig. 6.37 Film forming test of mud

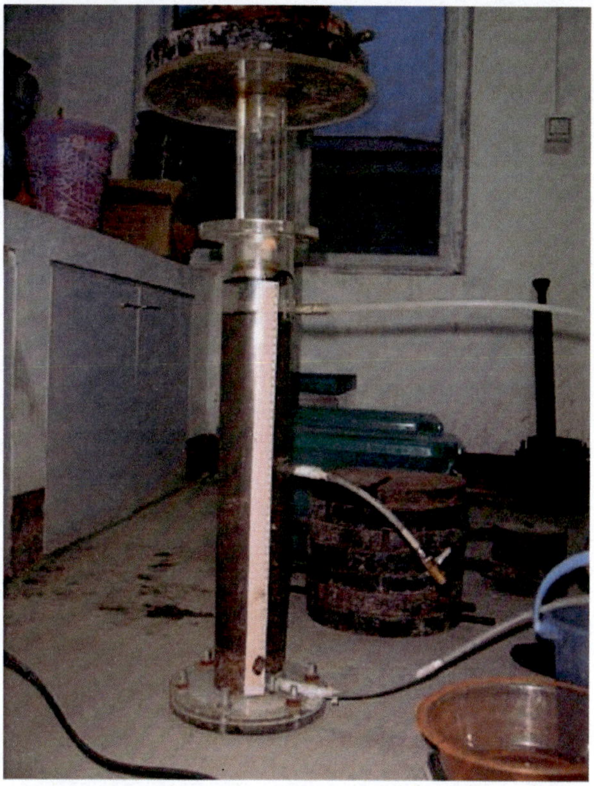

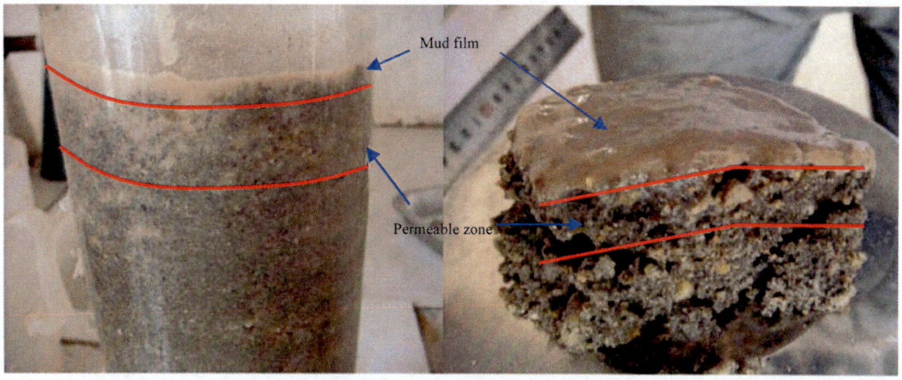

Fig. 6.38 Film formation of bentonite slurry with 12% concentration

passes through silty clay stratum) are mixed to configure the slurry with high physical stability with specific gravity of 1.15–1.20 and viscosity of more than 25 s.

(1) Physical stability of slurry. It is represented by the water evolution rate of 24 h slurry. The smaller the mud dewatering rate, the better its physical stability. The slurry was loaded into a measuring tube with a capacity of 100 mL, and

the volume of water released after 24 h was 1 mL, i. e., the water released rate after 24 h was 1%. The physical stability of the slurry was good.

(2) Mud film test. The test was carried out in the mixed layer of fine sand, gravel sand and pebble. The test method and procedure were the same as those of the above mud film test, but this test was graded pressure, and the variation curve of mud filtration water was plotted. The mud forms a mud film, as shown in Fig. 6.39, and the relationship between the amount of filtration water and time and mud pressure is shown in Fig. 6.40.

It can be seen from Fig. 6.40 that under the action of each stage of pressure, the filtration volume of slurry is very small, and the filtration volume per minute is

Fig. 6.39 Peat mud film

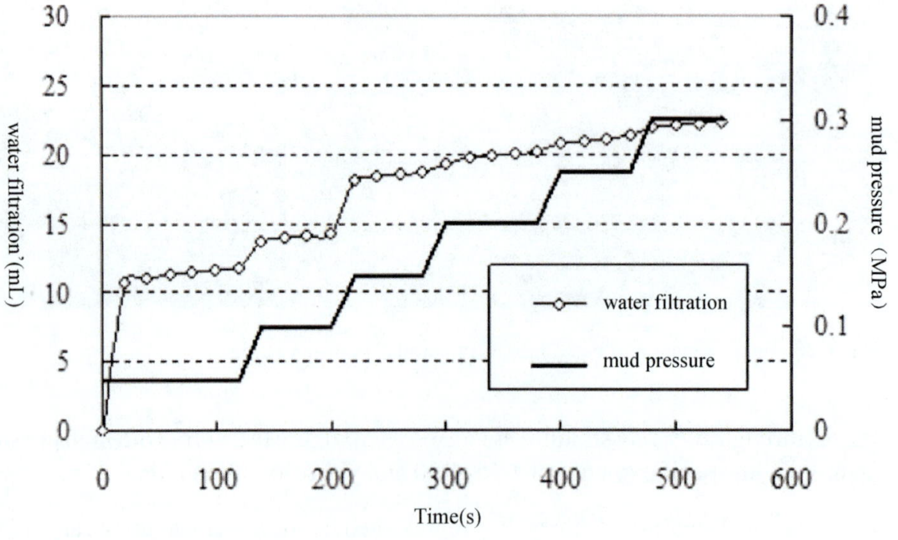

Fig. 6.40 Variation curve of filter water

about 0.6 mL. The filtration volume on the real excavation surface is about 3 m^3/2 h (meeting the empirical value of filtration volume: 10–20 m^3/2 h).

The results of the above experiments show that the bentonite slurry with the concentration of 8%–12% can quickly form a mud film containing permeability zone on the excavation surface. The mixed mud of bentonite and natural clay with specific gravity of 1.15–1.20 and viscosity of more than 25 s has good properties and high physical stability, which can quickly form micro permeable mud film on the excavation surface.

Therefore, when the Nanjing Yangtze River tunnel is opened in gravel stratum, the mud of the pressure chamber is replaced in two steps.

① Firstly, 8–12% bentonite mud is used as seepage mud to form a certain distance seepage zone in the excavation surface stratum and increase the cohesion of the stratum.

② The mud with specific gravity of 1.15–1.20 and viscosity of more than 25 s is used to form a mud film with mud skin type of micro-permeable on the excavation surface, which effectively acts on the excavation surface and reduces the loss of mud pressure.

3) Air tightness experiment of mud film

In order to facilitate the work of changing cutters and repairing cutter heads in the pressure chamber, it is necessary to appropriately reduce the mud slurry level in the excavation chamber and the pressure chamber according to the geological conditions, and support the excavation face with air pressure. The key to the pressurized cabin operation is to maintain the stability of the air pressure in the soil cabin. If the air tightness of the mud film is not good, the leakage will lead to sudden changes in air pressure, groundwater infiltration, and a large amount of flow sand, resulting in instability and collapse of the excavation face, which will endanger the safety of the cabin construction personnel. Therefore, before the construction personnel enter the cabin, the air tightness of the pressure cabin must be tested first.

Testing the air tightness of the pressure chamber is divided into two parts.

(1) Indoor simulation test: In the self-made mud film-forming device, the air tightness test was carried out on the formed mud film to test whether the quality of the mud film can effectively prevent gas leakage and meet the air tightness requirements of the pressure chamber, and then the mud formula was determined.

(2) Field test: on the basis of laboratory test, the air tightness of pressure chamber after grouting plugging is directly tested on the spot. The purpose is to observe the air supply interval of the air compressor, the speed of gas leakage in the pressure chamber, the change of pressure chamber pressure, the amount of residual soil and ground monitoring after the pressure is maintained, so as to analyze whether the excavation face is unstable and whether the pressure in the chamber can remain stable, so as to determine whether it can enter the cabin.

During the test, if the air pressure in the pressure cabin is stable at the set value, the starting frequency of the air compressor is stable, the air supply is normal, and there is no deformation and air leakage on the ground, which proves that the formation can achieve gas pressure stabilization. If there is no excess slag and soil taken out after the cabin cleaning, it indicates that the soil layer on the excavation surface is stable and no collapse occurs.

6.3.2.2 Pressure and Decompression Process

1) Pressing process

In order to pass the pressurized working area, people and tools need to enter the auxiliary cabin through the gate, where the air pressure should be increased first. The pressure process of shield opening in Nanjing Yangtze River Tunnel is introduced in the form of schematic diagram.

The shield diagram is shown in Fig. 6.41. It is defined that P_0 is the atmospheric pressure, P_1 is the incision pressure, namely the working cabin pressure, and P_2 is the pressure value between P_0 and P_1. In the initial state, only the working chamber (excavation chamber and pressure chamber) is under pressure, the pressure is P_1. If all pressure and decompression conditions are available, you can start working under pressure.

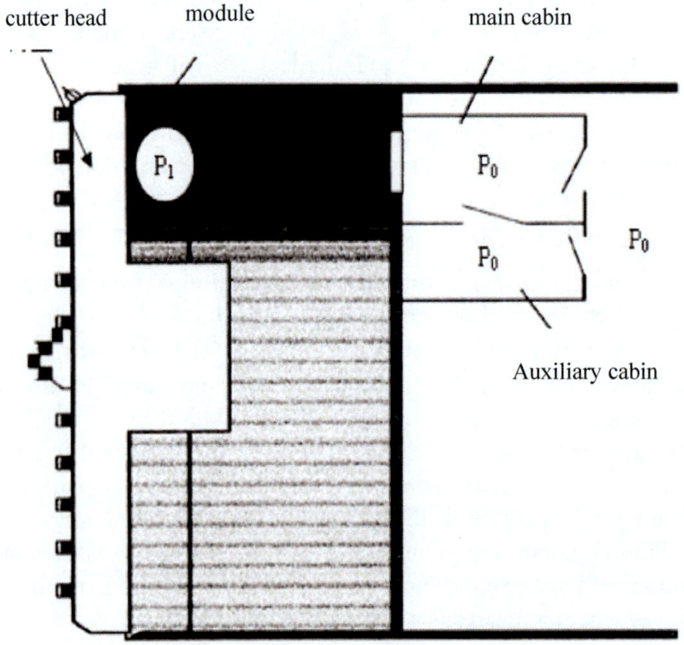

Fig. 6.41 Initial state of shield before opening

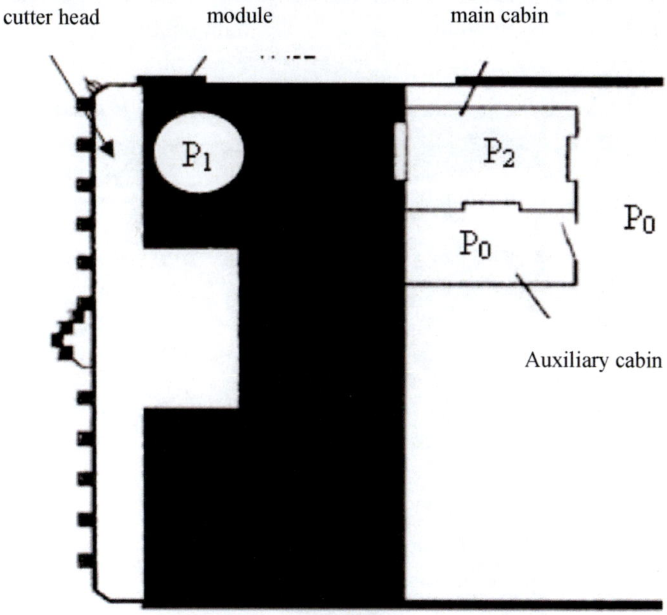

Fig. 6.42 Pressure schematic diagram of shield main cabin

First step, the main cabin is pressurized. As shown in Fig. 6.42, the main chamber pressure is increased from P_0 to P_2 by closing the main chamber gate and opening the main chamber valve, and then to the working chamber pressure P_1.

The second step, cabin pressure. The secondary cabin pressurization procedure is similar to the main cabin pressurization procedure as shown in Fig. 6.43. When the pressure of the main cabin is balanced with the pressure of the working cabin, the personnel enter the auxiliary cabin and begin to pressure the auxiliary cabin. In general, the main cabin is always pressurized first, and the auxiliary cabin always keeps the pressure relief state. Incoming personnel are pressurized in the secondary cabin to accommodate pressure in the main cabin and working cabin. The operation of adding personnel to the pressure operation area can be carried out by the pressure of the auxiliary cabin.

The third step, personnel into the cabin. When the pressure of the auxiliary cabin is balanced with the pressure of the main cabin and the working cabin, as shown in Fig. 6.44, and there is no adverse reaction for the personnel entering the cabin, they can enter the working cabin to repair the cutterhead and other equipment.

2) Decompression process

The entry staff must return within the prescribed time. The pressure time depends on the person who finally reaches the pressure compensation in the auxiliary cabin, and the time required to enter the auxiliary cabin is also calculated within the working time of the site. If the pressure decreases very slowly during decompression, the released gas can be discharged through the blood circulation and lung. If the pressure

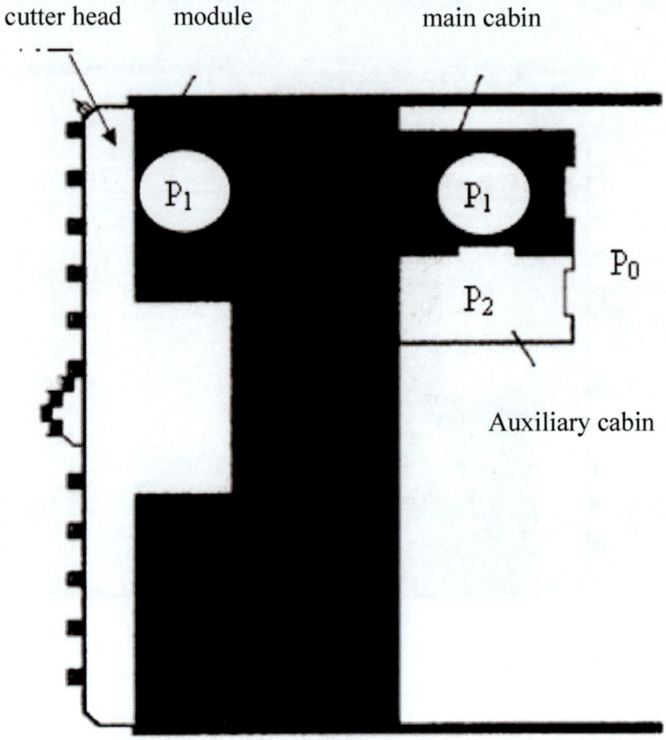

Fig. 6.43 Pressure schematic diagram of shield auxiliary cabin

is rapidly reduced, the bubbles will form water carbonate effect in body fluids and tissues. These can cause a series of decompression symptoms, releasing gas can cause temporary or permanent tissue damage, causing serious bodily harm to staff. Therefore, the decompression process and precautions should be strictly observed.

Similar to the pressure process, P_0 is defined as atmospheric pressure, P_1 is the working chamber pressure, and P_2 is the pressure value between P_0 and P_1.

In the initial state, as shown in Fig. 6.45, the working cabin and the main cabin are under pressure, the pressure is P_1, and the auxiliary cabin is under normal pressure. When all the conditions for staff and decompression are ready, decompression can be started. Normally, decompression is performed only through the main cabin. At the same time, the cabin is always in normal pressure.

As shown in Figs. 6.46 and 6.47, the main cabin pressure is reduced from P_1 to P_2, and then to P_0, and the decompression operation is completed.

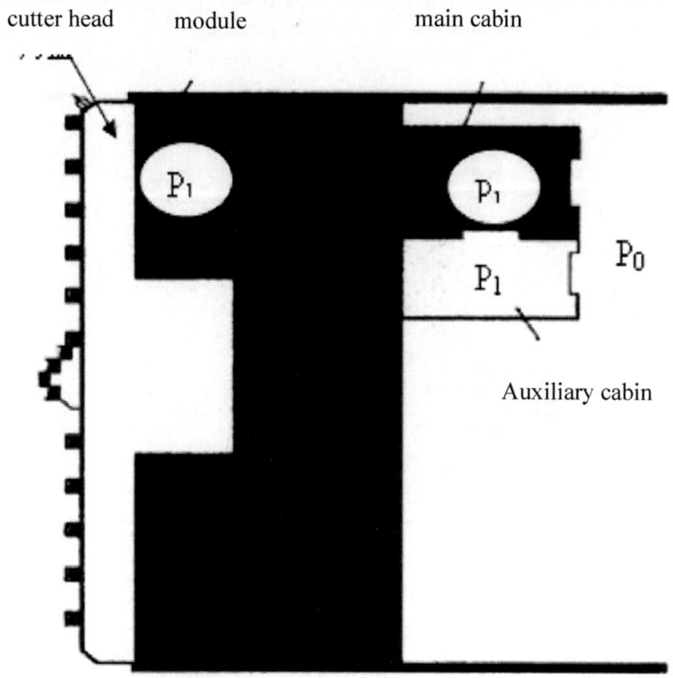

Fig. 6.44 Pressure completion of shield auxiliary cabin

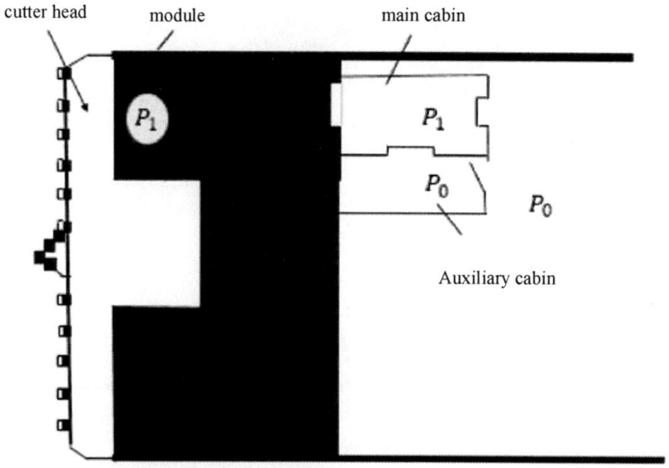

Fig. 6.45 State of shield before decompression operation

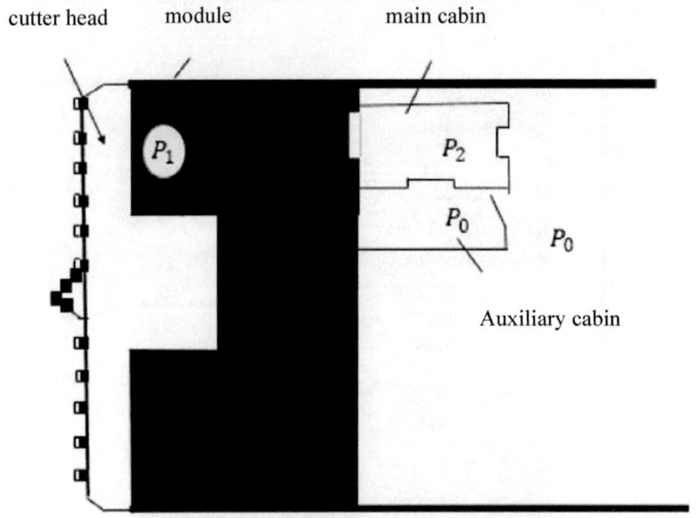

Fig. 6.46 Decompression process diagram of shield main cabin

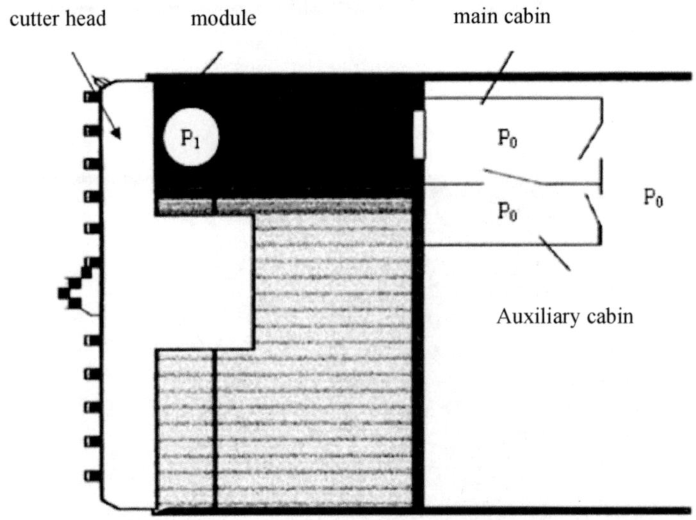

Fig. 6.47 End of shield decompression process

6.3.3 Replacement Technology of Shield Cutter Under Pressure

The earth pressure of the excavation surface is balanced by the air pressure instead of the earth pressure, and the stability of the air pressure on the excavation surface is used to provide the conditions for the construction personnel to enter the shield soil cabin safely. Air pressure stability can be roughly divided into three kinds: can prevent water gushing from the excavation face to prevent the collapse of the excavation face. Because the pressure on the palm surface can strengthen the stability of the palm surface. Due to the effect of compressed air on the surrounding rock gap to drain water, silt and clay layers are enhanced, the soil gap is reduced, and the formation stability is strengthened. The process of adding and reducing pressure is an important part of the tool changing process in the pressurized cabin. If the operation is improper, it is easy to cause the collapse of the excavation surface and the safety of the operators. Under the condition of compressed air, the personnel with pressure changer breathe compressed air, the general steps are as follows.

First of all, the pressure in the cabin is adjusted by the sequential pressure of the main and auxiliary cabins to prepare the personnel to enter the cabin. Secondly, when the pressure of the auxiliary cabin is balanced with the pressure of the main cabin and the working cabin, the operator enters the repair operation of the cutterhead and other equipment. Finally, after the completion of the operation, the decompression operation can be carried out. Generally, the decompression is only carried out through the main cabin, and the auxiliary cabin is always in the normal pressure state. After the completion of the decompression, the operator returns to the ground to rest under normal pressure.

This section takes Nanjing Yangtze River Tunnel as an example to illustrate the repair process of cutter head. The repair work of cutter head can be divided into two parts.

1) Preparation of cutterhead repair

(1) Template production. Before starting repair work, a template matching the original shovel seat should be made, including all bolt holes on the original shovel seat. The template will serve as a reference for the next repair of the tool holder. The cutter head repair is carried out by welding the blade holder, and the blade holder must be consistent with the wear shape of the outer edge region. In addition, the template will also be used for the lip welding of the edge blade. Templates are made of steel.

(2) Tool seat production. In order to repair the blade, two steel plates with right angles to the lip of the new blade are used as the blade seat (Fig. 6.48). The machining of tool holder must be carried out according to the shape of tool wear. To do this, you must first use templates. The welding seam should be reserved in the complete production process of the cutter seat, and the next step will be used in the repair work in the excavation chamber.

(3) Materials preparation. All materials used on TBM in each repair step should be prepared in advance. The materials include tools, welding equipment, blade holder, blade lip, screws, gaskets and all other necessities.

(4) Manufacture of outer edge shovel. The outer edge blade (blade) is connected with the adjacent blade by welding, as shown in Fig. 6.49. The purpose is to make full use of the remaining original screw holes to

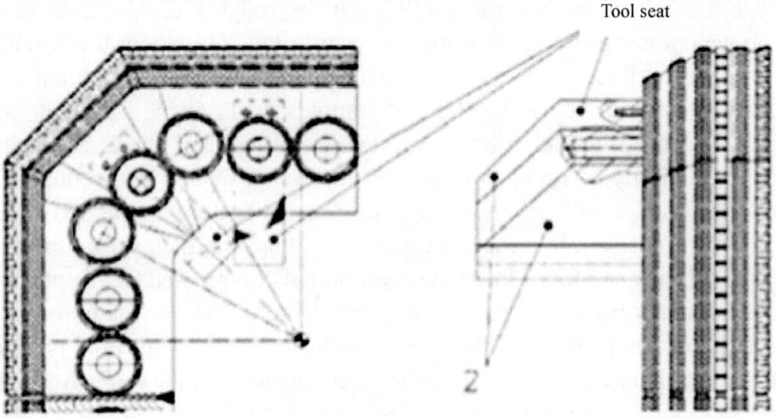

Fig. 6.48 The sketch of new shovel seat

Fig. 6.49 Welding blade

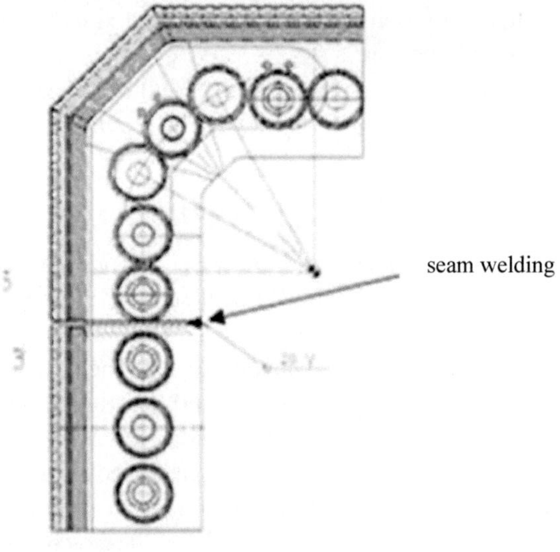

make the lip of the repaired blade more stable, and to make the positioning of the repaired blade base more accurate.

Before welding blade, the blade must be fixed to the template with screws. Welding operation should be carried out outside the excavation chamber, so that the weld can achieve the best welding quality. In the welding process, deformation control must be carried out continuously to avoid problems due to limited operating conditions when installing blade lips.

2) Disc repair process

Fig. 6.50 Adjustment schematic of cutter seat

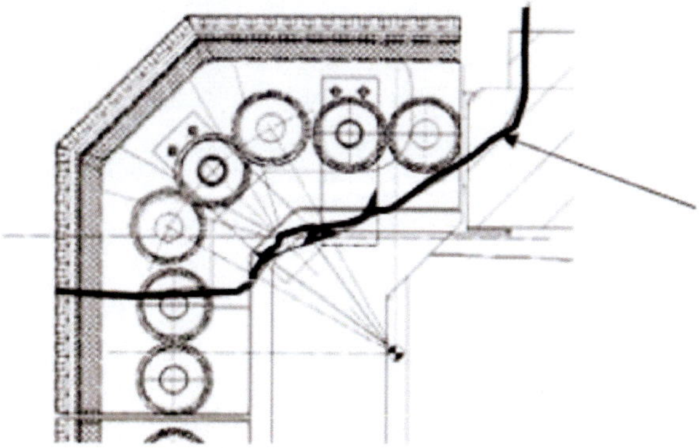

(1) Blade positioning. The repair of the blade holder will be carried out step by step, and only one blade in the blade arm will be repaired in one operation. Therefore, before entering the excavation cabin, the cutter head must be fixed at the position above the required repaired cutter arm. After the cutterhead is located, transmission facilities such as platforms and hangers should be installed to ensure operation safety.

(2) Demolition of worn blade. Before repair, the old shovel must be removed. The demolition part shall be marked in accordance with the installation location and must be transported out of the mud tank.

(3) Installation of shovel seat. When locating the new blade holder, the template should be fixed to the remaining screw hole of the original blade holder, and then the blade holder should be fixed to the template (Fig. 6.50). At this time, the tool holder should be finely adjusted locally until the tool holder is completely consistent with the worn steel structure. The precise alignment between the new tool holder and the template is crucial.

After adjustment, weld the tool holder to the cutter head. Then take off the template, free up the welding position, the new tool holder welding.

(4) Installation of new blade. After the welding of the shovel holder is completed, the new shovel must be installed in place and connected with bolts. Finally, the wear resistant block next to the tool arm must be rewelded.

Chapter 7
Construction Technology of Shield Tunnel Crossing the Middle Channel Section of Yangtze River

Due to the continuous development and application of large diameter slurry shield technology, the stability of excavation face in large-scale construction has aroused widespread concern in the engineering field. It is worth noting that the large-diameter slurry pressurized shield faces the following risks and problems in the construction of water area (river, lake and sea) crossing the channel section: the stability control of the excavation face is difficult, the splitting instability of the mud breakdown soil layer, the sand gushing and water gushing, the shield overexcavation, the tunnel floating and the segment damage. This makes maintaining the stability of excavation face become the top priority to ensure the safe and smooth construction of the tunnel under the conditions of special sections (punching section). The stability of the excavation face is achieved by balancing the water pressure and soil pressure of the stratum by the slurry pressure acting on the impermeable cement film. Therefore, the properties of the slurry and the mud film should be strictly controlled in the construction. This chapter compares and selects two construction schemes of shield crossing the middle channel section of the Yangtze River, analyzes the stability of excavation face through theoretical calculation and experimental simulation, and puts forward the method of selecting tunneling parameters and risk management countermeasures in engineering, which has important reference value for guiding practical construction.

7.1 Technical Difficulties and Schemes for Shield Tunneling Through River Channel

At present, the slurry pressurized shield method is almost the required method for river-crossing and sea-crossing tunnels. When the large diameter slurry pressurized shield machine is used to cross the river, it will inevitably cross some shallow soil sections (the thickness of the soil is less than twice the diameter of the shield). In the shallow overburden section of the middle Yangtze River, due to the high water

J. Chen et al., *Construction Technology of Large Diameter Underwater Shield Tunnel*, Key Technologies for Tunnel Construction under Complex Geological and Environmental Conditions, https://doi.org/10.1007/978-981-16-5896-9_7

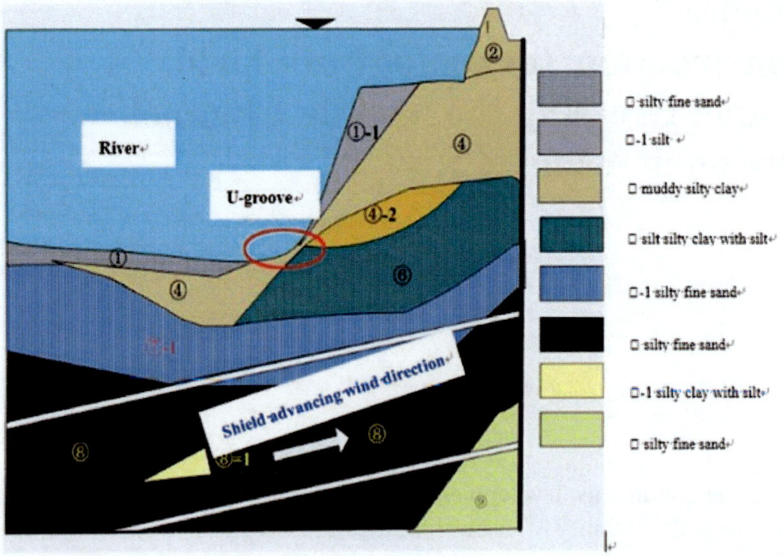

Fig. 7.1 Schematic diagram of the middle channel section

pressure, part of the overburden is scoured by high water pressure for years, thus forming a flushing groove. Therefore, this section is called the middle channel section of the Yangtze River, which is often U-shaped landform and the terrain fluctuates greatly, as shown in Fig. 7.1.

7.1.1 Technical Difficulties

In the process of slurry pressurized shield advancing in the middle channel section of the river, due to the shallow overburden soil and poor soil quality, the stability of the shield excavation face is difficult to control, and it is easy to fall and collapse. In order to prevent roof fall and collapse, the support pressure amplitude must be accurately controlled. If the amplitude is not properly controlled, it is easy to cause a small range of variation between the minimum limit and the maximum water pressure, making the support pressure of the excavation face difficult to control. At the same time, because of the high water pressure and large permeability in the channel section, it is difficult to set up the upper pressure due to the influence of tide and other factors in the river, which will cause the difficulty of mud pressure setting. Slurry pressure setting is too large, prone to pressure mud breakdown of shallow soil layer, to the formation of split penetration, mud to shield tail gap flow channeling and other mud loss risk. When the slurry pressure is set to be small, it is easy to occur the accident that the river water pours into the excavation face and causes the instability of the excavation face. When the shield passes through the sandy silt layer of shallow overburden, the stratum passed by the tunnel is mostly saturated muddy clay, silty clay and sandy silt, and the stratum is mostly soft plastic. In this stratum, there may be a slight liquefied soil layer, which is prone to flow sand and water gushing under certain hydrodynamic conditions, and seriously will cause the collapse of the soil ahead of the excavation. In addition, the lining segment is affected by the buoyancy of surrounding groundwater and shield tail grouting slurry. When the upper soil pressure and the weight of the segment cannot resist the buoyancy of the segment, the segment will float and rupture. When the thrust section in the middle of the Yangtze River is advancing, the force of the shield is unbalanced, and the axis is difficult to control. The asynchronous steering between the shield and the segment in the curve section will make the shield gap difficult to control, and it is prone to uneven phenomenon of segment gap, resulting in water balance, slurry leakage or segment shell.

In summary, the main technical difficulties of shield tunneling through the middle channel section of the Yangtze River are: ① The thickness of the overburden is small, the water pressure is large, the stratum mutation, the stability control of the excavation face is difficult, and it is easy to fall and collapse. ② It is difficult to set the mud pressure. If the mud pressure is too large, the soil layer will be split. If the mud pressure is too small, the river water will flow into the excavation surface and cause instability. ③ The soft plastic soil layer is easy to be liquefied under hydrodynamic conditions, resulting in sand gushing and water gushing. At the same time, when the shield tail sealing effect is poor or the grouting pipe is cleaned, water gushing and sand gushing may also occur. ④ It is inevitable to cause over-excavation during shield tunneling, and it is easy to cause water balance at shield tail, slurry leakage or segment shell sticking due to uneven segment gap. ⑤ When shield tunneling, the upper and lower forces are unbalanced, and the axis is difficult to control, which is prone to tunnel floating and segment damage.

7.1.2 Crossing Scheme

At present, there are two main construction schemes for shield crossing the channel section in the middle of the river: the reinforcement scheme of overburden at the bottom of the river and the non-filling construction scheme at the bottom of the river. After the preliminary scheme is selected, the specific content of the scheme must also be determined by numerical simulation, theoretical calculation and field test, as shown in Figs. 7.2 and 7.3.

1) Reinforcement scheme of river bottom soil

For the reinforcement scheme of covering soil, it is mainly by increasing the thickness of covering soil or using auxiliary work method to strengthen the covering soil, so that the covering soil can meet the requirements of shield construction. According to the shield construction practice at the bottom of the river at home and abroad, when the soil covering at the top of the shield cannot meet the safety requirements of the shield construction and the molding tunnel, the dumping clay is generally used

(a) 3D Data Simulation Analysis (b) Large Simulation Computer

Fig. 7.2 3D numerical simulation and analysis of key points in tunnel construction

(a)60cm diameter shield model (b) Model parameter debugging

Fig. 7.3 Simulation and optimization of tunneling parameters by shield model

in the deep water channel to increase the thickness of the overlying soil layer on the tunnel. In shallow water channel, due to the limitation of shipping and drainage, grouting reinforcement method is generally used to form an impervious hard shell layer, which is beneficial to the stability of soil layer and limit the floating of tunnel. When the overburden is too shallow, uplift piles and anti-floating slab structures are generally applied on the basis of grouting reinforcement to prevent roof fall caused by slurry pressure penetrating the overburden during shield tunneling.

2) Non-filling construction scheme of river bottom

When the buried depth of shallow overburden soil is greater than 0.7D and less than 1D, the dumping and filling measures are generally not adopted. Instead, the excavation surface is maintained by using high viscosity slurry, and the slurry pressure fluctuation and slurry flow of the excavation surface are controlled. The shield posture is strictly controlled in the tunneling process to ensure that the grouting is uniform and sufficient, and the safe and stable passage through the middle of the Yangtze River is achieved. When the shield passes through the shallow overburden section of the middle channel of the Yangtze River, due to the thin overburden thickness, high water pressure and high construction risk of the middle channel section of the Yangtze River, it is required to pass the shield at one time and avoid the long-term shutdown caused by equipment failure in the middle. Therefore, when the shield is close to the shallow soil section, the shield cutter should be stopped for inspection and replacement, a comprehensive maintenance of the shield equipment and the replacement of some parts, do a good job in inventory procurement of vulnerable spare parts, a comprehensive inspection and maintenance of ground slurry separation equipment, air compressor station, sewage pump and other equipment, and do a good job in equipment support for crossing the trough section. During the construction of the channel section in the middle of the river, high slurry pressure is prone to mud breakdown of shallow soil layer, and low pressure is easy to cause construction risks such as river bottom settlement and excavation face collapse. According to the characteristics of different strata, the tunneling and slurry parameters should be selected and adjusted timely and accurately so that the shield can pass through the middle channel section of the river smoothly. At the same time, the risks that may be faced during shield tunneling should be comprehensively evaluated and corresponding countermeasures should be formulated.

3) Construction scheme comparison

Because the flow velocity in the river is generally large and the clay particles are fine, the filling clay will cause local scour around the soil to disturb the stability of the original soil layer, so the bagged clay is needed to fill. However, this method is difficult to form a uniform overburden layer at the bottom of the river, and will lead to loose part of the soil filled with clay, which cannot become a whole with the soil under the original riverbed. The stability of soil pressure is small, the effect is negligible, and the economy is not suitable. In addition, the filling operation in the river, a huge amount of filling, not only time-consuming, but also affect the normal passage and safety of the waterway.

Therefore, compared with the dumping clay scheme with less adaptability and more disadvantages, the tunneling and slurry parameters should be adjusted in time, and the slurry quality should be optimized to stabilize the excavation surface so as to make the scheme of shield passing through the groove section more reasonable. At present, the Nanjing Yangtze River Tunnel project is a typical example of crossing the middle channel section of the Yangtze River in China. Through repeated analysis and demonstration, the project finally determines the construction scheme of "optimizing mud quality, fine control of pressure, strict control of posture, strengthening parameter matching, smooth control and propulsion, rapid segment assembly", and achieves the purpose of safe and rapid crossing the middle channel section.

7.2 Stability Analysis of Excavation Face and Calculation of Excavation Parameters Passing Through the Middle Channel

The key to the success of slurry shield construction is to ensure the stability of the excavation face. When large slurry shield is used in the bottom of the water, due to the slope control of the line selection and the influence of the riverbed topography, it is difficult to avoid the shallow overburden in some mileages, which increases the construction risk and technical difficulty. When the shield passes through the ultra-shallow overburden section, due to the thin overburden, on the one hand, in order to prevent collapse and roof fall, it is necessary to accurately control the amplitude of support pressure, resulting in a small variation range between the minimum limit and the maximum water pressure, which makes the support pressure of the excavation face difficult to control. On the other hand, the lining segment is affected by the buoyancy of surrounding groundwater and shield tail grouting slurry. When the upper soil pressure and the weight of the segment cannot resist the buoyancy of the segment, it will cause the segment floating, the longitudinal bolt of the segment brittle rupture, and cause engineering accidents. Therefore, how to ensure the stability of the excavation face, reasonably set the tunneling parameters and construction measures is an important topic in the construction of underwater shield crossing shallow soil section.

7.2.1 Stability Analysis of Excavation Face

1) Influence factors of excavation face stability of slurry pressurized shield

In the process of slurry pressurized balanced shield tunneling, the stability of excavation face depends on the following three factors:

(1) Mud pressure balances water pressure and soil pressure.
(2) The mud forms an impermeable mud film on the excavation surface, so that the mud pressure can be fully exerted.
(3) The slurry penetrates into the stratum from the excavation face, increasing the cohesion of the stratum in a certain range in front of the excavation.

In order to ensure the stability of the excavation face during the advancement of slurry shield, the appropriate slurry pressure should be set according to the stratum and soil water pressure of the excavation face. When the set mud pressure is too small, the risk of collapse of the excavation surface will increase. If the pressure is too large, the possibility of mud gushing will occur. Slurry pressure can be set according to the active earth pressure or static earth pressure generated by the soil. The horizontal displacement of excavation face and ground settlement in front of excavation should be considered comprehensively when setting slurry pressure.

In order to transmit and maintain mud pressure to the excavation face, the formation of mud film is crucial. Due to the existence of slurry pressure, the slurry continuously penetrates into the soil of the excavation face. Reducing the porosity of the soil of the excavation face makes the excavation face form an approximately impermeable mud film. Due to the formation of the mud film on the excavation surface, the effective mud pressure acting on the excavation surface increases, thus effectively maintaining the stability of the excavation surface. If the mud film is not formed or the quality of the mud film is not good, the mud will penetrate into the stratum under pressure. Not only the mud pressure cannot effectively act on the excavation face, but also the pore water pressure of the soil will rise and the effective stress will decrease, which is unfavorable to the stability of the excavation face. Therefore, effective measures must be taken to accelerate the formation of mud film in sandy soil and sandy gravel stratum with large loose permeability coefficient, so as to prevent collapse under the action of stratum pressure and cause instability of excavation face.

It can be seen that the formation of the mud film plays a vital role in the slurry balance, so it is very important to manage the mud indexes such as the specific gravity, viscosity, yield value and filtration characteristics of the mud. In addition, the stability of the tunnel excavation face is also related to the buried depth of the tunnel, the diameter of the tunnel and the stratum conditions of the tunnel.

2) Stability analysis of slurry shield advancing in the middle of Yangtze River

In the process of slurry pressurized shield tunneling, the cutterhead destroys the original balance of the excavation face, and the excavation face will produce active earth pressure, static earth pressure or passive earth pressure acting on the cutterhead. When the slurry shield is advancing, if the slurry pressure acting on the excavation surface is too small, the excavation surface soil will experience the displacement towards the cutterhead. In order to prevent the slip of the soil, the shear stress in the soil will continue to increase, and then the lateral stress of the soil will gradually reduce to the minimum. At this time, the state of the soil is in the active limit equilibrium state, and the relative soil pressure is called the active soil pressure. If the slurry pressure acting on the excavation surface is too large, the excavation surface soil will experience the displacement of the movement behind the cutterhead. In order to prevent the upward slip of the soil, the shear stress in the soil will increase, and then the lateral stress of the soil will gradually decrease to the minimum. At this time, the state of the soil is in a passive limit equilibrium state, and the relative soil pressure is called passive soil pressure.

Therefore, when the shield is advancing in the saturated soil layer at the river bottom, the slurry pressure of the slurry tank is equal to the active earth pressure, the slurry pressure of the slurry tank is equal to the static earth pressure, and the slurry pressure of the slurry tank is equal to the passive earth pressure. The three critical

Fig. 7.4 Force analysis chart when slurry pressure equals static earth pressure ($P_g = P_0$)

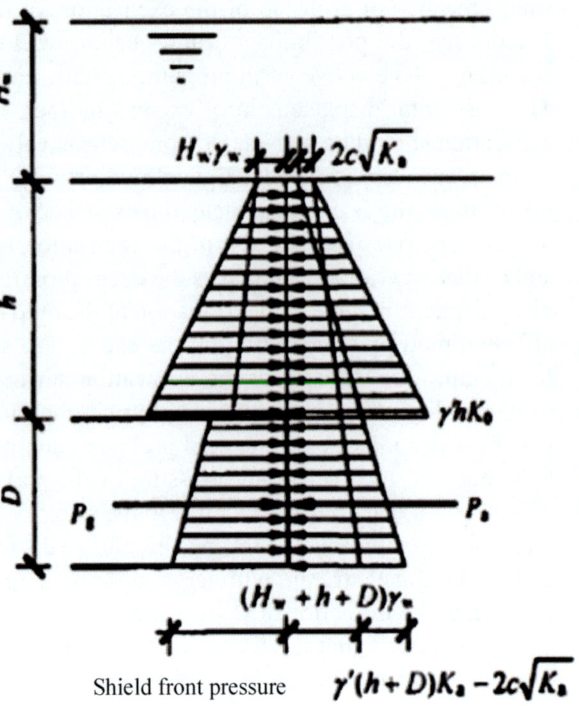

stress equilibrium states are shown in Figs. 7.4, 7.5 and 7.6 (when $c = 0$, the stratum is clay-free).

$$P_g = F_t - F_f \tag{7.1}$$

$$F_r = \pi DLP_r\mu \tag{7.2}$$

$$\left.\begin{aligned} P_a &= [(H_w + h + D/2)\gamma_w + (h + D/2)\gamma' K_a - 2c\sqrt{K_a}]\pi D^2/4 \\ P_0 &= [(H_w + h + D/2)\gamma_w + (h + D/2)\gamma' K_a]\pi D^2/4 \\ P_p &= [(H_w + h + D/2)\gamma_w + (h + D/2)\gamma' K_p - 2c\sqrt{K_p}]\pi D^2/4 \end{aligned}\right\} \tag{7.3}$$

$$\left.\begin{aligned} K_a &= \tan^2(45^\circ - \phi/2) \\ K_0 &= 1 - \sin\phi' \\ K &= \tan^2(45^\circ + \phi/2) \end{aligned}\right\} \tag{7.4}$$

P_g—Shield slurry tank slurry pressure.
F_t—Measured value of shield thrust.
F_f—Calculation value of shield friction.
L—Shield length.
D—Exterior diameter of shield.

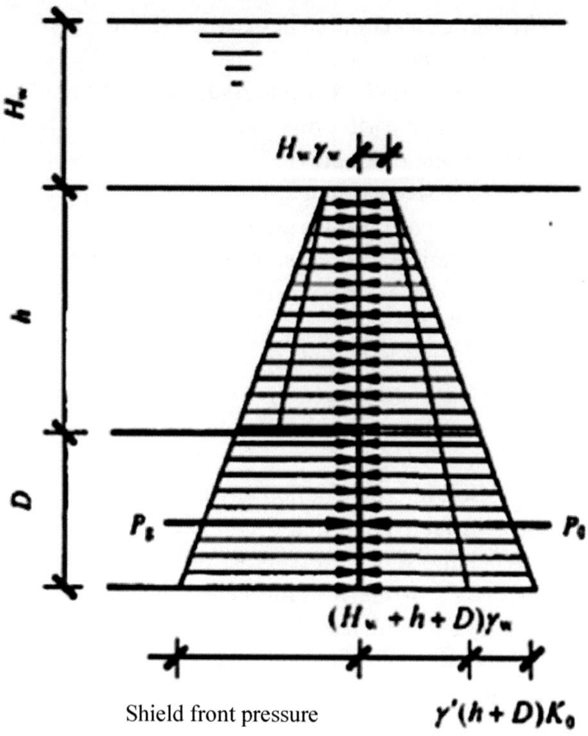

Fig. 7.5 Stress analysis diagram of active earth pressure equal to slurry pressure ($P_g = P_a$)

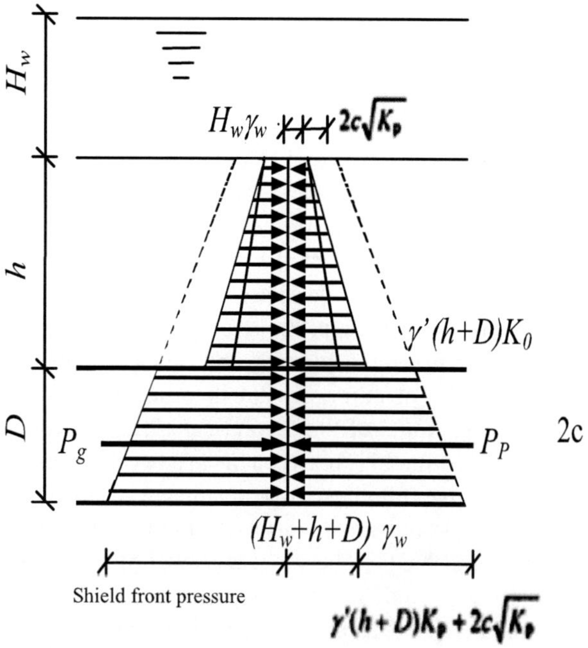

Fig. 7.6 Force analysis chart when slurry pressure equals passive earth pressure ($P_g = P_p$)

μ—Friction coefficient between shield periphery and soil, $\mu = 0.4$ in subclay, $\mu = 0.3$ in muddy clay.

P_r—Mean radial earth pressure around shield.

P_a—Active earth pressure.

P_0—Static earth pressure.

P_p—Passive earth pressure.

K_a—Active earth pressure coefficient.

K_0—Static earth pressure coefficient.

K_p—Passive earth pressure coefficient.

H_w—Water depth.

γ_w—Water bulk density.

h—Cover thickness.

γ'—Gravity of soil floating.

c—Soil cohesion.

φ—Soil internal friction angle.

φ'—Effective internal friction angle.

When the slurry pressure P_g of the shield slurry tank is less than the active earth pressure P_a of the working face, the soil of the excavation face will collapse into the soil tank, which will cause the overburden collapse and the ground subsidence. The river will enter the soil tank along the collapse crack and cause the water inrush accident. When the slurry pressure P_g of the shield slurry tank is between the active earth pressure P_a and the static earth pressure P_0 ($P_a < P_g < P_0$), there will be a certain settlement in the overburden layer ahead of the shield. The superposition of the settlement caused by the subsequent process will not be conducive to the control of the total settlement of the tunnel. At the same time, due to the disturbance of the front soil, it is difficult to control the shield, and it is difficult to coincide with the front earth pressure acting point on a straight line. The working condition suitable for the normal advance of shield is that the front extrusion pressure P_g is between the static earth pressure P_0 and the passive earth pressure P_p, so that the front overburden layer has a certain uplift to compensate for the settlement caused by the subsequent process and not to crack the overburden layer. When the positive extrusion pressure P_g is greater than the passive earth pressure P_p, the front covering soil will be cracked to produce cracks, and the river water will flow into the shield along the cracks to cause engineering accidents, and even cause the whole tunnel to be scrapped.

The value of slurry pressure in the actual construction process is between the upper and lower limits of the theoretical calculation value, that is, the positive extrusion pressure of shield is between the active earth pressure and the static earth pressure. In order to give full play to the advantages of good rock strata independence, the under-pressure propulsion mode is adopted, that is, under the condition of maintaining the overall balance between the level of the excavation chamber and the flow rate of the inlet and outlet slurry, the support pressure in the construction is slightly lower than the theoretical calculation value. The construction control method of rapid excavation, rapid assembly and high-quality slurry supplement after the excavation is adopted, which can well stabilize the working face. The slag discharge and slag discharge of each ring of the slurry field and the change of the slurry tank level are monitored at all times, and the extrusion pressure of the shield front is adjusted according to the actual monitoring data. However, in practical engineering, the control of support pressure may not be accurate due to the mutation of geology ahead, the inconsistency between excavation speed and slurry flow or the mismatch of slurry index. When the pressure of the excavation chamber reaches the limit pressure of the slurry, it may lead to the instability of the excavation face. The common methods to judge the stability of the excavation face are: ① Observing the stability of the support pressure, if the support pressure rises suddenly, it shows that the excavation face is most likely to have a large area of soil collapse, otherwise, it shows that the excavation face has a roof fall. ② Based on the observation of the support pressure, the stability of the level of the excavation chamber and the change of the inlet and outlet slurry flow are observed. If the level fluctuates greatly, it can be considered that the excavation that the excavation surface has been unstable.

3) Minimum thickness of shallow overburden

According to the above analysis, the earth chamber pressure should be kept between the static earth pressure and the passive earth pressure during the earth pressure balance shield construction, namely:

$$P_g < P_{\mathrm{p}} \tag{7.5}$$

From the formula (7.1), formula (7.4) and formula (7.5), the formula for calculating the minimum covering thickness h required for safe shield propulsion can be obtained.

$$h > \frac{\frac{4P_g}{\pi D^2} - 2c\sqrt{K_p} - H_w\gamma_w}{\gamma_w + \gamma' K_p} - \frac{D}{2} \tag{7.6}$$

7.2.2 Calculation of Tunneling Parameters

Due to the rapid advance, rapid assembly and rapid passage of the advancing scheme adopted in the process of shield tunneling in the middle reaches of the Yangtze River, in order to ensure the smooth and safe implementation of the construction scheme, it is necessary to calculate and set the various tunneling parameters in advance during the shield tunneling, including tunneling speed, cutterhead speed, jack thrust, incision pressure, post-wall grouting parameters, mud parameters, mud flow rate and mud pipe flow rate. Therefore, how to establish a benign and dynamic slurry cycle balance model in a short period of time is the key to determine the smooth progress.

The tunneling speed of the shield directly determines the excavation rate of the machine (i.e., the excavation amount per unit time). The faster the tunneling speed is, the higher the excavation rate is, which leads to the increase of the accumulation speed of the slag and soil in the excavation chamber. Under the condition of constant inflow and outflow of slurry, the slurry level in the excavation chamber increases, and ultimately leads to the increase of the incision water pressure, namely the support pressure in front of the shield, which is extremely unfavorable for the stability of the excavation face. How to establish a benign cycle between these three main parameters should be carried out in accordance with the following points: The first step, according to the geological characteristics of the rock ahead, the thickness of the tunnel, the height of the water level using the method of setting the water pressure of the incision mentioned above. The second step is to set the pressure of the bubble chamber according to the set water pressure of the incision and the liquid level on the shield 0 scale line. The third step, set reasonable penetration and cutter head speed, so as to obtain the shield propulsion speed. In the fourth step, when other factors (such as the uneven geology ahead) lead to significant changes in speed, the thrust of each group of jacks should be adjusted in time, or the speed should be stabilized by adjusting the penetration and rotational speed. The fifth step, when the speed changes, should actively adjust the amount of slurry in and out so as to stabilize the excavation tank level, adjust the amount of slurry in and out is mainly achieved by adjusting the speed of each pump.

Among the various tunneling parameters, the tunneling speed, cutterhead speed, post-wall grouting parameters, mud parameters, and mud pipe flow rate can be obtained according to experience or laboratory tests, while the incision pressure, mud flow rate and jack thrust need to be calculated. The calculation methods of three parameters, namely, incision pressure, mud flow rate and jack thrust, are introduced below.

When calculating the cutting water pressure of excavation face, the stress analysis is shown in Fig. 7.7.

(1) The upper limit formula of incision water pressure is as follows.

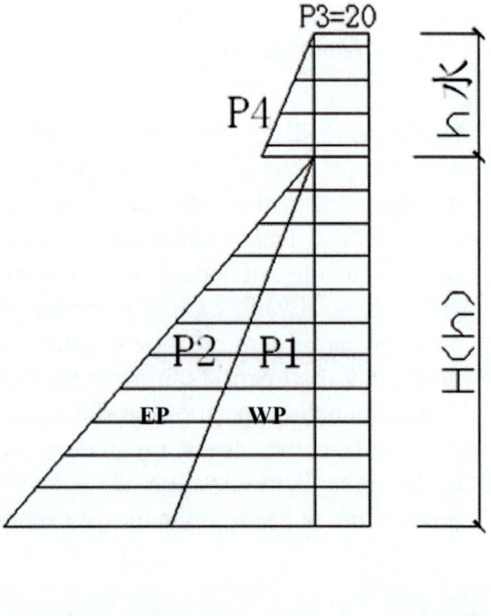

Fig. 7.7 Incision water pressure of excavation face (middle section of Yangtze River)

$$P_{fu} = P_1 + P_2 + P_3 + P_4 \tag{7.7}$$

$$\begin{aligned} P_{fu} &= P_1 + P_2 + P_3 + P_4 \\ &= \gamma_w \times h + K_0[(\gamma - \gamma_w) \times h + \gamma \times (H - h)] + 20 + \gamma_{water} h_{water} \end{aligned} \tag{7.8}$$

In the formula: P_{fu}—upper limit of incision pressure (kPa).

P_1—Groundwater pressure (kPa).

P_2—Static earth pressure (kPa).

P_3—Variable earth pressure, generally 20 kPa.

P_4—River water pressure is determined according to different water depths.

γ_w—Bulk density of water (kN/m^3).

H—Tunnel buried depth below groundwater level (calculated to tunnel center, m).

K_0—Static earth pressure coefficient.

γ—Soil bulk density (kN/m^3).

H—Tunnel buried depth (calculated to tunnel center, m).

γ_{water}—Bulk density of river water (kN/m^3).

h_{water}—Depth of river water (m), the actual water depth above the shield cut shall be determined according to the tidal gauge.

(2) The lower limit of incision water pressure is as follows.

$$\begin{aligned} P_{f1} &= P_1 + P_2' + P_3 + P_4 \\ &= \gamma_w \times h + K_a[(\gamma - \gamma_w) \times h + \gamma \times (H - h)] - 2 \times C_u \times \sqrt{K_a} + 20 + \gamma_w h_w \end{aligned} \tag{7.9}$$

In the formula: P_{f1}—cut water pressure lower limit (kPa).

P'_2—Active earth pressure (kPa).

K_a—Active earth pressure coefficient.

C_u—Soil cohesion (kPa).

2) Sludge flow rate

According to Duroude limit velocity formula:

$$v_L = F_L \sqrt{2gD \frac{d_s - d}{d}} \tag{7.10}$$

$$v_d = (1.2\text{–}1.25) v_L \tag{7.11}$$

v_L—Minimum limit velocity (m/s).

F_L—Depends on the constant of soil particle concentration and diameter, the value of feeding pipe is 0.7, the value of discharging pipe is 1.35.

d_s—The proportion of soil particles flowing through the mud pipe.

D—Pipe diameter.

d—The proportion of mud.

g—Gravity acceleration, 9.8 m/s^2.
v_d—Design minimum velocity.

3) The thrust calculation principle of shield jack.

The shield jack should have sufficient thrust to overcome the resistance encountered during shield tunneling. These propulsion resistance mainly include: slurry pressure acting on the shield reaction F_1. Soil friction to shield F_2. The traction force F_3 of the supporting system behind the shield. The component force of shield gravity in heading direction F_4. Cutting tool resistance F_5. Shield tail seal tension F_6.

The sum of the above various propulsion resistances is expressed by the following formula. In use, the specific situation of the shield machine should be considered, and a certain amount of surplus should be left, namely, the total thrust of the shield jack:

$$F = F_1 + F_2 + F_3 + F_4 + F_5 + F_6 \tag{7.12}$$

7.2.3 *Selection of Shield Construction Parameters and Numerical Calculation Examples*

Taking Nanjing Yangtze River Tunnel Project as an example, this section introduces the selection of relevant tunneling parameters and numerical simulation calculation during the construction of the middle channel section.

The distribution of strata in the middle channel section of Nanjing Yangtze River Tunnel Project is shown in Fig. 7.1. The specific geological conditions of the main strata under shield tunneling are as follows.

⑦-1 layer of silty sand: gray-greenish-grey, saturated, slightly dense-medium dense, rapid rocking reaction, microstratigraphy, containing mica debris, locally intercalated with silty soil thin layer, 1–3 mm thick, uneven.

⑧ Layered silty sand: grayish, saturated, medium dense, locally dense, poor particle gradation, and the main mineral components are quartz, feldspar, mica, etc., uneven and widely distributed.

⑨ Layered silty sand: blue-gray—blue-gray, saturated, dense, rapid rocking reaction, poor particle gradation. The main mineral components are quartz, feldspar, mica, and partially entrained with silt and silty clay. Couple gravel, particle size 2–15 mm. Mainly distributed in the north, south, missing in the middle.

Basic physical parameters of three formations are shown in Table 7.1.

7.2.3.1 Parameter Selection and Construction Control Measures

1) Mud pressure

Table 7.1 Table of physical parameters for shield tunneling through main strata in middle channel section of Yangtze River

Geotechnical layering	Name of rock and soil	Natural density	Water content	Void ratio	Shear test				Coefficient of compressibility	Modulus of compression	Modulus of deformation	Lateral pressure coefficient	Edgewise compressive strength
					Direct quick shear		Consolidated quick shear						
					Force of cohesion	Angle of internal friction	Force of cohesion	Angle of internal friction					
		ρ	w	e	c	q	c_c	c_q	$a_{0.1-0.2}$	E_{s1-2}	E	ξ	q_u
		g/cm^3	%		kPa	°	kPa	°	MPa^{-1}	MPa	MPa		kPa
④	Muddy-silty clay	1.80	38.9	1.10	15	3.3	17	12.7	0.63	3.14		0.72	41.2
⑥	Silt silty clay with silt	1.84	32.2	0.95	20	5.0	24	15.4	0.49	3.95		0.68	56.8
⑦-1	Silt-sand layer	1.92	28.8	0.80	5	28.3	0.0	30.0	0.15	13.2		0.43	
⑧	Fine sand layer	1.93	24.9	0.74	5	30.6	2	32.3	0.13	13.8		0.40	
⑨	Fine sand layer	1.94	23.6	0.72	6	30.9	2	34.1	0.13	13.9		0.37	

Since the slope section with a slope of 30° appears after the mileage K6+96.2, there is a slope effect, the slurry pressure of the whole shallow overburden section is set to be divided into two parts: the conventional pressure before section 3 and the slope pressure after section 3.

Conventional incision water pressure setting: $P_{conventional} = (P_a + P_0)/2 + 0.2$.

Slope section incision pressure setting: $P_{slope} = (P_a + P_0)/2 + P_{fix}$.

Among them, P_a is the main dynamic earth pressure, P_0 is the static earth pressure, and 0.35 bar is selected for the revision. The water depth in the calculation of slurry pressure is based on the actual water level of the Yangtze River during construction, and is adjusted according to the tidal water level of the Yangtze River when the shield passes through the shallow overburden. The mud pressure should be strictly controlled according to the technical bottom, and the deviation is between ±0.1 bar.

2) Velocity of slurry pipe

According to the experience, the slurry density of the discharge pipe is 1.30–1.40 t/m^3, and the dry density of the gravel sand layer is 1.72 t/m^3. According to the formula of the minimum limit flow rate of Durangde, the minimum sludge discharge of the discharge pipe in the gravel sand layer is $Q = 1731$ m^3/h. Considering the matching between the inlet and outlet slurry volume and the tunneling speed, the inlet slurry volume of fine sand and gravel sand layer is set to 1900 m^3/h, and the outlet slurry volume is set to about 2150 m^3/h.

3) Driving speed and cutter speed

Driving speed and cutter head speed are set according to geological conditions and construction experience, as shown in Table 7.2.

4) Grouting behind segment wall

(1) Selection of slurry type. In the early slurry shield tunnels, the synchronous grouting is generally used with double slurry. In the recent large slurry shield tunnels, the single slurry has replaced the double slurry, and has been successfully applied, such as the Shangzhong Road Cross River Tunnel and the Shanghai Yangtze River Tunnel. Compared with double slurry, single slurry has the following characteristics:

① Wide material, no pollution and low price.

Table 7.2 Excavation parameters of shallow overburden

	Speed of cutter head (rpm)	Tunnelling meterage (mm/min)	Cone petetration (mm/r)	Stratigraphic characteristics	Remark
Ultra-shallow overburden mileage K5+988~K6+104	0.65–0.8	25–30	35–45	⑦-1 silty fine sand, ⑧ silty fine sand and ⑧-1 silty clay with silt	Reduce excavation surface disturbance

② Good pumping, slurry filling effect, low grouting. However, it is difficult to control the setting time of the double-liquid slurry, which is easy to cause pipe plugging and slurry filling inaccuracy. Moreover, the slurry is easy to split the soil layer, resulting in the slurry being pressed into the soil with previous gaps, resulting in the grouting amount far exceeding the theoretical value, and the disturbance to the soil is large, which is easy to cause surface subsidence.

③ When the yield strength of the slurry exceeds 800 Pa for 20 h, the slurry has good resistance to tunnel floating, and the dislocation between the tunnel rings is generally within 0–6 mm. The strength of double slurry solid is uneven, and the tunnel settlement and floating are large during construction.

④ The combination of bentonite and admixture not only reduces the water demand of slurry, but also makes the slurry have good slump retention ability.

⑤ It has the performance of effectively preventing the back channeling of slurry, not damaging the shield tail device, and not being diluted or dispersed by slurry and groundwater. The double slurry is vulnerable to the dilution and penetration of groundwater and slurry of shield front support, and the setting time and strength of slurry are difficult to control.

In summary, Nanjing Yangtze River tunnel synchronous grouting using single slurry. Considering that the single liquid slurry has a long initial setting time and good fluidity, it will have a negative impact on the settlement that needs to be quickly controlled during shield construction. Therefore, when passing through the section, if the settlement cannot be effectively controlled, the rapid grouting filling is carried out with the tested double liquid speed cement to stabilize the soil.

(2) Setting slurry parameters. According to the geological conditions, the particle gradation of fine sand layer in the shallow overburden section of the middle Yangtze River is poor, and the gravel sand layer filling is easy to be washed. In order to prevent shield tail slurry leakage, tunnel floating and stratum instability, it is necessary to strengthen the post-grouting control of segment wall to ensure the quality of synchronous slurry. The slurry density was set as 1.96 g/cm^3, and the slurry slump was controlled at 18–22 cm. The grouting volume is controlled at 150%–200% of the theoretical void volume to ensure that the grouting behind the wall is dense and effective. At the same time, the grouting pressure is controlled to prevent the breakdown of the shallow soil layer. The grouting pressure is set to 95%–105% (fluctuation ± 0.1 bar, matching with the slurry pressure) of the slurry pressure at the position of the grouting pipe, and the pressure loss from the pump head to the outlet of the slurry pipe.

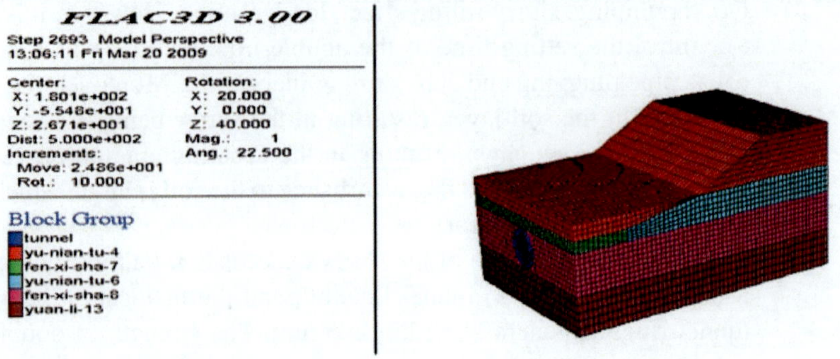

Fig. 7.8 Topographic map of gully section when vertical and vertical scales are consistent

7.2.3.2 Numerical Simulation Analysis of Slurry Pressure Setting in Left-Line Tunnel Flushing Section

The static earth pressure and instability pressure of the shield incision with the minimum overburden mileage are simulated by numerical simulation. The numerical calculation results are compared with the calculation results of the classical formula, and finally a more reasonable set value of slurry pressure is obtained.

1) Model calculation range

The actual slope of the groove section is about 30°, and the geological longitudinal section is illusioned due to the inconsistent vertical and vertical ratios. The topographic conditions of the groove section are redrawn to obtain the topographic map of the groove section when the vertical and vertical ratios are consistent, and the numerical modeling is carried out accordingly.

As shown in Fig. 7.8, the slope mileage is K6+098–K6+118. According to the actual construction situation, when the shield passes through the groove section, in order to ensure the safe passage of this section, the slurry pressure setting value in the slurry shield pressure cabin is constantly changing. The specific change process is as follows: from K6+090 to K6+098 and then to P_{slope}, from K6+115 to K6+118 and then to $P_{\text{conventional}}$.

2) Establishment of calculation model

The finite difference software FLAC^{3D} is used to simulate the actual working conditions. The numerical model takes K6+000–K6+175 strata and tunnels for modeling analysis. The model takes the most unfavorable working conditions, and takes the minimum overburden mileage as the starting mileage of slope toe. The minimum overburden thickness is 11.3 m, and the water level elevation is 5 m. The model is shown in Fig. 7.9. Figure 7.9a is a complete tunnel model. In order to shorten the calculation time, one-half model Fig. 7.9b is used and inside the tunnel. The lining shell element is added, and the stress perpendicular to the excavation surface is applied on the excavation surface to simulate the slurry pressure of the shield slurry tank applied on the excavation surface.

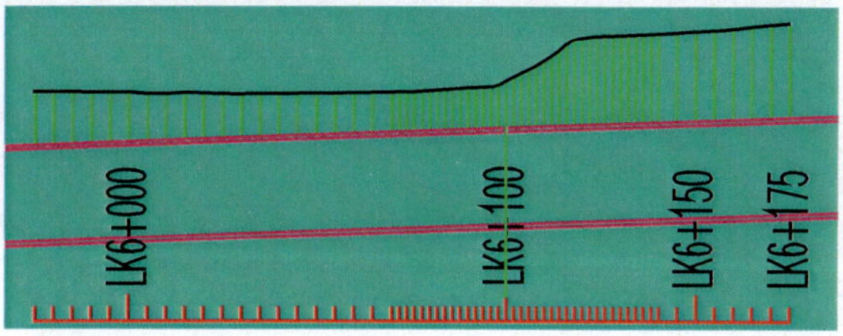

(a) Calculations established

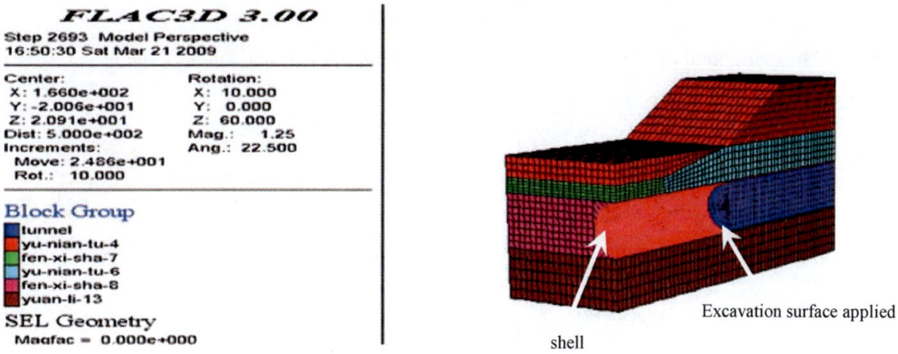

(b) The model used for calculation

Fig. 7.9 FLAC^{3D} numerical simulation

3) Calculation and analysis of calculation results

   (1) Calculation results. The formula and FLAC^{3D} are used to calculate the pressure of the incision. The results are shown in Table 7.3.
   (2) Analysis of calculation results. According to the calculation results obtained by the finite difference software FLAC^{3D}, it is analyzed to obtain a more reasonable pressure value and guide the construction. The results are shown in Figs. 7.10, 7.11 and 7.12.

The actual values of active soil pressure and static soil pressure in the slope section should be greater than the formula calculation results, but the numerical solution of active soil pressure in the current calculation results is less than the formula solution, which is not desirable. The reason is that the software is difficult to simulate the slip failure of soil, and the horizontal displacement of soil has lost practical significance when instability occurs.

Table 7.3 Comparison of calculation results of incision pressure

	Active earth pressure(bar)	Earth pressure at rest(bar)
Formula solution	3.37	3.41
Numerical solution	2.90	4.10

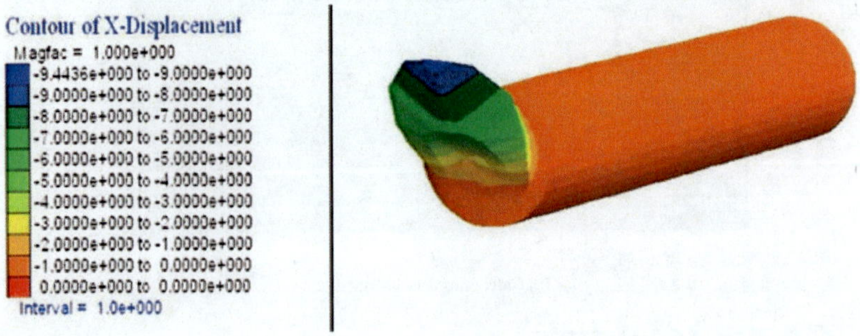

Fig. 7.10 Horizontal displacement of excavation face at 2.9 bar incision pressure

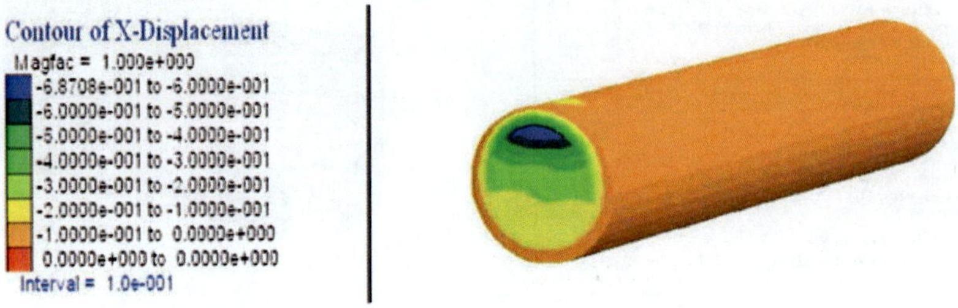

Fig. 7.11 Horizontal displacement of excavation face at 3.0 bar incision pressure

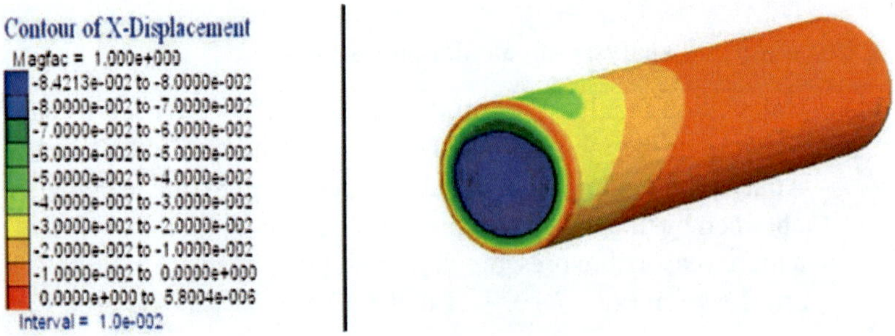

Fig. 7.12 Horizontal displacement of excavation face at 3.4 bar incision pressure

Standard for numerical calculation of static earth pressure: when horizontal displacement of excavation face is almost zero, support pressure is static earth pressure. The numerical solution of static earth pressure is 4.1 bar, and the formula solution is 3.41 bar. The difference between them is about 0.7 bar.

Therefore, the recommended value of $P_{\text{amendment}}$ can be initially taken as half of the difference between the numerical solution and the formula solution of the static earth pressure, $P_{\text{amendment}} = 0.35$ bar, adjusted according to the surface subsidence or other monitoring feedback results when the actual pressure is set.

7.3 Study on Slurry Ratio and Film Formation of Shield Tunneling Across the Middle of Yangtze River

Mud not only undertakes the heavy responsibility of maintaining the stability of excavation face, but also undertakes the important task of carrying slag soil. The selection and control of mud parameters are of great significance to the normal and safe construction of slurry shield. Especially in the construction of the punching section, due to the shallow overburden thickness, the slurry parameters must be strictly controlled to maintain the stability of the excavation face.

7.3.1 Experimental Study on Slurry Ratio and Film Formation of Fine Sand Stratum in the Channel Section

The strata crossing the channel section of Nanjing Yangtze River Tunnel are mainly silty sand strata. When the shield punching section passes through the silty sand layer, due to the single particle size and poor gradation of the silty sand layer, the slightly disturbed stratum is easy to be liquefied. The dense mud film is formed on the excavation surface to maintain the stability of the excavation surface and ensure the safety of the excavation surface. However, the particle size of this kind of fine sand stratum is small, and the mud is not easy to penetrate into the formation to form a permeable zone, which can only form mud skin in the surface siltation. Since the mud is prone to damage under disturbance, in order to ensure the stability of the excavation face in such strata, it is necessary to ensure the formation of mud with high strength and strong anti-disturbance ability on the surface of excavation face. This paper focuses on how to form dense mud skin on the surface of fine sand.

1) Test mud

Slurry used in this experiment: the old slurry + pulping agent in the field conditioning tank, the slurry ratio and the property parameters are shown in Table 7.4, and the particle curve of the old slurry is shown in Fig. 7.13. Due to the flocculation phenomenon of No. 7 slurry, the first six kinds of slurry were selected for film forming test.

2) Test strata

In this experiment, the fine sand with particle size of 0.075–0.25 mm was selected to simulate the fine sand layer in practical engineering. The dry density and permeability coefficient of the stratum used in the experiment were consistent with those in practical engineering. The dry density $\rho_d = 1.73$ g/cm^3, and the permeability coefficient $K = 10^{-3}$ cm/s.

3) Film forming test results

Table 7.4 Proportion and property parameters of mud used in laboratory test research for mud proportioning in fine sand stratum

Number	Slurry ratio	Density	Funnel viscosity (s)	Φ600 (mPa s)	Φ300 (mPa s)	Plastic viscosity (mPa s)	24 h Water bled rate (%)
1	Old slurry	1.21	17.3	10	5	5	41
2	Old slurry + 1‰ 1#	1.21	21.4	16	8	8	40
3	Old slurry + 2‰ 1#	1.21	22.8	20	10	10	20
4	Old slurry + 1‰ 3#	1.22	22.5	17	8	9	39
5	Old slurry + 2‰ 3#	1.21	24.4	23	11	12	14
6	Old slurry + 1‰ 5#	1.21	21	14	7	7	40
7	Old slurry + 2‰ 5#	Slurry produces flocculation, sediment is similar to silt, mud water is serious					

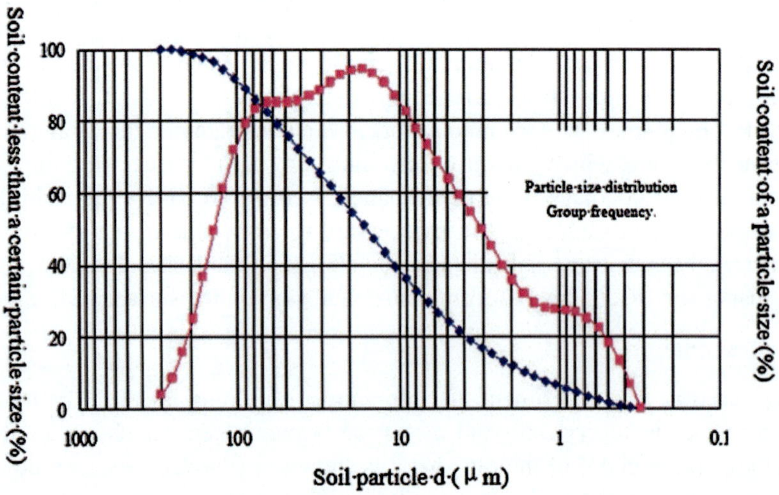

Fig. 7.13 Particle size distribution curves of old slurry

The test results of slurry film-forming test in each group are as follows:

The test results of No. 1 mud are shown in Figs. 7.14 and 7.15.

The test results of No. 2 mud are shown in Figs. 7.16 and 7.17.

The test results of No. 3 mud are shown in Figs. 7.18 and 7.19.

The test results of No. 4 mud are shown in Figs. 7.20 and 7.21.

The test results of No. 5 mud are shown in Figs. 7.22 and 7.23.

The test results of No. 6 mud are shown in Figs. 7.24 and 7.25.

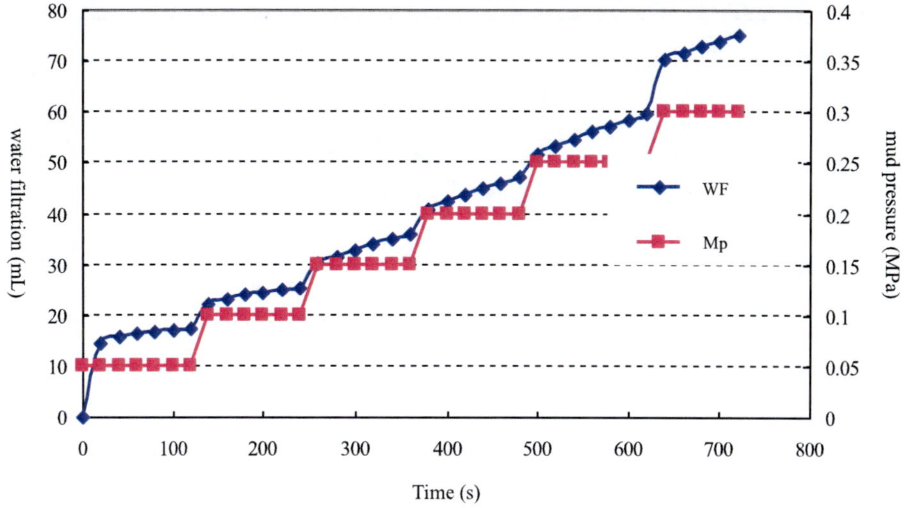

Fig. 7.14 Filtration curve of No. 1 mud film forming test

Fig. 7.15 Mud film formed in No. 1 mud film forming test

4) Analysis of test results

From the above results, it can be seen that the filtration curve changes step by step with the pressure gradient. The greater the viscosity of the mud, the smaller the 24 h precipitation rate, the smaller the amount of filter water formed by the mud, the thinner the mud, and the denser the mud.

According to the test results, it is recommended to use No. 3 or No. 5 formula should be used when the project crosses the silty sand stratum of shallow overburden section of Jiangzhongchongcao, that is, the slurry is prepared by adding NSHS-1 or NSHS-3 slurry preparation agent to the old slurry (mixed slurry of bentonite and

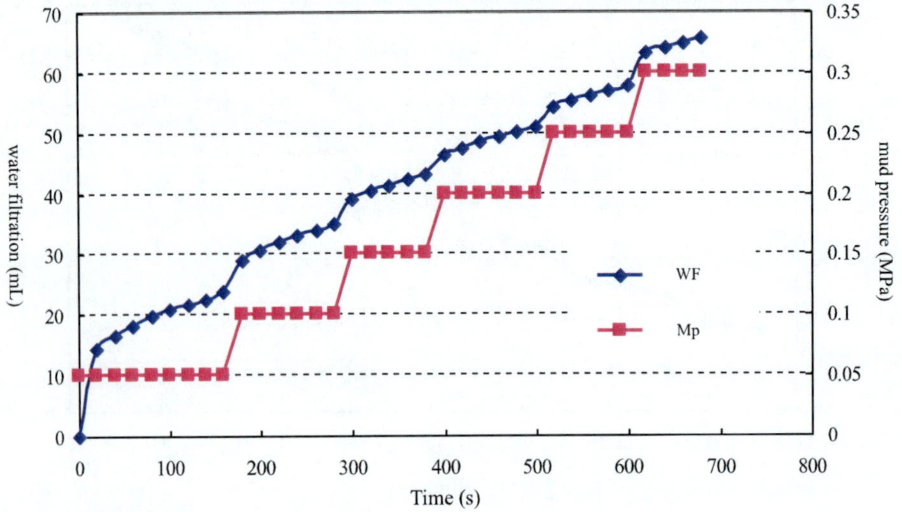

Fig. 7.16 The variation curve of filter water in No. 2 mud film forming test

Fig. 7.17 Mud film formed in No. 2 mud film forming test

clay) in the slurry mixing tank. The density of the old slurry is controlled above 1.12 g/cm^3, the addition amount of NSHS-1 or NSHS-3 slurry preparation agent is about 2%, and the viscosity is controlled within 23–25 s.

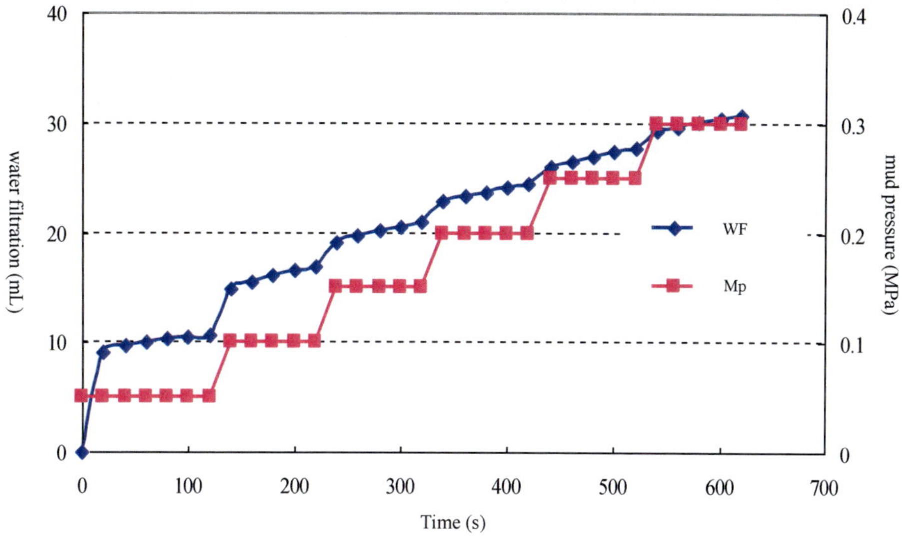

Fig. 7.18 Filtration curve of No. 3 mud film forming test

Fig. 7.19 Mud film formed in No. 3 mud film forming test

7.3.2 Determination of Mud Ratio and Index During Construction of Trench Section

Taking Nanjing Yangtze River Tunnel as an example, combined with the actual construction situation, the determination method of mud ratio and corresponding indexes is introduced.

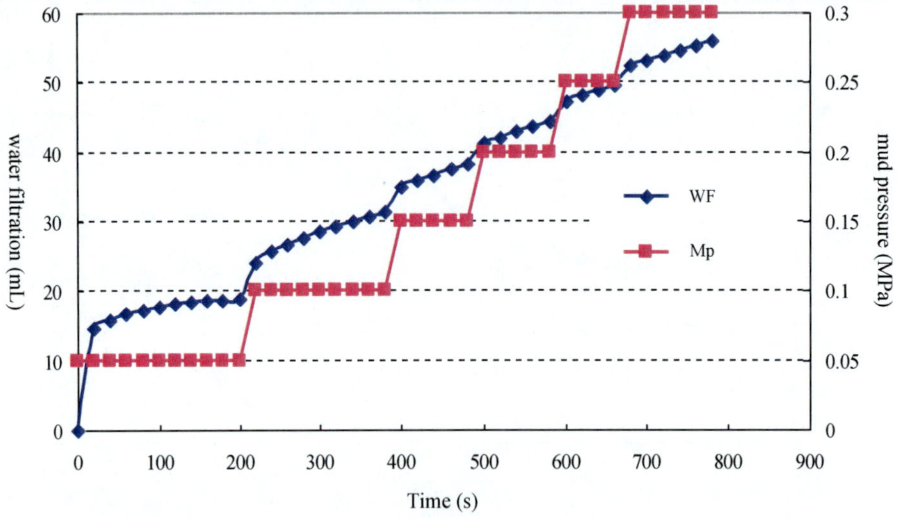

Fig. 7.20 Filtration curve of No. 4 mud film forming test

Fig. 7.21 Mud film formed in No. 4 mud film forming test

1) Density

Large mud density is conducive to stabilizing the excavation face, but too large mud density increases the burden of mud circulation equipment and affects the slag discharge efficiency. In order to maintain the stable grouting density of excavation face at 1.23–1.26 g/cm^3 and the grouting density at 1.30–1.40 g/cm^3.

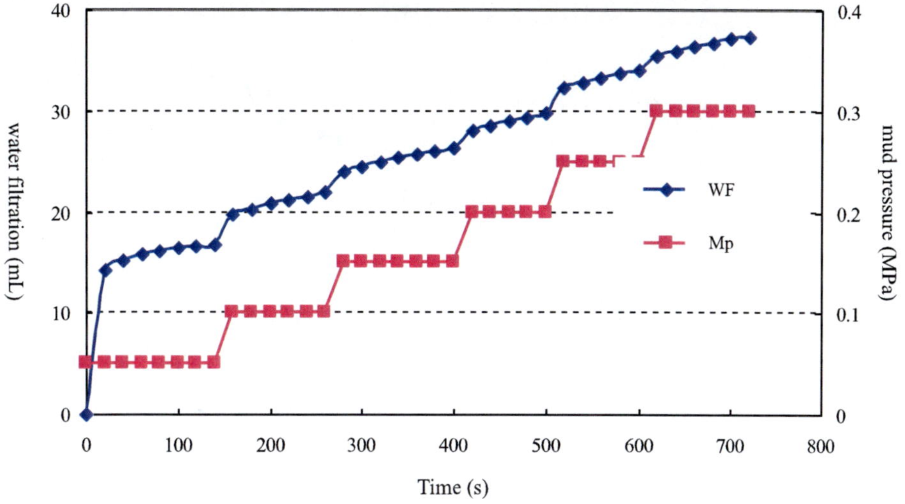

Fig. 7.22 Filtration curve of No. 5 mud film forming test

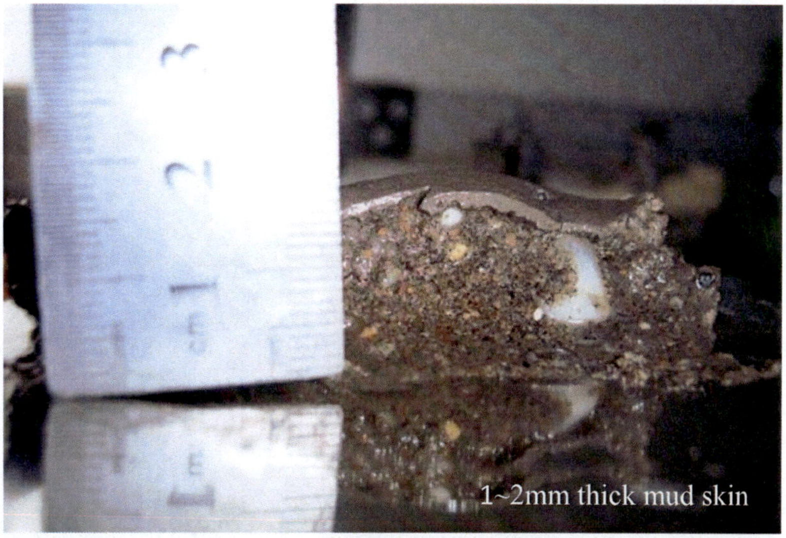

Fig. 7.23 Mud film formed in No. 5 mud film forming test

2) Viscosity

In order to maintain the stability of the excavation face and form an effective mud film on the palm surface, it is necessary to improve the quality of circulating mud when preparing new mud in the mud treatment field. The mud viscosity of the slurry regulating tank is controlled at 23–25 s, while ensuring that the mud leakage is less than 10 m^3/h and the water separation rate is not more than 5%. The mud parameters are determined by the construction experience of similar projects at home and abroad

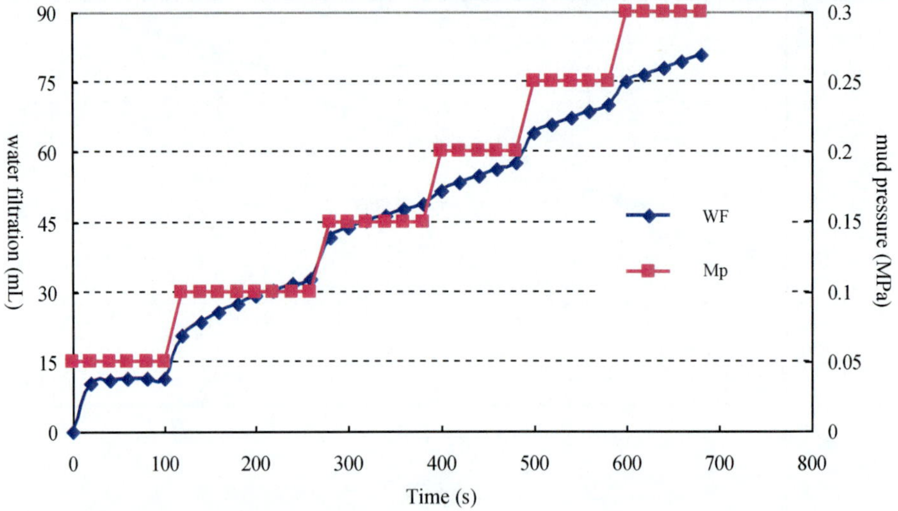

Fig. 7.24 Filtration curve of No. 6 mud film forming test

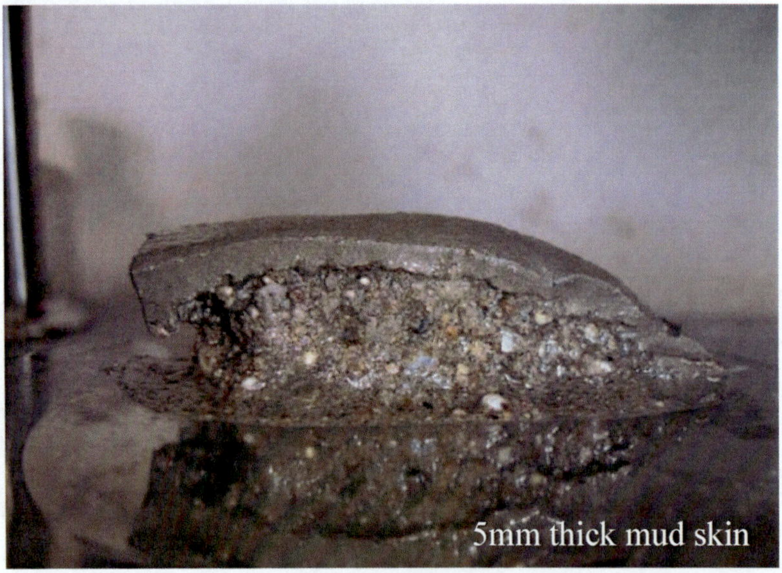

Fig. 7.25 Mud film formed in No. 6 mud film forming test

and the tests carried out by relevant scientific research units. In the construction of shallow soil section in the middle of the river, 80–100 m^3 new mud is added to each propulsion ring (because the mud leakage probability in the middle of the river is greater) to adjust the viscosity and density of the circulating mud. The concrete implementation also needs to be adjusted according to the actual situation to ensure the mud quality.

Table 7.5 Mud parameters in shallow overburden

Item	Slurry proportion (g/cm^{-3})	Pulp flow (m^{-3}/)h	Proportion of pulp discharge (g/cm^3)	Discharge flow (m^{-3}/)h	Water bled rate	Seepage loss (m^{-3}/)h
Shallow overburden in river	1.23–1.26	1800–2000	1.30–1.40	2050–2250	<5%	<10

3) Sand content

In the gravel layer with large porosity and sand layer with poor gradation, the sand particles in the mud have a blocking effect on the formation pores, so the formation of the mud film is closely related to the particle size and content of the sand in the mud. Through the mud treatment field screening precipitation, retain useful clay particles, remove more than 74 μm most of the sand particles and more than 45 μm part of the silt particles, the formation of appropriate solid particle gradation, ensure the formation of mud film in the excavation face. Therefore, the control of sand content in mud is also an important index in mud treatment, and the sand content in most sections of Nanjing Yangtze River tunnel project is controlled at 15%–25%.

4) Pump pressure control of slurry pump

In order to avoid the failure of equipment overload operation caused by high pump pressure of inlet and outlet slurry pump in slurry circulation system, it is necessary to control the pump pressure of inlet and outlet slurry pump, and the maximum pump pressure is less than 0.9 MPa. The mud parameters of shallow overburden are shown in Table 7.5.

5) Preparation of slurry in punching section

In order to ensure the stability of the excavation face in the middle-shallow overburden section of the Yangtze River, it is necessary to allocate 3000 m^3 of the basic slurry according to the determined slurry parameters before entering the middle-shallow overburden construction. The slurry viscosity reaches 25 s, and the density is controlled at 1.23–1.26 g/cm^3. According to the tunneling experience of Nanjing Yangtze River Tunnel Project in ⑧ layer of silty fine sand and ⑩ layer of gravel sand, the mud loss in construction mainly includes the leakage of working face during tunneling and assembly, the carrying of slag and soil during the separation of mud and water, and the mud loss during the cleaning of sedimentation tank. The slurry loss of each ring is 110–130 m^3 when tunneling in the shallow silt fine sand layer in the middle of the river. According to the velocity and distance of the mud pipe, 80 m^3 of the new mud is added to the tunnel face 20 min before the completion of each ring excavation to maintain the stability of the tunnel face. In order to keep the mud volume of slurry circulation system stable, 30–50 m^3 new mud was added to the slurry tank.

7.4 Shield Attitude Control and Risk Management Technology for Shield Crossing the Middle Channel Section of Yangtze River

The excessive snake movement of shield will inevitably cause frequent rectification. The process of rectification is the process of uneven force on the segment ring surface. Therefore, it is necessary to control the posture of shield in the tunneling process and make it do a small amount of snake movement along the tunnel axis as far as possible. According to the specification, the allowable deviation of the plane position and elevation of the central axis of the assembled segment in shield tunneling is ±100 mm. The deviation should be gradually corrected to avoid sudden correction, so as not to cause serious uneven stress on the segment torus.

When the shield passes through the ultra-shallow overburden section, due to the thin overburden, on the one hand, in order to prevent collapse and roof fall, it is necessary to accurately control the amplitude of support pressure, resulting in a small variation range between the minimum limit and the maximum water pressure, which makes the support pressure of the excavation face difficult to control. On the other hand, the lining segment is affected by the buoyancy of surrounding groundwater and shield tail grouting slurry. When the upper soil pressure and the weight of the segment cannot resist the buoyancy of the segment, it will cause the segment floating, the longitudinal bolt of the segment brittle rupture, and cause engineering accidents. Therefore, how to safely pass through the shallow soil section is the key problem of underwater shield tunnel construction. In order to ensure the safety of construction, it is necessary to analyze the construction risks and formulate corresponding countermeasures.

7.4.1 Shield Attitude Control in Shield Tunnel Construction

The attitude control of shield in shield tunnel construction includes the rolling control of the body and the control of the forward direction. In the tunneling process, the shield operator adjusts the attitude of the shield according to the data displayed on the computer screen by the laser automatic guidance system. The operation principle of shield attitude control is as follows:

(1) The roll angle of the body should be appropriate. The shield roll angle is too large, and the shield cannot maintain the correct attitude, which affects the assembly quality of the segment. At this time, the roll angle can be reduced by reversing the cutterhead.

(2) The thrust of the right jack partition needs to be increased if the shield moves horizontally to the right. Conversely, the thrust of the left jack partition needs to be increased. If the shield head is downward, the thrust of the lower jack partition needs to be increased, and vice versa.

1) General rules of shield attitude control

In general, the direction correction of shield should be controlled between ±20 mm. In the transition curve and circular curve section, the direction correction of shield machine should be controlled between ±30 mm. Try to keep the shield axis parallel to the tunnel design axis, otherwise it may cause too small shield tail clearance and segment dislocation cracks due to bad posture.

When the soil on the excavation face is uniform, the attitude control of the shield is easy, and the direction deviation angle is usually controlled in the ±5 mm/m.

When the left and right soft and hard strata in the excavation face are uneven and are in the curve section, the shield attitude control is difficult. At this time, the tunneling speed can be reduced and the jack thrust of each partition can be reasonably adjusted. If necessary, the overbreak cutter (shield equipped with overbreak cutter) can be considered for overbreak in hard rock area.

When the shield encounters the soft and hard soil layer, in order to prevent the shield from "raising", keep the downward attitude. On the contrary, we should maintain the upward posture. Attention should be paid to the jack travel difference between the upper and lower ends and the left and right sides can not be too large, generally controlled between ±20 mm.

In the curve section tunneling, generally according to the curve radius of the shield to the curve inside a small amount of offset, offset generally take 10–30 mm.

In the shield attitude control, the stroke control of the propulsion cylinder is the key point. For 1.5 m wide segment, in principle, the stroke is controlled in 1700–1800 mm, the stroke difference is controlled in 0–40 mm, the stroke is too large, the shield tail brush is easy to expose, the segment is more out of the shield tail, and the deformation is large. If the stroke difference is too large, it is easy to increase the angle between the shield and the segment, which is easy to cause the damage and dislocation of the segment.

2) Control technology of shield tunneling posture in different geological environments

(1) Control of shield tunneling posture in muddy soil.

When shield tunneling in soft soil layer, due to the extremely poor self-stability of stratum, in order to control the horizontal and vertical deviation of shield within the allowable range and avoid excessive disturbance to stratum caused by excessive shield snake, it is appropriate to control the shield tunneling speed at 30–40 mm/min and the cutter head speed at about 1.5 r/min. During tunneling in this section of strata, the thrust of the four groups of jacks should be relatively balanced, so as to avoid the large stroke difference of jacks during tunneling. Otherwise, it may cause that the thrust axis and the segment center axis are not in the same straight line. In the tunneling process, a certain amount of additives should be added according to the actual situation in order to keep the excavation smoothly and keep the continuous tunneling of the shield as far as possible. At the same time, the synchronous grouting amount should be strictly controlled to ensure the effective filling of the pipe back gap.

(2) Control of shield tunneling posture in sand.

Due to the poor self-stability and high water content of water-bearing sand layer, shield tunneling in full-section water-rich sand layer is prone to "bump" phenomenon, and the shield is also prone to floating in water-bearing sand layer.

In order to avoid the "bump" phenomenon of shield tunneling in water-bearing sand layer, the shield machine should maintain the upward trend in the process of advancing. If the "bump" trend is found, the upper and lower pressure should be adjusted immediately to maintain the upward trend of shield.

In order to avoid the floating phenomenon of shield tunneling in water-bearing sand layer, the rotational speed of cutterhead should be reduced and the disturbance to surrounding sand layer should be reduced in shield tunneling. If the buried depth of the tunnel is less than 2/3 times of the shield diameter, measures such as geological improvement and ground surcharge should be implemented in water-bearing sand layer.

(3) Control of shield tunneling posture in strata with large fluctuation of strata.

Rock layer fluctuation will lead to uneven soft and hard rock in tunnel excavation face. The shield tunneling in this stratum is difficult to control the attitude of the shield, and it is easy to cause excessive hunting in the vertical direction of the shield, resulting in dislocation and cracking of the segment. Taking the soft and hard strata above as an example, under such geological conditions, the force of shield cutterhead is uneven, and the tunneling speed is unbalanced. This requires that during the tunneling process, the shield attitude data provided by the measurement system must be observed at all times, and the thrust and total thrust of and total thrust of the jacks in each partition must be continuously adjusted in combination with the travel difference between the propulsion jack and the hinged jack to maintain the stability of the shield attitude. If you don't pay attention to adjusting the travel difference of the jack, it will cause large changes in the selection of the segment, and even cause too small shield tail clearance to make the segment not smoothly out of the shield tail. Therefore, in the process of advancing, we should not only pay attention to the shield attitude provided by the measurement system to guide tunneling, but also take into account the travel difference of jacks in each partition.

(4) Control of shield tunneling posture in the hard rock stratum of the whole section.

The whole cross-section hard rock stratum belongs to uniform rock stratum. The axis attitude of shield tunneling in this kind of stratum can be well controlled. When tunneling, the thrust of jacks in each partition is uniform, the total thrust and tunneling speed are uniform, and the good attitude of shield can be maintained.

3) Correction measures for shield tunneling

Shield always deviates from the design axis during tunneling, and rectification must be carried out in a planned and step by step. Correction measures are as follows:

(1) Attention should be paid to the change of roll angle during tunneling, and the rotation direction of cutterhead should be adjusted according to the roll angle of shield in time.
(2) The tunneling parameters should be adjusted according to the geological conditions of each section.
(3) In the process of deviation correction, the tunneling speed should be slowed down, and attention should be paid to avoiding the damage to the segment caused by the excessive force on the unilateral jack during deviation correction.
(4) The reasonable segment type should be selected as far as possible to avoid the excessive influence of human factors on shield attitude, and the segment assembly quality should be strictly controlled to avoid the adjustment of shield attitude.
(5) When correcting the deviation, it is necessary to pay close attention to the attitude of the shield, the selection of the segment and the gap between the shield tail and the segment. The gap between the shield tail and the should be uniform.
(6) When the shield deviates from the design axis, it shall not be adjusted violently to avoid excessive deviation in the opposite direction.

4) Shield attitude control requirements of Nanjing Yangtze River Tunnel

According to the construction experience of Nanjing Yangtze River Tunnel, the shield posture and formed segments are stable in the sand layer, and prone to floating in the shallow overburden. The shield posture in the shallow soil section of the middle reaches of the Yangtze River is vertically controlled at −30 to +10 mm, and the shield posture is controlled at +10 to +30 mm inside the curve due to the advance of the curve section on the plane.

7.4.2 Float Control and Treatment Measures

After the assembled tunnel ring is removed from the tail of the shield, due to the buoyancy of the surrounding groundwater or grouting slurry at the tail of the shield, when the earth pressure on the upper part of the segment and the weight of the segment cannot resist the buoyancy, the tunnel will float and cause passive damage to the soil above the segment. There are five control methods for the floating phenomenon.

1) Synchronous grouting control

In water-bearing fine sand stratum, solving the problem of segment floating is essentially a competition between synchronous grouting and stabilizing segment floating in time. The ideal grouting method should be that the shield tunneling along the axis, the grouting slurry fully fills the construction gap and rapidly solidifies to form a consolidation body with high early strength, so that the tunnel and the surrounding soil form an overall structure to achieve stability.

According to the relationship between the regular value of segment floating and shield advancing posture, the grouting hole position and grouting quantity are reasonably selected. According to the construction experience of Nanjing Yangtze River Tunnel, the grouting volume ratio of the upper, middle and lower rows of six grouting holes in the upper and lower rows of shield tail is about 5:3:2. According to the diameter of shield excavation and the gap of shield tail, the grouting volume of each ring segment (2 m wide) is 26–30 m^3.

2) Control shield posture

Poor attitude control during shield tunneling will inevitably lead to frequent deviation correction. The process of deviation correction is the process of uneven stress on the segment torus. Therefore, in the tunneling process, the attitude of the shield should be strictly controlled, and the shield should be pushed along the design axis of the tunnel as much as possible to avoid the deviation correction snake. When the deviation is found, it should be gradually corrected, and sudden correction should be prohibited, so as not to cause large dislocation between segments, and the stress on the segment torus is seriously uneven.

3) Controlling tunneling speed

If the slurry cannot consolidate and stabilize the segment timely and effectively in the synchronous grouting process, the shield tunneling speed should be properly controlled. Since the synchronous grouting amount is 13–15 m^3/m and the grouting speed is about 0.5 m^3/min, the advancing speed should not be greater than 30 mm/min to ensure the balance between the gap formed when the segment is out of the shield tail and the grouting amount. If the advancing speed is too fast, the slurry cannot fill the gap between the segment and the tunnel in time. The slurry is concentrated at the bottom of the segment by gravity, and the segment is buoyed by the slurry buoyancy.

4) Reasonable control of shield advancing elevation

According to the shield construction experience of Nanjing Yangtze River Tunnel, the empirical values of segment floating after assembly in various strata are counted, and the shield posture is controlled to advance according to the axis elevation minus the segment floating during tunneling. In the construction of the shallow soil section in the middle of the river, the elevation of the shield propulsion axis is reduced to 20 mm below the design axis, so as to offset the buoyancy of the segment lining in the later stage. Practice has proved that the central axis of the formed tunnel is basically consistent with the design axis by controlling tunneling.

5) Treatment after tube floating

It is difficult to treat the segment after floating. Generally, the grouting hole can be opened at the bottom of the tunnel to release the grouting slurry at the bottom of the segment. According to the similar engineering construction experience, the effect of this method is not ideal, and the tunnel is polluted, and the construction risk is large. If it is found that the floating segment exceeds the limit, the shield tunneling must be stopped immediately, and the floating segment is grouted twice through the grouting hole. The grouting material is the best with instantaneous setting double liquid slurry, and the grouting sequence should follow the direction of tunnel slope, from the tunnel vault to the two waists, and finally the arch bottom is pressed. The principle of stopping grouting is to open the grouting hole at the arch bottom without water seepage, so as to prevent the segment from floating after the shield is restored to tunneling.

7.4.3 Construction Risk Analysis and Countermeasures of River Channel Section

When the shield tunneling in the shallow soil area of the channel section under the water area, due to the unbalanced force on the upper and lower parts of the shield, the attitude of the shield is upward, the crushing is difficult, the tunnel is floating and the axis is difficult to control. After the assembled tunnel ring is removed from the tail of the shield, when the earth pressure on the upper part of the segment and the weight of the segment cannot resist the buoyancy of the surrounding groundwater or the grouting slurry at the tail of the shield, the tunnel will float and cause passive damage to the soil above the segment. Therefore, it is necessary to analyze the possible construction risks of shield tunneling in the middle of the Yangtze River, and formulate corresponding control measures and emergency plans.

1) Risk analysis and countermeasures

Taking the shield construction of Nanjing Yangtze River Tunnel crossing the middle channel section as an example.

(1) The overburden layer of the punching section is very shallow. The thickness of the overburden layer of 72 m punching section is less than twice the shield diameter, and the minimum thickness is 11.31–12.34 m, only 0.78 times the shield diameter. The stratum has hydraulic connection with the river water, which is prone to roof fall and collapse.
(2) The vertical direction of the punching segment shield is 4.5% upward slope, and the excavation face has 4.5% inclination, which is easy to lose stability.
(3) The horizontal direction of the trench shield is $R = 2500$ m horizontal curve segment, the curve segment shield will inevitably cause overbreak, which is not conducive to the stability of the excavation face. In addition, due to the non-synchronization of shield and segment steering in the curve section, the shield tail clearance is difficult to control, and the segment gap is prone to unevenness, resulting in water leakage, slurry leakage or segment shelling at the shield tail.
(4) The upper soil of the punching section is silty sand layer, and the liquefaction grade is mild liquefaction to medium liquefaction. When it is disturbed by shield tunneling, it is prone to liquefaction and collapse. When the shield tail

sealing effect is not good or the grouting pipe is cleaned, water gushing and sand gushing may occur.

In view of the above four points, the following countermeasures are adopted for construction control:

(1) In order to prevent the formation splitting caused by long-term shutdown of the shield in this section, a comprehensive inspection of the shield is carried out before the shield enters this section. All the spare parts with hidden dangers are replaced, and the comprehensive overhaul and replacement of the assembly system are completed. A high-voltage cable is replaced before entering the groove section, so as to ensure that the groove section is promoted in accordance with the principles of high viscosity and heavy slurry, reasonable low pressure, smooth propulsion, rapid assembly, prohibition of shutdown and one-pass, and strive to shorten the crossing time. The high quality slurry was prepared in the slurry tank, and 400 m^3 new slurry was kept in the storage tank for maintenance of excavation face and slurry adjustment at the end of each ring. Reserve 3100 m^3 slurry in the mixing tank to meet the requirements of the index, and reserve 5000 m^3 slurry in the waste slurry tank. On-site storage room reserves not less than 20t NSHS-1, 10t NSHS-2, 20t NSHS-3, 10t NSHS-4 and other polymer materials and 50t bentonite, storage room reserves 100t NSHS-1, 30t NSHS-2, 100t NSHS-3, 50t NSHS-4 and other polymer materials and 200t bentonite. Drivage using high density, high viscosity mud to form dense mud film, closed excavation face. The mud pressure is calculated according to the parameters of weak strata, providing sufficient support pressure for the excavation face. At the same time, the mud pressure and grouting pressure (fluctuation ± 0.1 bar) are strictly controlled to prevent the pressure from breaking through the overburden. In order to ensure the accuracy of tunneling parameters, a field measurement of the topographic section of the bottom of the river was carried out before passing through the gully section. During the crossing process, the water level was measured by special person before each ring tunneling, and the tunneling parameters were corrected. Strengthen the condition monitoring and maintenance of equipment in the process of crossing, equipped with necessary vulnerable parts.
(2) 4.5% of the slope of the excavation face is difficult to maintain stability when the excavation stops, so the mud must have the ability to rapidly film. According to the empirical data, combined with the calculation pressure, some mud pressure should be appropriately increased to offset the instability of the excavation face caused by the influence of the slope. Therefore, 80 m^3 high concentration new slurry must be injected into the excavation face in time before the end of each ring excavation to maintain the stability of the excavation face.
(3) In order to reduce the influence of the plane curve section on the construction of the chute section, the uniform steering is carried out according to the design offset of each ring during tunneling, so as to prevent the phenomenon of rapid turning. In order to ensure that the shield tail clearance does not deteriorate, each ring is divided into 0.5 m, 1 m, 1.5 m, 2 m four mileage segments of the

shield tail clearance selected 10 points for measurement, and according to the measured data of the shield tail clearance, in the process of excavation, timely slow correction, strictly prohibit excessive correction. The segment selection is mainly based on machine selection and manual review, and the segment installation form that is most matched with the shield machine attitude is selected. When slurry leakage is caused by uneven shield tail clearance, the water-stop sponge can be pasted on the outer arc surface of the segment in order to block the water absorption expansion when slurry leakage occurs at shield tail. At the same time, the oil injection amount can be increased with the increase of synchronous grouting amount.

(4) For the liquefaction of sand layer, the disturbance is minimized by reducing the tunneling speed and the rotational speed of the cutterhead in the tunneling parameters. The dense mud film was formed by using high density and high viscosity mud, and the excavation surface was closed. At the same time, through the mud film formation process, the viscous particles are penetrated into the silty sand layer, which can also improve the liquefaction phenomenon of the silty sand layer and eliminate the liquefaction effect of the soil contacted with the cutterhead. According to the parameters of fine sand layer, the slurry pressure is calculated to provide sufficient support pressure for the excavation face. In view of the possible water and sand gushing phenomenon, on the one hand, it is necessary to ensure that the shield tail gap is uniform. On the other hand, it is necessary to ensure that the injection amount of back wall grouting and shield tail sealing oil is sufficient. At the end of each grouting, the mixed slurry of fly ash and bentonite is used to keep pressure on the shield tail grouting pipe to prevent the inflow of external mud.

2) Grouting measures at river bottom

The main countermeasures adopted in the occurrence of river bottom slurry fall are as follows:

(1) When the river bottom slurry is found, it is necessary to closely monitor the liquid level display of the main control room and the variation of the slurry flow rate and dry sand volume. If the slurry is slight and can be pushed forward without reducing the water pressure of the excavation surface, it will continue to move forward. At the same time, it is necessary to accelerate the propulsion speed appropriately. NSHS-2 plugging material is injected from the emergency grouting mouth. When the machine is stopped and assembled, the slurry pressure is appropriately reduced, and the plugging material is injected. At the same time, the jack thrust is appropriately increased to maintain the supporting effect of the cutterhead on the excavation surface, improve the assembly efficiency and shorten the assembly time, so that the shield can pass through the slurry area as soon as possible.

(2) When the slurry burst is serious and cannot be pushed forward, the slurry pressure is gradually reduced to +0.4 bar of groundwater pressure, maintaining the balance of excavation face, further improving the density and viscosity of

slurry, and adding 2%–4% NSHS-2 plugging agent to the slurry to block the position of slurry burst. The slurry flow is moderately increased by using the reserved slurry, and the smooth propulsion is adopted to shorten the assembly time and quickly pass through the slurry burst area.

(3) When it is found that the river water flows into the tunnel from the tail of the shield, the current situation should be first analyzed and the force should be organized for emergency. In the emergency process, the cement mortar and high quality shield tail oil (WR90) with short setting time (3–4 h) were injected at the leakage site. At the same time, the existing sewage pump (200 m^3/h + 200 m^3/h) is used to do the drainage work, and the pump is added according to the needs to ensure the smooth discharge of the river into the shield tunnel.

3) Emergency plan

Due to the great risks in the construction of shield crossing the middle channel section of the Yangtze River, the following engineering safety plans are formulated in combination with the actual situation of Nanjing Yangtze River Tunnel.

(1) Technical measures.

This paper discusses and analyzes all kinds of dangerous sources that may occur when shield passes through the middle channel of the river, and formulates corresponding countermeasures, as shown in Table 7.6.

The specific operation control points are as follows:

① Ground monitoring should be timely feedback to the duty leader, technical personnel and shield, slurry field operators.

② If the river surface slurry burst occurs, it indicates that the mud pressure has broken through the overburden. At this time, the hydraulic connection between the river water and the mud tank has occurred, and a large number of mud

Table 7.6 Technical measures

Number	Possible situations	Countermeasures	Remark
1	Grouting on river surface	Adjusting mud pressure, increasing mud density, shield tunneling through the slurry area	
2	Water gushing into the tunnel	Starting emergency sealing device of shield machine, strong drainage and plugging in shield machine, and increasing injection amount of grease and synchronous grouting slurry	
3	Excavation face instability	Replace the slurry of excavation face with high concentration and high density slurry, increase the pressure of support face, and make the shield machine pass through the collapsed stratum	

leakage and soil erosion may occur. At this time, it is necessary to appropriately reduce the mud pressure, increase the mud density, and increase the tunneling speed, so that the shield can quickly cross the slurry burst area. After the shield passes through the grouting area, the machine is stopped to supplement grouting and the through joint is closed.

③ If the river water pours into the tunnel, it indicates that there is a partial failure of the shield tail seal or the segment water stop. At this time, it is necessary to carry out strong drainage to prevent the shield components from being soaked in water, and to identify the leakage position. For example, the segment leakage can be treated by conventional concrete plugging technology such as polyurethane injection. For example, the shield tail leakage can start the shield tail emergency sealing device, increase the amount of oil and synchronous grouting, so as to play the role of waterproof and waterproof.

④ If the excavation face is unstable or collapsed, it indicates that the mud film on the excavation face has failed, and the mud support pressure is less than the external water and soil pressure. In this case, the ground prefabricated high viscosity and high density new slurry should be used to replace the slurry on the excavation face immediately, and the support pressure should be increased to restore the stability of the excavation face. Because the excavation face of the collapsed stratum is not complete, it is necessary to make the shield tunneling slowly through the collapsed area and reestablish the balance after the slurry replacement is completed.

⑤ In order to ensure the smooth implementation of the above measures, a comprehensive inspection of all equipment, materials and personnel is arranged before the shield passes through the river channel section.

(2) Rescue measures.

① When a sudden accident occurs, the on-duty leader, technical personnel or security personnel on the spot should immediately organize unrelated personnel to evacuate the dangerous area quickly, and the machinery that cannot be evacuated immediately should not be evacuated and people are the main force to ensure the safety of the construction personnel.

② After the evacuation of dangerous places (generally evacuated to the hole), immediately count the number of site construction personnel, check whether the personnel have not escaped from the scene, and immediately report the relevant situation to the project department leader.

③ After receiving the notification, the leaders of the project department should immediately start the emergency relief program, organize manpower and material resources to make full efforts to rescue and relief, and reduce disaster losses.

④ When casualties occur, according to the emergency rescue plan timely rescue work. Rescue should be carried out to ensure that rescue workers have no life and safety threats. If they have no rescue ability, they should report to the local government or relevant departments in time for rescue, and cooperate with the rescue work.

⑤ When rescuing the wounded, according to the number of wounded and the degree of injury, the medical staff should take corresponding emergency measures at the scene. According to the principle of "first heavy and then light," the wounded were sent to the hospital for rescue and treatment in time.

⑥ On-site isolation measures such as safety warning lines are adopted to prevent other personnel from entering dangerous areas and avoid the expansion of disaster losses.

⑦ According to the situation of disaster losses, in accordance with the relevant laws, regulations and provisions of the national and local governments at all levels, timely reporting to higher authorities or relevant departments, pending further investigation and processing.

(3) Logistics support measures.

Field commanders should be equipped with intercoms to ensure smooth communication and communication.

A sufficient amount of emergency materials such as straw bags, woven bags, polyurethanes and cotton quilts, and equipment such as pumps, excavators and slag trucks, including stretchers, were prepared on the site before the shield passed through the middle reaches of the Yangtze River. The pump is installed in place, and the pipeline is connected. Once the water gushes, it can be discharged immediately, and the slag truck is equipped with clay for standby. Site communications, lighting, drainage, ventilation, fire and other equipment to be configured to prevent the occurrence of dangerous scene chaos. In addition, all personnel are required to wear safety hats, water shoes, life jackets and other labor protection products to protect the life safety of workers.

References

Anagnostou, G., and K. Kovari. 1994. The face stability of slurry-shield-driven tunnel. *Tunneling and Underground Space Technology* 9 (2): 165–174.
Broere, W. 2001. *Tunnel face stability and new CPT applications.* Delft: Delft University of Technology.
Chen, Jian. 2014. Summary of key construction technologies for super large diameter slurry shield crossing the Yangtze River. *Modern Tunnel Technology* 51 (S1): 81–86.
China Standardization Committee. 2007. *Bentonite: GB/T 20973-2007.* Beijing: China Quality Inspection Press.
Fengxiang, Zhang, Zhu Hehuan and Fu Deming. 2004. *Shield tunnel.* Beijing: People's Transportation Press.
Fritz, P., R. Hermanns Stengele, and A. Heinz. 2002. Modified bentonite slurries for slurry shields in highly permeable soils. In *4th international symposium geotechnical aspects of underground construction in soft ground*, Toulouse.
Guo, Xinjun, and Hongwei Dai. 2013. *Super-large diameter shield tunnel crossing the Yangtze River construction technology.* Beijing: China Construction Industry Press.
Guo, Xinjun, Fanlu Min, Xiaochun Zhong, and Wei Zhu. 2012. Nanjing Yangtze River tunnel engineering difficulties analysis and key technology summary. *Rock Mechanics and Engineering Journal* 31 (10): 2154–2160.
Han, Xiaorui. 2009. *Research on the formation law of slurry shield slurry film based on the matching between slurry and stratum.* Nanjing: Hehai University.
Han, Yuewang. 2009. *The influence mechanism of consolidation-cement characteristics of grouting body behind shield wall on formation deformation.* Nanjing: Hehai University.
Heinz, A. 2006. *Modifizierte Bentonitsuspensionen fir geotechnische Bauverfahren in Boden hoher Durchlassigkeit.* Zurich: Swiss Federal Institute of Technology ETH.
Kim, S.H., and F. Tonon. 2010. Face stability and required support pressure for TBM driven tunnels with ideal face membrane-drained case. *Tunnelling and Underground Space Technology* 25 (5): 526–542.
Li, Liang. 2010. *Study on the mechanism and countermeasures of embankment damage under high strength mining.* China University of Mining and Technology.
Li, Yun, and Zixin Zhang. 2007. Research on the influence of slurry penetration on the stability of shield excavation face. *Underground Space and Engineering* (04): 720–725+750.
Liu, Xueyan, and Dajun Yuan. 2014. Sludge pressure setting to prevent slurry fracturing during shield tunneling. *Journal of Civil Engineering* 47 (05): 128–132.
Minfan, Road. 2012. *Study on permeability law of slurry in stratum and slurry film formation of slurry pressure shield.* Nanjing: Hehai University.

J. Chen et al., *Construction Technology of Large Diameter Underwater Shield Tunnel*, Key Technologies for Tunnel Construction under Complex Geological and Environmental Conditions, https://doi.org/10.1007/978-981-16-5896-9

Minfan, Road, Teng Jiang, Daiwei Wei, and Asia Zhang. 2014. Experimental study on mud preparation and mud film formation of slurry shield under pressure. *Tunnel Construction* 34 (09): 857–861.

Minfan, Road, Jingbo Xu, Jiarui Du, and Huitang Xing. 2015. Experimental study on slurry preparation and film formation of large diameter slurry shield gravel sand stratum. *Modern Tunnel Technology* 52 (06): 141–146.

Troops, Xuemei Xu, and Hongwei Xia. 2017. Analysis of hidden dangers of trenchless dike crossing technology in dike reinforcement. *Flood Control and Drought Relief in China* 27 (02): 97–98.

Wang, Fei. 2012. *Study on slope stability analysis and reinforcement measures of embankment engineering*. Hefei University of Technology.

Wang, Mingsheng. 2015. Research on the key construction technology of slurry shield of Nanchang subway crossing Ganjiang shallow covering permeable layer. *Tunnel Construction Set* 35 (11): 1222–1228.

Wang, Hai, and Xiaochun Zhong. 2013. Influence of slag removal efficiency of slurry shield slurry-water separation system on slurry. *Tunnel Construction* 33 (11): 928–932.

Wang, Song, and Yong Wang. 2014. Construction technology of shield tunnel in water. *Shanxi Building* 40 (05): 171–172.

Watanabe, T., and H. Yamazaki. 1981. Giant size slurry shield is a success in Tokyo. *Tunnels and Tunnelling* 13 (1): 13–17.

Wei, Liangwen, Qinghe Zhang, and Zhongyi Deng. 2007. Stability mechanism and application of large slurry shield tunnel excavation face. *Underground Space and Engineering* (01): 87–91.

Wu, Shiming, Cungang Lin, Zhongmiao Zhang, and Ning Wang. 2011. Risk analysis and control of slurry shield undercrossing embankment. *Journal of Rock Mechanics and Engineering* 30 (05): 1034–1042.

Xinghuitang. 2010. Nanjing Yangtze River tunnel construction technology of slurry shield crossing the super shallow Earth section of the Yangtze River. *Modern Tunnel Technology* 47 (02): 68–73.

Yao, Xianli. 2010. The application of casing drilling method in pile pulling construction of Nanjing Metro Line 2. *Tunnel Construction Set* 30 (01): 48–52+62.

Yin, Luchao, Zhenhong Zhu, and Yuzhen Li. 1999. *Japan tunnel shield new technology*. Wuhan: Huazhong University of Technology Press.

Yuan, Dajun, Fan Yin, Huawei Wang, Qingfei Huang, and Heng Xiao. 2009. Ultra-large diameter slurry shield tunneling on soil disturbance research. *Rock Mechanics and Engineering* 28 (10): 2074–2080.

Yuan, Dajun, Qingfei Huang, Xinggao Li, et al. 2010. Study on the phenomenon of slurry splitting and stretching in clay stratum under shield tunneling. *Journal of Geotechnical Engineering* 32 (05): 712–717.

Zhang, Asia. 2016. *Surface subsidence and collapse caused by large-diameter slurry shield tunnel construction*. Nanjing: Hehai University.

Zhang, Zhongan. 2016. Application of steel sleeve in shield reception of subway tunnel. *Railway Standard Design* 60 (02): 110–113.

Zhang, Qinghe, Shentang Wang, Changzheng Yan, and Wei Zhang. 2004. The construction technology countermeasures of shield tunnel crossing shallow overburden under water. *Rock Mechanics and Engineering* (05): 857–861.

Zhou, Wenbo. 2004. *Shield tunnel construction technology and application*. Beijing: China Construction Industry Press.

2014. *Calculation and analysis of shallow soil thickness of shield tunnel crossing Inland River*. Tianjin University.

2014. Mud management in slurry shield construction of subway tunnel. *Construction* (05): 58–59.

2017. Discussion on construction technology of slurry shield shallow overburden. *Construction Mechanization* 38 (09): 45–47.